Quantum Groups, Quantum Foundations, and Quantum Information: a Festschrift for Tony Sudbery

Journal of Physics: Conference Series Volume 254

York, United Kingdom
29-30 September 2008

Editors:

Stefan Weigert

ISBN: 978-1-61782-244-5
ISSN: 1742-6588

Printed from e-media with permission by:

Curran Associates, Inc.
57 Morehouse Lane
Red Hook, NY 12571

Some format issues inherent in the e-media version may also appear in this print version.

Printed by Curran Associates, Inc. (2011)

For permission requests, please contact the Institute of Physics
at the address below.

Institute of Physics
Dirac House, Temple Back
Bristol BS1 6BE UK

Phone: 44 1 17 929 7481
Fax: 44 1 17 920 0979

techtracking@iop.org

Additional copies of this publication are available from:

Curran Associates, Inc.
57 Morehouse Lane
Red Hook, NY 12571 USA
Phone: 845-758-0400
Fax: 845-758-2634
Email: curran@proceedings.com
Web: www.proceedings.com

TABLE OF CONTENTS

PREFACE

On 29 July 2008, Professor Anthony Thomas Sudbery – known as Tony to his friends and colleagues – celebrated his 65th birthday. To mark this occasion and to honour Tony's scientific achievements, a 2-day Symposion was held at the University of York on 29–30 September 2008 under the sponsorship of the *Institute of Physics* and the *London Mathematical Society*. The breadth of Tony's research interests was reflected in the twelve invited lectures by A Beige, I Bengtsson, K Brown, N Cerf, E Corrigan, J Ladyman, A J Macfarlane, S Majid, C Manogue, S Popescu, J Ryan and R W Tucker. This *Festschrift*, also made possible by the generosity of the IOP and the LMS, reproduces the majority of these contributions together with other invited papers.

Tony obtained his PhD from the University of Cambridge in 1970. His thesis, written under the guidance of Alan Macfarlane, is entitled *Some aspects of chiral su(3) × su(3) symmetry in hadron dynamics*. He arrived in York in 1971 with his wife Rodie, two young daughters, a lively mind and a very contemporary shock of hair. He was at that stage interested in mathematical physics and so was classed as an applied mathematician in the departmental division in place at that time. But luckily Tony did not fit into this category. His curiosity is combined with a good nose for problems and his capacity for knocking off conjectures impressed us all.

Within a short time of his arrival he was writing papers on group theory, complex analysis and combinatorics, while continuing to work on quantum mechanics. His important paper on quaternionic analysis is an example of the imagination and elegance of his ideas. By developing a derivative, he replaced the relatively obscure analytical theory of quaternions by one informed by modern complex analysis. Other interests emerged, centred round the quantum: quantum mechanics and its foundations, quantum groups and quantum information. He didn't just dabble in these areas but mastered them, gaining a national and international reputation; for instance he joined Roger Penrose in a discussion on 'The Physics of Reality' in Melvyn Bragg's radio series 'In our time'. He was much in demand for contributions to the 'News and Views' section in *Nature* and has written numerous book reviews in scientific and semi-popular journals as well as newspaper commentaries on important scientific developments.

Quantum foundations and quantum information have remained Tony's main professional preoccupations. As a contribution to the conceptual problems surrounding quantum measurement, he undertook a detailed analysis of the observation of decay, introducing the crucial distinction between *continuous measurement* and *continual observation*. A red thread through his work in quantum mechanics has been the critical scrutiny of different interpretations of quantum mechanics and the question of their experimental testability. As a result he has become an eloquent proponent of a version of what is commonly known as the 'many-worlds' interpretation in his most recent paper (arXiv:1009.3914), Tony explains why he considers the name 'Everett–Wheeler interpretation' as most appropriate and 'many worlds' unnecessary.

Tony started directing his research effort to quantum information theory in the late 1990s. He quickly established himself in the quantum information community, putting York on the map in this field. He was among the first people to investigate entanglement properties of multipartite states, particularly the 3-qubit states. While studying the pure 4-qubit states he mysteriously came up with what could rightly be called the 'Sudbery state':

$$|S\rangle = \frac{1}{\sqrt{6}}\Big(|0011\rangle + |1100\rangle + \omega(|1010\rangle + |0101\rangle) + \omega^2(|1001\rangle + |0110\rangle)\Big)$$

(where ω is a primitive root of 1), which is the most entangled 4-qubit state in natural measures of entanglement. He continues to work in quantum information theory, particularly on quantum entanglement.

Not least there is also Tony the philosopher who wrote articles with such intriguing titles as 'The necessity of not doing otherwise' or 'Why am I me? and why is my world so classical?'.

Tony was a teacher of the old school. His lectures were inspiring and fun. He believed that the personal element is central to understanding and inspiration, shown by his whole-hearted commitment to small group teaching. Tony's experience of teaching quantum physics to mathematicians is reflected in his widely known text on quantum mechanics for mathematicians entitled *Quantum Mechanics and the Particles of Nature* (Cambridge 1986).

In 1994, Tony succeeded John Fountain as Head of Department. The 1990s had been a difficult decade for universities in Britain but John had managed to keep the Department on a sound footing, providing foundations which enabled Tony to develop a growth strategy and realise new opportunities when they arose. Despite these challenges Tony was not to be distracted from his scientific work but started to make his significant contributions to quantum information science.

Tony's interests extend beyond mathematics, physics and philosophy: to science fiction, music (both classical and jazz) and theatre – he exercised his thespian talents in his lectures and as a member of the *York Shakespeare Project*. And his lively mind is matched by a lively body: Tony still plays a mean game of squash and of tennis, and his enthusiastic and acrobatic dancing is most remarkable – it would be no surprise if he were to appear on the popular British TV show 'Strictly Come Dancing'.

A man of many parts...

On behalf of all contributors to the *Festschrift* it remains for us to wish Tony many productive and happy years to come in this new phase of his life that he himself characterises with the word 'freedom' (and that surely doesn't match the definition of 'retirement'). There is no doubt that he will utilise this newly-gained freedom to continue to inspire and challenge his fellow scientists with his inquisitive mind and cheerful spirit.

York, October 2010

Paul Busch, Maurice Dodson and Atsushi Higuchi
Stefan Weigert (editor)

All papers published in this volume of *Journal of Physics: Conference Series* have been peer reviewed through processes administered by the proceedings Editors. Reviews were conducted by expert referees to the professional and scientific standards expected of a proceedings journal published by IOP Publishing.

Quantum – a Festschrift for Tony Sudbery IOP Publishing
Journal of Physics: Conference Series **254** (2010) 012001 doi:10.1088/1742-6596/254/1/012001

Exotic bialgebras from 9×9 braid matrices

B Abdesselam[1], A Chakrabarti[2], V K Dobrev[3,4] and S G Mihov[3]

[1] Laboratoire de Physique Quantique de la Matière et de Modélisations Mathématiques, Centre Universitaire de Mascara, 29000-Mascara, Algérie
and
Laboratoire de Physique Théorique, Université d'Oran Es-Sénia, 31100-Oran, Algérie
[2] Centre de Physique Théorique, Ecole Polytechnique, 91128 Palaiseau Cedex, France
[3] Institute of Nuclear Research and Nuclear Energy Bulgarian Academy of Sciences 72 Tsarigradsko Chaussee, 1784 Sofia, Bulgaria
[4] Abdus Salam International Center for Theoretical Physics Strada Costiera 11, 34100 Trieste, Italy

E-mail: boucif@yahoo.fr, chakra@cpht.polytechnique.fr, dobrev@inrne.bas.bg, smikhov@inrne.bas.bg

Abstract. We construct exotic bialgebras that arise from multiparameter 9×9 R-matrices, some of which are new. We also construct the dual bialgebras of two of these exotic bialgebras.

Received: 22 November 2009

In honour of Tony Sudbery's 65th birthday

1. Introduction

For several years [1–4] our initial collaboration studied the algebraic structures coming from 4x4 R-matrices (solutions of the Yang–Baxter equation) that are not deformations of classical ones (i.e., the identity up to signs). In parallel, higher dimensional ($N^2 \times N^2$ matrices for all $N > 2$) exotic braid matrices have been presented and studied in [5, 6]. More recently, our follow-up collaboration (with Boucif Abdesselam replacing our deceased friend and co-author Daniel Arnaudon) constructed $N^2 \times N^2$ unitary braid matrices $\widehat{R}$ for $N > 2$ generalizing the class known for $N = 2$ [7, 8]. Some of these results were applied to model-building [9, 10].

In the present paper we start the systematic study of the bialgebras that arise from these higher dimensional unitary braid matrices. We start with the simplest possible case $N = 3$ in order to get the necessary expertise. However, even this case is complicated enough.

The paper is organized as follows. In Section 2 we give a the overall general setting. Section 3 is devoted to the explicit construction of exotic bialgebras starting from a 9×9 braid matrix that we constructed in [7]. In Section 4 we find the dual bialgebra of the most promising exotic bialgebra constructed in Section 3. In Section 5 we take a much more general 9×9 R-matrix and find many more examples of exotic bialgebras. In Section 6 we find the dual bialgebra of the most promising exotic bialgebra constructed in Section 5.

2. Preliminaries

Our starting point is the following 9×9 braid matrix from [7]:

$$\widehat{R}(\theta) = \begin{vmatrix} a_+ & 0 & 0 & 0 & 0 & 0 & 0 & 0 & a_- \\ 0 & b_+ & 0 & 0 & 0 & 0 & 0 & b_- & 0 \\ 0 & 0 & a_+ & 0 & 0 & 0 & a_- & 0 & 0 \\ 0 & 0 & 0 & c_+ & 0 & c_- & 0 & 0 & 0 \\ 0 & 0 & 0 & 0 & 1 & 0 & 0 & 0 & 0 \\ 0 & 0 & 0 & c_- & 0 & c_+ & 0 & 0 & 0 \\ 0 & 0 & a_- & 0 & 0 & 0 & a+ & 0 & 0 \\ 0 & b_- & 0 & 0 & 0 & 0 & 0 & b_+ & 0 \\ a_- & 0 & 0 & 0 & 0 & 0 & 0 & 0 & a_+ \end{vmatrix} \tag{2.1}$$

where

$$a_\pm = \tfrac{1}{2}(e^{m^+_{11}\theta} \pm e^{m^-_{11}\theta}), \quad b_\pm = \tfrac{1}{2}(e^{m^+_{21}\theta} \pm e^{m^-_{21}\theta}), \quad c_\pm = \tfrac{1}{2}(e^{m^+_{22}\theta} \pm e^{m^-_{22}\theta}), \tag{2.2}$$

and $m^\pm_{ij}$ are parameters. The above braid matrix satisfies the baxterized braid equation:

$$\widehat{R}_{12}(\theta)\widehat{R}_{23}(\theta+\theta')\widehat{R}_{12}(\theta') = \widehat{R}_{23}(\theta')\widehat{R}_{12}(\theta+\theta')\widehat{R}_{23}(\theta). \tag{2.3}$$

For the RTT relations of Faddeev-Reshetikhin-Takhtajan [11] we need the corresponding R-matrix, $R = P\hat{R}$, (P is the permutation matrix):

$$R(\theta) = \begin{pmatrix} a_+ & 0 & 0 & 0 & 0 & 0 & 0 & 0 & a_- \\ 0 & 0 & 0 & c_+ & 0 & c_- & 0 & 0 & 0 \\ 0 & 0 & a_- & 0 & 0 & 0 & a_+ & 0 & 0 \\ 0 & b_+ & 0 & 0 & 0 & 0 & 0 & b_- & 0 \\ 0 & 0 & 0 & 0 & 1 & 0 & 0 & 0 & 0 \\ 0 & b_- & 0 & 0 & 0 & 0 & 0 & b_+ & 0 \\ 0 & 0 & a_+ & 0 & 0 & 0 & a_- & 0 & 0 \\ 0 & 0 & 0 & c_- & 0 & c_+ & 0 & 0 & 0 \\ a_- & 0 & 0 & 0 & 0 & 0 & 0 & 0 & a_+ \end{pmatrix} \tag{2.4}$$

which satisfies the baxterized Yang-Baxter equation:

$$R_{12}(\theta)R_{13}(\theta+\theta')R_{23}(\theta') = R_{23}(\theta')R_{13}(\theta+\theta')R_{12}(\theta) \tag{2.5}$$

In fact, we need the solutions of the constant YBE which are as follows:

$$a_+ = b_+ = c_+ = 1/2, \quad a_+ = \pm a_-, \quad b_+ = \pm b_-, \quad c_+ = \pm c_- \tag{2.6}$$

In view of (2.2) we see that for $a_+ = a_-$ the proper limit is obtained, for example, by taking the following limits: first $m^-_{11} = -\infty$, and then $\theta = 0$, while for $a_+ = -a_-$ the limit may be obtained for $m^+_{11} = -\infty$ first, and then $\theta = 0$. Similarly are obtained the limits for $b_\pm$ and $c_\pm$.

So we have eight R matrices satisfying the constant YBE :

$$(+,+,+),\ (-,+,+),\ (+,-,+),\ (+,+,-),\ (+,-,-),\ (-,+,-),\ (-,-,+),\ (-,-,-) \tag{2.7}$$

where the $\pm$ signs denote respectively the signs of $a_+ = \pm a_-$, $b_+ = \pm b_-$ and $c_+ = \pm c_-$.

For the elements of the 3×3 T matrix we introduce the notation:

$$T = \begin{pmatrix} k & p & l \\ q & r & s \\ m & t & n \end{pmatrix} \tag{2.8}$$

Journal of Physics: Conference Series **254** (2010) 012001 doi:10.1088/1742-6596/254/1/012001

3. Solutions of the RTT equations and exotic bialgebras

We consider matrix bialgebras which are unital associative algebras generated by the nine elements from (2.8) $k, l, m, n, p, q, r, s, t$. The co-product and co-unit relations are the classical ones:

$$\delta(T) = T \otimes T \quad (3.1a)$$
$$\varepsilon(T) = \mathbf{1}_3 \quad (3.1b)$$

We expect the bialgebras under consideration not to be Hopf algebras which, as in the $S03$ case [2], would be easier to check after we find the dual bialgebras.

In the next subsections we obtain the desired bialgebras by applying the RTT relations of [11]:

$$R\,T_1\,T_2 \;=\; T_2\,T_1\,R\ , \quad (3.2)$$

where $T_1 = T \otimes \mathbf{1}_2$, $T_2 = \mathbf{1}_2 \otimes T$, for $R = R(\theta)$, (2.4), and the parameters are the constants in (2.6) following the eight cases of (2.7).

3.1. Algebraic relations

I) Relations, which do not depend on the parameters $a_\pm$, $b_\pm, c_\pm$. We have the set of relations

$$\begin{aligned}(N) = \{k^2 = n^2,\ \ kn = nk,\ \ l^2 = m^2,\ \ lm = ml\\ km = nl,\ \ kl = nm,\ \ lk = mn,\ \ mk = ln\\ r(k-n) = (k-n)r = 0,\ \ r(l-m) = (l-m)r = 0\}\end{aligned} \quad (3.3)$$

The last two relations suggest to introduce the generators:

$$\begin{aligned}k = \tilde{k} + \tilde{n},\ \ n = \tilde{k} - \tilde{n};\ \ l = \tilde{l} + \tilde{m},\ \ m = \tilde{l} - \tilde{m},\\ p = \tilde{p} + \tilde{t},\ \ t = \tilde{p} - \tilde{t};\ \ q = \tilde{q} + \tilde{s},\ \ s = \tilde{q} - \tilde{s},\end{aligned} \quad (3.4)$$

In terms of these generators we have:

$$\begin{aligned}(N) = \{\tilde{k}\tilde{n} = \tilde{n}\tilde{k} = 0;\ \ \tilde{l}\tilde{m} = \tilde{m}\tilde{l} = 0;\\ \tilde{k}\tilde{m} = \tilde{n}\tilde{l} = 0;\ \ \tilde{l}\tilde{n} = \tilde{m}\tilde{k} = 0;\\ r\tilde{m} = r\tilde{n} = 0,\ \ \tilde{m}r = \tilde{n}r = 0\}\end{aligned} \quad (3.5)$$

II) Relations that do not depend on the relative signs of (a_-, b_-), (a_-, c_-) and (b_-, c_-). In that case we have

IIa) $a_+ = \pm a_-$

If $a_+ = a_-$ we have the set of relations

$$A_+ = \{p^2 = t^2,\ \ pt = tp;\ \ q^2 = s^2,\ \ qs = sq\} \quad (3.6)$$

or in terms of the alternative generators we have:

$$A_+ = \{\tilde{p}\tilde{t} = \tilde{t}\tilde{p} = 0;\ \ \tilde{q}\tilde{s} = \tilde{s}\tilde{q} = 0\} \quad (3.7)$$

If $a+ = -a_-$ the set of relations is

$$A_- = \{p^2 = -t^2,\ \ pt = -tp;\ \ q^2 = -s^2,\ \ qs = -sq\} \quad (3.8)$$

or alternatively

$$A_- = \{\tilde{p}^2 = \tilde{t}^2 = 0;\quad \tilde{q}^2 = \tilde{s}^2 = 0\} \tag{3.9}$$

IIb) $b_+ = \pm b_-$
If $b_+ = b_-$ we have the set of relations:

$$B_+ = \{rp = rt;\quad rq = rs\} \tag{3.10}$$

Alternatively

$$B_+ = \{r\tilde{t} = 0;\quad r\tilde{s} = 0\} \tag{3.11}$$

If $b+ = -b_-$

$$B_- = \{rp = -rt;\quad rq = -rs\} \tag{3.12}$$

Alternatively

$$B_- = \{r\tilde{p} = 0;\quad r\tilde{q} = 0\} \tag{3.13}$$

IIc) $c_+ = c_\pm$
If $c_+ = c_-$ we have

$$C_+ = \{pr = tr;\quad qr = sr\} \tag{3.14}$$

Alternatively

$$C_+ = \{\tilde{t}r = 0;\quad \tilde{s}r = 0\} \tag{3.15}$$

If $c_+ = -c_-$ we have

$$C_- = \{pr = -tr;\quad qr = -sr\} \tag{3.16}$$

Alternatively

$$C_- = \{\tilde{p}r = 0;\quad \tilde{q}r = 0\} \tag{3.17}$$

III) Relations depending on the relative signs of (a_-, b_-), (a_-, b_-0) and (b_-, c_-)
IIIa) $a_- = \pm b_-$
If $a_- = b_-$ we have the set of relations

$$\begin{aligned}(AB)_+ = \{&pk = tn,\ tk = pn;\ pl = tm,\ tl = pm;\\ &qk = sn,\ qn = sk;\ ql = sm,\ sl = qm\}\end{aligned} \tag{3.18}$$

Alternatively

$$(AB)_+ = \{\tilde{p}\tilde{m} = \tilde{p}\tilde{n} = \tilde{t}\tilde{k} = \tilde{t}\tilde{l} = 0;\quad \tilde{q}\tilde{m} = \tilde{q}\tilde{n} = \tilde{s}\tilde{k} = \tilde{s}\tilde{l} = 0\} \tag{3.19}$$

If $a_- = -b_-$ we have

$$\begin{aligned}(AB)_- = \{&pk = -tn,\ tk = -pn;\ pl = -tm,\ tl = -pm;\\ &qk = -sn,\ qn = -sk;\ ql = -sm,\ sl = -qm\}\end{aligned} \tag{3.20}$$

Alternatively

$$(AB)_- = \{\tilde{p}\tilde{k} = \tilde{p}\tilde{l} = \tilde{t}\tilde{m} = \tilde{t}\tilde{n} = 0;\quad \tilde{q}\tilde{k} = \tilde{q}\tilde{l} = \tilde{s}\tilde{m} = \tilde{s}\tilde{n} = 0\} \tag{3.21}$$

IIIb) $a_- = \pm c_-$
If $a_- = c_-$ we have

$$\begin{aligned}(AC)_+ = \{&kp = nt,\ kt = np;\ lp = mt,\ lt = mp;\\ &kq = ns,\ nq = ks;\ lq = ms,\ ls = mq\}\end{aligned} \tag{3.22}$$

Alternatively

$$(AC)_+ = \{\tilde{k}\tilde{t} = \tilde{l}\tilde{t} = \tilde{m}\tilde{p} = \tilde{n}\tilde{p} = 0;\ \ \tilde{k}\tilde{s} = \tilde{l}\tilde{s} = \tilde{m}\tilde{q} = \tilde{n}\tilde{q} = 0\} \tag{3.23}$$

If $a_- = -c_-$ we have

$$\begin{aligned}(AC)_- = \{&kp = -nt,\ kt = -np;\ lp = -mt,\ lt = -mp;\\ &kq = -ns,\ nq = -ks;\ lq = -ms,\ ls = -mq\}\end{aligned} \tag{3.24}$$

Alternatively

$$(AC)_- = \{\tilde{k}\tilde{p} = \tilde{l}\tilde{p} = \tilde{m}\tilde{t} = \tilde{n}\tilde{t} = 0;\ \ \tilde{k}\tilde{q} = \tilde{l}\tilde{q} = \tilde{m}\tilde{s} = \tilde{n}\tilde{s} = 0\} \tag{3.25}$$

IIIc) $b_- = \pm c_-$
If $b- = c_-$ we have

$$(BC)_+ = \{pq = ts,\ tq = ps;\ \ qp = st,\ qt = sp\} \tag{3.26}$$

Alternatively

$$(BC)_+ = \{\tilde{p}\tilde{s} = \tilde{t}\tilde{q} = 0;\ \ \tilde{s}\tilde{p} = \tilde{q}\tilde{t} = 0\} \tag{3.27}$$

If $b_- = -c_-$ we have

$$(BC)_- = \{pq = -ts,\ tq = -ps;\ \ qp = -st,\ qt = -sp\} \tag{3.28}$$

Alternatively

$$(BC)_- = \{\tilde{p}\tilde{q} = \tilde{t}\tilde{s} = 0;\ \ \tilde{q}\tilde{p} = \tilde{s}\tilde{t} = 0\} \tag{3.29}$$

3.2. Classification of bialgebras
Thus we have the following solutions

- for $a_+ = a_- = b_- = c_-$ we have the set of relations

$$(+,+,+) = \{N \cup A_+ \cup B_+ \cup C_+ \cup (AB)_+ \cup (AC)_+ \cup (BC)_+\} \tag{3.30}$$

Explicitly we have:

$$\begin{aligned}\tilde{k}\tilde{m} = \tilde{m}\tilde{k} = 0;\ \ \tilde{k}\tilde{n} = \tilde{n}\tilde{k} = 0;\ \ \tilde{k}\tilde{t} = \tilde{t}\tilde{k} = 0;\ \ \tilde{k}\tilde{s} = \tilde{s}\tilde{k} = 0;\\ \tilde{l}\tilde{m} = \tilde{m}\tilde{l} = 0;\ \ \tilde{l}\tilde{n} = \tilde{n}\tilde{l} = 0;\ \ \tilde{l}\tilde{t} = \tilde{t}\tilde{l} = 0;\ \ \tilde{l}\tilde{s} = \tilde{s}\tilde{l} = 0;\\ \tilde{p}\tilde{m} = \tilde{m}\tilde{p} = 0;\ \ \tilde{p}\tilde{n} = \tilde{n}\tilde{p} = 0;\ \ \tilde{p}\tilde{t} = \tilde{t}\tilde{p} = 0;\ \ \tilde{p}\tilde{s} = \tilde{s}\tilde{p} = 0;\\ \tilde{q}\tilde{m} = \tilde{m}\tilde{q} = 0;\ \ \tilde{q}\tilde{n} = \tilde{n}\tilde{q} = 0;\ \ \tilde{q}\tilde{t} = \tilde{t}\tilde{q} = 0;\ \ \tilde{q}\tilde{s} = \tilde{s}\tilde{q} = 0;\\ r\tilde{m} = \tilde{m}r = 0;\ \ r\tilde{n} = \tilde{n}r = 0;\ \ r\tilde{t} = \tilde{t}r = 0;\ \ r\tilde{s} = \tilde{s}r = 0.\end{aligned} \tag{3.31}$$

From (3.31) we see that the algebra $\mathcal{A}_{+++}$ is a direct sum of two subalgebras: $\mathcal{A}^1_{+++}$ with generators $\tilde{k}, \tilde{l}, \tilde{p}, \tilde{q}, r$, and $\mathcal{A}^2_{+++}$ with generators $\tilde{m}, \tilde{n}, \tilde{s}, \tilde{t}$. Both subalgebras are free, with no relations, and thus, no PBW bases.

- for $a_+ = -a_- = b_- = c_-$ we have the set of relations

$$(-,+,+) = \{N \cup A_- \cup B_+ \cup C_+ \cup (AB)_- \cup (AC)_- \cup (BC)_+\} \tag{3.32}$$

Journal of Physics: Conference Series **254** (2010) 012001 doi:10.1088/1742-6596/254/1/012001

Explicitly we have:

$$\begin{aligned} \tilde{k}\tilde{m} = \tilde{m}\tilde{k} = 0;\quad \tilde{k}\tilde{n} = \tilde{n}\tilde{k} = 0;\quad \tilde{k}\tilde{p} = \tilde{p}\tilde{k} = 0;\quad \tilde{k}\tilde{q} = \tilde{q}\tilde{k} = 0;\\ \tilde{l}\tilde{m} = \tilde{m}\tilde{l} = 0;\quad \tilde{l}\tilde{n} = \tilde{n}\tilde{l} = 0;\quad \tilde{l}\tilde{p} = \tilde{p}\tilde{l} = 0;\quad \tilde{l}\tilde{q} = \tilde{q}\tilde{l} = 0;\\ r\tilde{m} = \tilde{m}r = 0;\quad r\tilde{n} = \tilde{n}r = 0;\quad r\tilde{t} = \tilde{t}r = 0;\quad r\tilde{s} = \tilde{s}r = 0;\\ \tilde{t}\tilde{m} = \tilde{m}\tilde{t} = 0;\quad \tilde{t}\tilde{n} = \tilde{n}\tilde{t} = 0;\quad \tilde{t}\tilde{q} = \tilde{q}\tilde{t} = 0;\\ \tilde{s}\tilde{m} = \tilde{m}\tilde{s} = 0;\quad \tilde{s}\tilde{n} = \tilde{n}\tilde{s} = 0;\quad \tilde{s}\tilde{p} = \tilde{p}\tilde{s} = 0;\\ \tilde{p}^2 = \tilde{q}^2 = \tilde{s}^2 = \tilde{t}^2 = 0. \end{aligned} \tag{3.33}$$

The structure of this algebra, denoted $\mathcal{A}_{-++}$, is more complicated. There are two quasi-free subalgebras: $\mathcal{A}^1_{-++}$ with generators $\tilde{k}, \tilde{l}, \tilde{t}, \tilde{s}$, and $\mathcal{A}^2_{-++}$ with generators $\tilde{m}, \tilde{n}, \tilde{p}, \tilde{q}$. They are quasi-free due to the last line of (3.33). They do not form a direct sum due to the existence of the following 12 two-letter building blocks of the basis $\mathcal{A}_{-++}$: $r\tilde{k}\,, r\tilde{l}\,, r\tilde{p}\,, r\tilde{q}\,, \tilde{p}\tilde{t}\,, \tilde{q}\tilde{s}$ plus the reverse order.

- for $a_+ = a_- = -b_- = c_-$ we have

$$(+,-,+) = \{N \cup A_+ \cup B_- \cup C_+ \cup (AB)_- \cup (AC)_+ \cup (BC)_-\} \tag{3.34}$$

Explicitly we have:

$$\begin{aligned} \tilde{k}\tilde{m} = \tilde{m}\tilde{k} = 0;\quad \tilde{k}\tilde{n} = \tilde{n}\tilde{k} = 0;\quad \tilde{k}\tilde{t} = \tilde{k}\tilde{s} = 0;\quad \tilde{p}\tilde{k} = \tilde{q}\tilde{k} = 0;\\ \tilde{l}\tilde{m} = \tilde{m}\tilde{l} = 0;\quad \tilde{l}\tilde{n} = \tilde{n}\tilde{l} = 0;\quad \tilde{l}\tilde{t} = \tilde{l}\tilde{s} = 0;\quad \tilde{p}\tilde{l} = \tilde{q}\tilde{l} = 0;\\ r\tilde{m} = \tilde{m}r = 0;\quad r\tilde{n} = \tilde{n}r = 0;\quad r\tilde{p} = r\tilde{q} = 0;\quad \tilde{t}r = \tilde{s}r = 0;\\ \tilde{t}\tilde{p} = \tilde{p}\tilde{t} = 0;\quad \tilde{t}\tilde{s} = \tilde{s}\tilde{t} = 0;\quad \tilde{t}\tilde{m} = \tilde{t}\tilde{n} = 0;\\ \tilde{q}\tilde{p} = \tilde{p}\tilde{q} = 0;\quad \tilde{q}\tilde{s} = \tilde{s}\tilde{q} = 0;\quad \tilde{m}\tilde{q} = \tilde{n}\tilde{q} = 0;\\ \tilde{s}\tilde{m} = \tilde{s}\tilde{n} = 0;\quad \tilde{m}\tilde{p} = \tilde{n}\tilde{p} = 0. \end{aligned} \tag{3.35}$$

The structure of this algebra, denoted $\mathcal{A}_{+-+}$, is also complicated. There are four free subalgebras: $\mathcal{A}^1_{+-+}$ with generators $\tilde{k}, \tilde{l}, r$, $\mathcal{A}^2_{+-+}$ with generators $\tilde{m}, \tilde{n}$, $\mathcal{A}^3_{+-+}$ with generators $\tilde{p}, \tilde{s}$, $\mathcal{A}^4_{+-+}$ with generators $\tilde{q}, \tilde{t}$. Only the first two are in direct sum, otherwise all are related by the following 20 two-letter building blocks: $\tilde{k}\tilde{p}\,, \tilde{k}\tilde{q}\,, \tilde{s}\tilde{k}$, $\tilde{t}\tilde{k}$, $\tilde{l}\tilde{p}\,, \tilde{l}\tilde{q}\,, \tilde{s}\tilde{l}$, $\tilde{t}\tilde{l}$, $r\tilde{s}\,, r\tilde{t}\,, \tilde{p}r\,, \tilde{q}r$, $\tilde{m}\tilde{s}\,, \tilde{m}\tilde{t}\,, \tilde{p}\tilde{m}\,, \tilde{q}\tilde{m}$, $\tilde{n}\tilde{s}\,, \tilde{n}\tilde{t}\,, \tilde{p}\tilde{n}\,, \tilde{q}\tilde{n}$.

There is no overall ordering. There is some partial order if we consider the subalgebra formed by the generators of the latter three subalgebras: $\tilde{m}, \tilde{n}$, $\tilde{p}, \tilde{s}$, $\tilde{q}, \tilde{t}$, namely, we have:

$$\tilde{p}, \tilde{q} \;>\; \tilde{m}, \tilde{n} \;>\; \tilde{s}, \tilde{t} \tag{3.36}$$

But for the natural subalgebra formed by generators $\tilde{k}, \tilde{l}, r$, $\tilde{p}, \tilde{s}$, $\tilde{q}, \tilde{t}$, we have cyclic ordering:

$$\tilde{p}, \tilde{q} \;>\; r \;>\; \tilde{s}, \tilde{t} \;>\; \tilde{k}, \tilde{l} \;>\; \tilde{p}, \tilde{q}\;, \tag{3.37}$$

i.e., no ordering. We have seen this phenomenon in the simpler exotic bialgebra $S03$, [2].

- for $a_+ = a_- = b_- = -c_-$ we have the set of relations

$$(+,+,-) = \{N \cup A_+ \cup B_+ \cup C_- \cup (AB)_+ \cup (AC)_- \cup (BC)_-\} \tag{3.38}$$

Journal of Physics: Conference Series **254** (2010) 012001 doi:10.1088/1742-6596/254/1/012001

Explicitly we have:

$$\begin{aligned}\tilde{k}\tilde{m} = \tilde{m}\tilde{k} = 0;\ \ \tilde{k}\tilde{n} = \tilde{n}\tilde{k} = 0;\ \ \tilde{k}\tilde{p} = \tilde{k}\tilde{q} = 0;\ \ \tilde{s}\tilde{k} = \tilde{t}\tilde{k} = 0;\\ \tilde{l}\tilde{m} = \tilde{m}\tilde{l} = 0;\ \ \tilde{l}\tilde{n} = \tilde{n}\tilde{l} = 0;\ \ \tilde{l}\tilde{p} = \tilde{l}\tilde{q} = 0;\ \ \tilde{s}\tilde{l} = \tilde{t}\tilde{l} = 0;\\ r\tilde{m} = \tilde{m}r = 0;\ \ r\tilde{n} = \tilde{n}r = 0;\ \ r\tilde{t} = r\tilde{s} = 0;\ \ \tilde{p}r = \tilde{q}r = 0;\\ \tilde{t}\tilde{p} = \tilde{p}\tilde{t} = 0;\ \ \tilde{t}\tilde{s} = \tilde{s}\tilde{t} = 0;\ \ \tilde{m}\tilde{t} = \tilde{n}\tilde{t} = 0;\\ \tilde{q}\tilde{p} = \tilde{p}\tilde{q} = 0;\ \ \tilde{q}\tilde{s} = \tilde{s}\tilde{q} = 0;\ \ \tilde{q}\tilde{m} = \tilde{q}\tilde{n} = 0;\\ \tilde{p}\tilde{m} = \tilde{p}\tilde{n} = 0;\ \ \tilde{m}\tilde{s} = \tilde{n}\tilde{s} = 0.\end{aligned} \tag{3.39}$$

This algebra, denoted $\mathcal{A}_{++-}$, is a conjugate of the previous one. It has the same four free algebras, and the only difference is that the subalgebras are related by 20 two-letter building blocks which are in reverse order w.r.t. the previous case: $\tilde{k}\tilde{s}$, $\tilde{k}\tilde{t}$, $\tilde{p}\tilde{k}$, $\tilde{q}\tilde{k}$, $\tilde{l}\tilde{s}$, $\tilde{l}\tilde{t}$, $\tilde{p}\tilde{l}$, $\tilde{q}\tilde{l}$, $r\tilde{p}$, $r\tilde{q}$, $\tilde{s}r$, $\tilde{s}r$, $\tilde{m}\tilde{p}$, $\tilde{m}\tilde{q}$, $\tilde{s}\tilde{m}$, $\tilde{t}\tilde{m}$, $\tilde{n}\tilde{p}$, $\tilde{n}\tilde{q}$, $\tilde{s}\tilde{n}$, $\tilde{t}\tilde{n}$.

- for $a_+ = a_- = -b_- = -c_-$ we have the set of relations

$$(+,-,-) = \{N \cup A_+ \cup B_- \cup C_- \cup (AB)_- \cup (AC)_- \cup (BC)_+\} \tag{3.40}$$

Explicitly we have:

$$\begin{aligned}\tilde{k}\tilde{m} = \tilde{m}\tilde{k} = 0;\ \ \tilde{k}\tilde{n} = \tilde{n}\tilde{k} = 0;\ \ \tilde{k}\tilde{p} = \tilde{p}\tilde{k} = 0;\ \ \tilde{k}\tilde{q} = \tilde{q}\tilde{k} = 0;\\ \tilde{l}\tilde{m} = \tilde{m}\tilde{l} = 0;\ \ \tilde{l}\tilde{n} = \tilde{n}\tilde{l} = 0;\ \ \tilde{l}\tilde{p} = \tilde{p}\tilde{l} = 0;\ \ \tilde{l}\tilde{q} = \tilde{q}\tilde{l} = 0;\\ r\tilde{m} = \tilde{m}r = 0;\ \ r\tilde{n} = \tilde{n}r = 0;\ \ r\tilde{p} = \tilde{p}r = 0;\ \ r\tilde{q} = \tilde{q}r = 0;\\ \tilde{s}\tilde{m} = \tilde{m}\tilde{s} = 0;\ \ \tilde{s}\tilde{n} = \tilde{n}\tilde{s} = 0;\ \ \tilde{s}\tilde{p} = \tilde{p}\tilde{s} = 0;\ \ \tilde{s}\tilde{q} = \tilde{q}\tilde{s} = 0;\\ \tilde{t}\tilde{m} = \tilde{m}\tilde{t} = 0;\ \ \tilde{t}\tilde{n} = \tilde{n}\tilde{t} = 0;\ \ \tilde{t}\tilde{p} = \tilde{p}\tilde{t} = 0;\ \ \tilde{t}\tilde{q} = \tilde{q}\tilde{t} = 0.\end{aligned} \tag{3.41}$$

This algebra, denoted $\mathcal{A}_{+--}$, is a conjugate of the algebra $\mathcal{A}_{+++}$, obtained by the exchange of the pairs of generators $(\tilde{p}, \tilde{q})$ and $(\tilde{s}, \tilde{t})$.

- for $a_+ = -a_- = b_- = -c_-$ we have the set of relations

$$(-,+,-) = \{N \cup A_- \cup B_+ \cup C_- \cup (AB)_- \cup (AC)_+ \cup (BC)_-\} \tag{3.42}$$

Explicitly we have:

$$\begin{aligned}\tilde{k}\tilde{m} = \tilde{m}\tilde{k} = 0;\ \ \tilde{k}\tilde{n} = \tilde{n}\tilde{k} = 0;\ \ \tilde{k}\tilde{t} = \tilde{k}\tilde{s} = 0;\ \ \tilde{p}\tilde{k} = \tilde{q}\tilde{k} = 0;\\ \tilde{l}\tilde{m} = \tilde{m}\tilde{l} = 0;\ \ \tilde{l}\tilde{n} = \tilde{n}\tilde{l} = 0;\ \ \tilde{l}\tilde{t} = \tilde{l}\tilde{s} = 0;\ \ \tilde{p}\tilde{l} = \tilde{q}\tilde{l} = 0;\\ r\tilde{m} = \tilde{m}r = 0;\ \ r\tilde{n} = \tilde{n}r = 0;\ \ r\tilde{t} = r\tilde{s} = 0;\ \ \tilde{p}r = \tilde{q}r = 0;\\ \tilde{s}\tilde{t} = \tilde{t}\tilde{s} = 0;\ \ \tilde{s}\tilde{m} = \tilde{s}\tilde{n} = 0;\ \ \tilde{t}\tilde{m} = \tilde{t}\tilde{n} = 0;\\ \tilde{p}\tilde{q} = \tilde{q}\tilde{p} = 0;\ \ \tilde{m}\tilde{p} = \tilde{n}\tilde{p} = 0;\ \ \tilde{m}\tilde{q} = \tilde{n}\tilde{q} = 0;\\ \tilde{p}^2 = \tilde{q}^2 = \tilde{s}^2 = \tilde{t}^2 = 0.\end{aligned} \tag{3.43}$$

The structure of this algebra, denoted $\mathcal{A}_{-+-}$, is very complicated. There are two free subalgebras: $\mathcal{A}^1_{-+-}$ with generators $\tilde{k}, \tilde{l}r$, $\mathcal{A}^2_{-+-}$ with generators $\tilde{m}, \tilde{n}$, and four quasi-free subalgebras: $\mathcal{A}^3_{-+-}$ with generators $\tilde{p}, \tilde{s}$, $\mathcal{A}^4_{-+-}$ with generators $\tilde{q}, \tilde{t}$, $\mathcal{A}^5_{-+-}$ with generators $\tilde{p}, \tilde{t}$, $\mathcal{A}^6_{-+-}$ with generators $\tilde{q}, \tilde{s}$. The first four subalgebras have generators as in the $\mathcal{A}_{+-+}$ case (but taking into account the last line of (3.43)). Only the first two subalgebras are in direct sum, and there are intersections between the last four. Furthermore, all are related by the following 20 two-letter building blocks: $\tilde{k}\tilde{p}\,,\tilde{k}\tilde{q}\,,\tilde{s}\tilde{k},\,\tilde{t}\tilde{k},\ \ \tilde{l}\tilde{p}\,,\tilde{l}\tilde{q}\,,\tilde{s}\tilde{l},\,\tilde{t}\tilde{l},\ \ r\tilde{p}\,,r\tilde{q}\,,\tilde{s}r\,,\tilde{t}r,\ \ \tilde{m}\tilde{s}\,,\tilde{m}\tilde{t}\,,\tilde{p}\tilde{m}\,,\tilde{q}\tilde{m},$ $\tilde{n}\tilde{s}\,,\tilde{n}\tilde{t}\,,\tilde{p}\tilde{n}\,,\tilde{q}\tilde{n}$, which are the same as in the $\mathcal{A}_{+-+}$ case, except those involving r.

The last difference makes things better. Indeed, there is no overall ordering, more precisely we have:

$$\tilde{p},\tilde{q}\ >\ \tilde{m},\tilde{n}\ >\ \tilde{s},\tilde{t}\ >\ \tilde{k},\tilde{l},r\ >\ \tilde{p},\tilde{q} \tag{3.44}$$

i.e., we have some cyclic order.

Thus, the bialgebra $\mathcal{A}_{-+-}$ may turn out to be the easiest to handle, as the exotic bialgebra $S03$, [2].

- for $a_+ = -a_- = -b_- = c_-$ we have the set of relations

$$(-,-,+) = \{N \cup A_- \cup B_- \cup C_+ \cup (AB)_+ \cup (AC)_- \cup (BC)_-\} \tag{3.45}$$

Explicitly we have:

$$\begin{aligned}
\tilde{k}\tilde{m} = \tilde{m}\tilde{k} = 0;\ \ \tilde{k}\tilde{n} = \tilde{n}\tilde{k} = 0;\ \ \tilde{k}\tilde{p} = \tilde{k}\tilde{q} = 0;\ \ \tilde{s}\tilde{k} = \tilde{t}\tilde{k} = 0;\\
\tilde{l}\tilde{m} = \tilde{m}\tilde{l} = 0;\ \ \tilde{l}\tilde{n} = \tilde{n}\tilde{l} = 0;\ \ \tilde{l}\tilde{p} = \tilde{l}\tilde{q} = 0;\ \ \tilde{s}\tilde{l} = \tilde{t}\tilde{l} = 0;\\
r\tilde{m} = \tilde{m}r = 0;\ \ r\tilde{n} = \tilde{n}r = 0;\ \ r\tilde{p} = r\tilde{q} = 0;\ \ \tilde{t}r = \tilde{s}r = 0;\\
\tilde{s}\tilde{t} = \tilde{t}\tilde{s} = 0;\ \ \tilde{m}\tilde{s} = \tilde{n}\tilde{s} = 0;\ \ \tilde{m}\tilde{t} = \tilde{n}\tilde{t} = 0;\\
\tilde{p}\tilde{q} = \tilde{q}\tilde{p} = 0;\ \ \tilde{p}\tilde{m} = \tilde{p}\tilde{n} = 0;\ \ \tilde{q}\tilde{m} = \tilde{q}\tilde{n} = 0;\\
\tilde{p}^2 = \tilde{q}^2 = \tilde{s}^2 = \tilde{t}^2 = 0.
\end{aligned} \tag{3.46}$$

This algebra, denoted $\mathcal{A}_{--+}$, is a conjugate of the previous algebra obtained by the exchange of the pairs of generators $(\tilde{p},\tilde{q})$ and $(\tilde{s},\tilde{t})$.

- for $a_+ = -a_- = -b_- = -c_-$ we have the set of relations

$$(-,-,-) = \{N \cup A_- \cup B_- \cup C_- \cup (AB)_+ \cup (AC)_+ \cup (BC)_+\} \tag{3.47}$$

Explicitly we have:

$$\begin{aligned}
\tilde{k}\tilde{m} = \tilde{m}\tilde{k} = 0;\ \ \tilde{k}\tilde{n} = \tilde{n}\tilde{k} = 0;\ \ \tilde{k}\tilde{s} = \tilde{s}\tilde{k} = 0;\ \ \tilde{k}\tilde{t} = \tilde{t}\tilde{k} = 0;\\
\tilde{l}\tilde{m} = \tilde{m}\tilde{l} = 0;\ \ \tilde{l}\tilde{n} = \tilde{n}\tilde{l} = 0;\ \ \tilde{l}\tilde{s} = \tilde{s}\tilde{l} = 0;\ \ \tilde{l}\tilde{t} = \tilde{t}\tilde{l} = 0;\\
r\tilde{m} = \tilde{m}r = 0;\ \ r\tilde{n} = \tilde{n}r = 0;\ \ r\tilde{p} = \tilde{p}r = 0;\ \ r\tilde{q} = \tilde{q}r = 0;\\
\tilde{p}\tilde{m} = \tilde{m}\tilde{p} = 0;\ \ \tilde{p}\tilde{n} = \tilde{n}\tilde{p} = 0;\ \ \tilde{p}\tilde{s} = \tilde{s}\tilde{p} = 0;\\
\tilde{q}\tilde{m} = \tilde{m}\tilde{q} = 0;\ \ \tilde{q}\tilde{n} = \tilde{n}\tilde{q} = 0;\ \ \tilde{q}\tilde{t} = \tilde{t}\tilde{q} = 0;\\
\tilde{p}^2 = \tilde{t}^2 = \tilde{q}^2 = \tilde{s}^2 = 0.
\end{aligned} \tag{3.48}$$

This algebra, denoted $\mathcal{A}_{---}$, is a conjugate of the algebra $\mathcal{A}_{-++}$, obtained by the exchanges of generators: $\tilde{p} \longleftrightarrow \tilde{s}$ and $\tilde{q} \longleftrightarrow \tilde{t}$.

3.3. Summary

Thus, taking into account conjugation, we have found *four* different bialgebras:

$$\mathcal{A}_{+++}\ \cong\ \mathcal{A}_{+--}\,,\quad \mathcal{A}_{---}\ \cong\ \mathcal{A}_{-++}\,,\quad \mathcal{A}_{+-+}\ \cong\ \mathcal{A}_{++-}\,,\quad \mathcal{A}_{-+-}\ \cong\ \mathcal{A}_{--+} \tag{3.49}$$

Journal of Physics: Conference Series **254** (2010) 012001 doi:10.1088/1742-6596/254/1/012001

Thus, for future use we shall use shorter notation:

$$\mathcal{A}_{++} \equiv \mathcal{A}_{+++}, \quad \mathcal{A}_{--} \equiv \mathcal{A}_{---}, \quad \mathcal{A}_{+-} \equiv \mathcal{A}_{+-+}, \quad \mathcal{A}_{-+} \equiv \mathcal{A}_{-+-} \tag{3.50}$$

The first two bialgebras have no ordering. The first one is simpler, since it is split in two subalgebras with five and four generators. The third bialgebra has partial ordering in one subalgebra. The last one, is the most promising since it has partial cyclic ordering.

The next task in this line of research is to find the dual bialgebras, analogously, as done for the four-element exotic bialgebras in [1, 2]. We do this in the next Section for the most interesting of the above: $\mathcal{A}_{-+} \equiv \mathcal{A}_{-+-}$.

4. The dual bialgebra of A_{-+}

To start with we begin with the coproducts of the elements of the T-matrix. We have:

$$\delta\begin{pmatrix} k & p & l \\ q & r & s \\ m & t & n \end{pmatrix} = \begin{pmatrix} k\otimes k + p\otimes q + l\otimes m & k\otimes p + p\otimes r + l\otimes t & k\otimes l + p\otimes s + l\otimes n \\ q\otimes k + r\otimes q + s\otimes m & q\otimes p + r\otimes r + s\otimes t & q\otimes l + r\otimes s + s\otimes n \\ m\otimes k + t\otimes q + n\otimes m & mo\times p + t\otimes r + n\otimes t & m\otimes l + t\otimes s + n\otimes n \end{pmatrix} \tag{4.1}$$

We have also

$$\epsilon\begin{pmatrix} k & p & l \\ q & r & s \\ m & t & n \end{pmatrix} = \begin{pmatrix} 1 & 0 & 0 \\ 0 & 1 & 0 \\ 0 & 0 & 1 \end{pmatrix} \tag{4.2}$$

Going to the set of "tilde" generators we see that the co-products are more complicated. Namely

$$\begin{aligned}
\delta(\tilde{k}) &= \tilde{k}\otimes\tilde{k} + \tilde{n}\otimes\tilde{n} + \tilde{l}\otimes\tilde{l} - \tilde{m}\otimes\tilde{m} + \tilde{p}\otimes\tilde{q} + \tilde{t}\otimes\tilde{s}, \\
\delta(\tilde{n}) &= \tilde{k}\otimes\tilde{n} + \tilde{n}\otimes\tilde{k} - \tilde{l}\otimes\tilde{m} + \tilde{m}\otimes\tilde{l} + \tilde{p}\otimes\tilde{s} + \tilde{t}\otimes\tilde{q}, \\
\delta(\tilde{l}) &= \tilde{k}\otimes\tilde{l} + \tilde{n}\otimes\tilde{m} + \tilde{l}\otimes\tilde{k} - \tilde{m}\otimes\tilde{n} + \tilde{p}\otimes\tilde{q} - \tilde{t}\otimes\tilde{s}, \\
\delta(\tilde{m}) &= \tilde{k}\otimes\tilde{m} + \tilde{n}\otimes\tilde{l} - \tilde{l}\otimes\tilde{n} + \tilde{m}\otimes\tilde{k}\tilde{p}\otimes\tilde{s} + \tilde{t}\otimes\tilde{q}, \\
\delta(r) &= r\otimes r + 2\tilde{q}\otimes\tilde{p} + 2\tilde{s}\otimes\tilde{t}, \\
\delta(\tilde{p}) &= \tilde{k}\otimes\tilde{p} + \tilde{n}\otimes\tilde{t} + \tilde{l}\otimes\tilde{p} - \tilde{m}\otimes\tilde{t} + \tilde{p}\otimes r, \\
\delta(\tilde{t}) &= \tilde{k}\otimes\tilde{t} + \tilde{n}\otimes\tilde{p} - \tilde{l}\otimes\tilde{t} + \tilde{m}\otimes\tilde{p} + \tilde{t}\otimes r, \\
\delta(\tilde{q}) &= \tilde{q}\otimes\tilde{k} + \tilde{q}\otimes\tilde{l} + \tilde{s}\otimes\tilde{n} + \tilde{s}\otimes\tilde{m} + r\otimes\tilde{q}, \\
\delta(\tilde{s}) &= \tilde{q}\otimes\tilde{n} - \tilde{q}\otimes\tilde{m} + \tilde{s}\otimes\tilde{k} - \tilde{s}\otimes\tilde{m} + r\otimes\tilde{s}.
\end{aligned} \tag{4.3}$$

However, looking at the last four equations we see that they can be essentially simplified we introduce the new "hat" generators:

$$\tilde{k} = \hat{k} + \hat{l}, \ \tilde{l} = \hat{k} - \hat{l}; \ \ \tilde{m} = \hat{m} - \hat{n}, \ \tilde{n} = \hat{m} + \hat{n}. \tag{4.4}$$

Journal of Physics: Conference Series **254** (2010) 012001 doi:10.1088/1742-6596/254/1/012001

Thus we have:

$$\begin{aligned}
\delta(\hat{k}) &= 2\hat{k} \otimes \hat{k} - 2\hat{m} \otimes \hat{n} + \tilde{p} \otimes \tilde{q},\\
\delta(\hat{l}) &= 2\hat{l} \otimes \hat{l} - 2\hat{n} \otimes \hat{m} + \tilde{t} \otimes \tilde{s},\\
\delta(\hat{m}) &= 2\hat{k} \otimes \hat{m} - 2\hat{m} \otimes \hat{l} - \tilde{p} \otimes \tilde{s},\\
\delta(\hat{n}) &= 2\hat{l} \otimes \hat{n} + 2\hat{n} \otimes \hat{k} + \tilde{t} \otimes \tilde{q},\\
\delta(\hat{p}) &= 2\hat{k} \otimes \hat{p} - 2\hat{m} \otimes \hat{t} + \tilde{p} \otimes r,\\
\delta(\hat{q}) &= 2\hat{q} \otimes \hat{k} + 2\hat{s} \otimes \hat{n} + r \otimes \tilde{q},\\
\delta(\hat{s}) &= 2\hat{s} \otimes \hat{l} - 2\hat{q} \otimes \hat{m} + r \otimes \tilde{s},\\
\delta(\hat{t}) &= 2\hat{l} \otimes \hat{t} + 2\hat{n} \otimes \hat{p} + \tilde{t} \otimes r,\\
\delta(r) &= r \otimes r + 2\tilde{q} \otimes \tilde{p} + 2\tilde{s} \otimes \tilde{t}.
\end{aligned} \tag{4.5}$$

With the "hat" generators we have

$$\begin{aligned}
\epsilon(\hat{k}) = \epsilon(\hat{l}) = 1/2,\quad \epsilon(r) = 1,\\
\epsilon(z) = 0,\ \text{for } z = (\hat{m}, \hat{n}, \tilde{p}, \tilde{q}, \tilde{s}, \tilde{t}).
\end{aligned} \tag{4.6}$$

The bialgebra relations are as follows:

$$\begin{aligned}
\hat{k}\hat{m} = \hat{m}\hat{k} = \hat{k}\hat{n} = \hat{n}\hat{k} = 0;\ \hat{k}\tilde{t} = \hat{k}\tilde{s} = \tilde{p}\hat{k} = \tilde{q}\hat{k} = 0;\\
\hat{l}\hat{m} = \hat{m}\hat{l} = \hat{l}\hat{n} = \hat{n}\hat{l} = 0;\ \hat{l}\tilde{t} = \hat{l}\tilde{s} = \tilde{p}\hat{l} = \tilde{q}\hat{l} = 0;\\
r\hat{m} = \hat{m}r = r\hat{n} = \hat{n}r = 0;\quad r\tilde{t} = r\tilde{s} = \tilde{p}r = \tilde{q}r = 0;\\
\tilde{s}\tilde{t} = \tilde{t}\tilde{s} = 0;\quad \tilde{s}\hat{m} = \tilde{s}\hat{n} = \tilde{t}\hat{m} = \tilde{t}\hat{n} = 0;\\
\tilde{p}\tilde{q} = \tilde{q}\tilde{p} = 0;\quad \hat{m}\tilde{p} = \hat{n}\tilde{p} = \hat{m}\tilde{q} = \hat{n}\tilde{q} = 0;\\
\tilde{p}^2 = \tilde{q}^2 = \tilde{s}^2 = \tilde{t}^2 = 0.
\end{aligned} \tag{4.7}$$

The dual elements are defined by the standard procedure of [12], namely:

$$\langle Z, f\rangle = \epsilon\left(\frac{\partial f}{\partial z}\right), \quad \text{where} \quad z = (\hat{k}, \hat{l}, \hat{m}, \hat{n}, \tilde{p}, \tilde{q}, \tilde{s}, \tilde{t}, r). \tag{4.8}$$

The basis we are working is essentially the following

$$\begin{aligned}
\hat{k}^{\kappa}\hat{l}^{\ell}r^{\tau}, &\quad \text{and all permutations of } (\hat{k}\hat{l}r),\\
\hat{k}^{\kappa}\hat{l}^{\ell}r^{\tau}\tilde{p}, &\quad \text{and all permutations of } (\hat{k}\hat{l}r),\\
\hat{k}\hat{l}r)\tilde{q}, &\quad \text{and all permutations of } (\hat{k}\hat{l}r),\\
\tilde{s}\hat{k}^{\kappa}\hat{l}^{\ell}r^{\tau}, &\quad \text{and all permutations of } (\hat{k}\hat{l}r),\\
\tilde{t}\hat{k}^{\kappa}\hat{l}^{\ell}r^{\tau} &\quad \text{and all permutations of } (\hat{k}\hat{l}r),\\
&\quad \hat{m},\ \hat{n}.
\end{aligned} \tag{4.9}$$

Thus the following dual bialgebra is obtained:

$$\begin{aligned}
\left[\hat{K}^{\kappa}, \hat{L}^{\ell}\right] &= 0, \\
\hat{K}^{\kappa}\hat{M} = 2^{\kappa}\hat{M}, \quad \hat{N}\hat{K}^{\kappa} = 2^{\kappa}\hat{N}, \quad \hat{M}\hat{K} = \hat{K}\hat{N} &= 0, \\
\hat{M}\hat{L}^{\ell} = 2^{\ell}\hat{M}, \quad \hat{L}^{\ell}\hat{N} = 2^{\ell}\hat{N}, \quad \hat{N}\hat{L} = \hat{L}\hat{M} &= 0, \\
\hat{M}^2 = \hat{N}^2 = 0, \quad \hat{M}\hat{N} = -2\hat{K}, \quad \hat{N}\hat{M} &= -2\hat{L}, \\
\left[\hat{K}, \tilde{P}\right] = 2\tilde{P}, \quad \left[\hat{L}, \tilde{P}\right] = 0, \quad \left[\hat{R}, \tilde{P}\right] &= -\tilde{P}, \\
\left[\hat{K}, \tilde{Q}\right] = -2\tilde{Q}, \quad \left[\hat{L}, \tilde{Q}\right] = 0, \quad \left[\hat{R}, \tilde{Q}\right] &= \tilde{Q}, \\
\left[\hat{K}, \tilde{S}\right] = 0, \quad \left[\hat{L}, \tilde{S}\right] = -2\tilde{S}, \quad \left[\hat{R}, \tilde{S}\right] &= \tilde{S}, \\
\left[\hat{K}, \tilde{T}\right] = 0, \quad \left[\hat{L}, \tilde{T}\right] = 2\tilde{T}, \quad \left[\hat{R}, \tilde{T}\right] &= -\tilde{T}, \\
\hat{M}\tilde{T} = -2\tilde{P}, \quad \tilde{T}\hat{M} = 0, \quad \tilde{Q}\hat{M} = -2\tilde{S}, \quad \hat{M}\tilde{Q} &= 0, \\
\hat{N}\tilde{P} = 2\tilde{T}, \quad \tilde{P}\hat{N} = 0, \quad \tilde{S}\hat{N} = 2\tilde{Q}, \quad \hat{N}\tilde{S} &= 0, \\
\hat{M}\tilde{P} = \tilde{P}\hat{M} = \hat{M}\tilde{S} = \tilde{S}\hat{M} &= 0, \\
\hat{N}\tilde{Q} = \tilde{Q}\hat{N} = \hat{N}\tilde{T} = \tilde{T}\hat{N} &= 0, \\
\left[\tilde{S}, \tilde{P}\right] = \hat{M}, \quad \left[\tilde{Q}, \tilde{T}\right] = \hat{N}, \quad \left[\tilde{Q}, \tilde{S}\right] = \left[\tilde{P}, \tilde{T}\right] &= 0, \\
\tilde{P}\tilde{Q} = \tilde{T}\tilde{S}, \quad \tilde{Q}\tilde{P} &= \tilde{S}\tilde{T}, \\
\tilde{P}^2 = \tilde{Q}^2 = \tilde{S}^2 = \tilde{T}^2 = 0, \quad \tilde{P}\tilde{T} = \tilde{T}\tilde{P} = \tilde{Q}\tilde{S} = \tilde{S}\tilde{Q} &= 0.
\end{aligned} \tag{4.10}$$

Finally we write down the co-products of the dual bialgebra:

$$\begin{aligned}
\delta(\hat{K}) &= \hat{K} \otimes 1_U + 1_U \otimes \hat{K}, \\
\delta(\hat{L}) &= \hat{L} \otimes 1_U + 1_U \otimes \hat{L}, \\
\delta(\hat{M}) &= \hat{M} \otimes 1_U + 1_U \otimes \hat{M}, \\
\delta(\hat{N}) &= \hat{N} \otimes 1_U + 1_U \otimes \hat{N}, \\
\delta(\tilde{P}) &= \tilde{P} \otimes 1_U + 1_U \otimes \tilde{P}, \\
\delta(\tilde{Q}) &= \tilde{Q} \otimes 1_U + 1_U \otimes \tilde{Q}, \\
\delta(\tilde{S}) &= \tilde{S} \otimes 1_U + 1_U \otimes \tilde{S}, \\
\delta(\tilde{T}) &= \tilde{T} \otimes 1_U + 1_U \otimes \tilde{T}, \\
\delta(R) &= R \otimes 1_U + 1_U \otimes R.
\end{aligned} \tag{4.11}$$

.

Journal of Physics: Conference Series **254** (2010) 012001 doi:10.1088/1742-6596/254/1/012001

5. Generalized R matrix

We propose now a more general 9×9 R matrix with 16 parameters:

$$R_{gen} = \begin{pmatrix} a_1^+ & 0 & 0 & 0 & 0 & 0 & 0 & 0 & a_1^- \\ 0 & 0 & 0 & a_2^+ & 0 & a_2^- & 0 & 0 & 0 \\ 0 & 0 & b_1^+ & 0 & 0 & 0 & b_1^- & 0 & 0 \\ 0 & b_2^+ & 0 & 0 & 0 & 0 & 0 & b_2^- & 0 \\ 0 & 0 & 0 & 0 & 1 & 0 & 0 & 0 & 0 \\ 0 & c_1^+ & 0 & 0 & 0 & 0 & 0 & c_1^- & 0 \\ 0 & 0 & c_2^+ & 0 & 0 & 0 & c_2^- & 0 & 0 \\ 0 & 0 & 0 & d_1^+ & 0 & d_1^- & 0 & 0 & 0 \\ d_2^+ & 0 & 0 & 0 & 0 & 0 & 0 & 0 & d_2^- \end{pmatrix}. \tag{5.1}$$

The eight-parameter R-matrix in (2.4) is obviously a special case of the above. Substituting (5.1) in the constant YBE we obtain the equation:

$$(a_1^+ - b_1^-)(b_1^+ - c_2^-) = 0 \tag{5.2}$$

We take the first possible solution

$$a_1^+ - b_1^- = 0 \tag{5.3}$$

The other possibility we leave for future investigations. With that choice we obtain the following unconditional relations:

$$a_1^+ = b_1^- = c_2^+ = d_2^- = a_2^+ = b_2^+ = c_1^- = d_1^- = 1/2. \tag{5.4}$$

These relations are just the same as for the corresponding elements of the previous R-matrix. However, for the remaining elements of (5.1) we obtain a large variety of possible relations. They read as follows:

$$1a)\ a_1^- = b_1^+ = c_2^- = d_2^+ = 1/2;\ \ a_2^- = b_2^- = c_1^+ = d_1^+ = \pm 1/2 \tag{5.5}$$
$$1b)\ a_1^- = b_1^+ = c_2^- = d_2^+ = 1/2;\ \ a_2^- = -b_2^- = -c_1^+ = d_1^+ = \pm 1/2 \tag{5.6}$$
$$1c)\ a_1^- = b_1^+ = c_2^- = d_2^+ = -1/2;\ \ a_2^- = b_2^- = c_1^+ = d_1^+ = \pm 1/2 \tag{5.7}$$
$$1d)\ a_1^- = b_1^+ = c_2^- = d_2^+ = -1/2;\ \ a_2^- = -b_2^- = -c_1^+ = d_1^+ = \pm 1/2 \tag{5.8}$$

$$2a)\ a_1^- = b_1^+ = c_2^- = d_2^+ = 1/2;\ \ a_2^- = b_2^- = c_1^+ = d_1^+ = \pm i/2 \tag{5.9}$$
$$2b)\ a_1^- = b_1^+ = c_2^- = d_2^+ = 1/2;\ \ a_2^- = -b_2^- = -c_1^+ = d_1^+ = \pm i/2 \tag{5.10}$$
$$2c)\ a_1^- = b_1^+ = c_2^- = d_2^+ = -1/2;\ \ a_2^- = b_2^- = c_1^+ = d_1^+ = \pm i/2 \tag{5.11}$$
$$2d)\ a_1^- = b_1^+ = c_2^- = d_2^+ = -1/2;\ \ a_2^- = -b_2^- = -c_1^+ = d_1^+ = \pm i/2 \tag{5.12}$$

$$3a)\ a_1^- = ib_1^+ = ic_2^- = -d_2^+ = i/2;\ \ a_2^- = b_2^- = ic_1^+ = id_1^+ = \pm 1/2q_+ \tag{5.13}$$
$$3b)\ a_1^- = ib_1^+ = ic_2^- = -d_2^+ = -i/2;\ \ a_2^- = b_2^- = ic_1^+ = id_1^+ = \pm 1/2q_+ \tag{5.14}$$
$$3c)\ a_1^- = ib_1^+ = ic_2^- = -d_2^+ = i/2;\ \ a_2^- = -b_2^- = -ic_1^+ = id_1^+ = \pm 1/2q_+ \tag{5.15}$$
$$3d)\ a_1^- = ib_1^+ = ic_2^- = -d_2^+ = -i/2;\ \ a_2^- = -b_2^- = -ic_1^+ = id_1^+ = \pm 1/2q_+ \tag{5.16}$$

Here we have denoted

$$q_+ = \sqrt{i} = \sqrt{2}/2(1+i),\ \ q_- = \sqrt{-i} = \sqrt{(2)}/2(1-i) \tag{5.17}$$

Next we have:

$$\begin{aligned}
4a)\ & a_1^- = -ib_1^+ = -ic_2^- = -d_2^+ = i/2;\ \ a_2^- = b_2^- = -ic_1^+ = -id_1^+ = \pm 1/2q_- && (5.18)\\
4b)\ & a_1^- = -ib_1^+ = -ic_2^- = -d_2^+ = i/2;\ \ a_2^- = b_2^- = -ic_1^+ = -id_1^+ = \pm 1/2q_- && (5.19)\\
4c)\ & a_1^- = -ib_1^+ = -ic_2^- = -d_2^+ = i/2;\ \ a_2^- = -b_2^- = ic_1^+ = -id_1^+ = \pm 1/2q_- && (5.20)\\
4d)\ & a_1^- = -ib_1^+ = -ic_2^- = -d_2^+ = -i/2;\ \ a_2^- = -b_2^- = ic_1^+ = -id_1^+ = \pm 1/2q_- && (5.21)
\end{aligned}$$

In fact the bialgebras obtained from the R matrices $1a - 1d$, $2a - 2d$, $3a - 3d$ and $4a - 4d$ are equivalent to those described in section 3 as follows:

$$\mathcal{A}_{++} \cong 1a+ \cong 1a- \cong 2c+ \cong 2c- \cong 3a+ \cong 3a- \cong 4b+ \cong 4b- \qquad (5.22)$$

$$\mathcal{A}_{+-} \cong 1b+ \cong 1b- \cong 2d+ \cong 2d- \cong 3c+ \cong 3c- \cong 4d+ \cong 4d- \qquad (5.23)$$

$$\mathcal{A}_{-+} \cong 1d+ \cong 1d- \cong 2b+ \cong 2b- \cong 3d+ \cong 3d- \cong 4c+ \cong 4c- \qquad (5.24)$$

$$\mathcal{A}_{--} \cong 1c+ \cong 1c- \cong 2a+ \cong 2a- \cong 3b+ \cong 3b- \cong 4a+ \cong 4a- \qquad (5.25)$$

Further we present the solutions which give different exotic types of algebras from the above. First of all we give the full list of solutions for the above R-matrix:

$$\begin{aligned}
5a)\ & a_1^- = -ib_1^+ = -ic_2^- = -d_2^+ = i/2;\ \ a_2^- = b_2^- = ic_1^+ = id_1^+ = \pm 1/2q_+\\
5b)\ & a_1^- = -ib_1^+ = -ic_2^- = -d_2^+ = -i/2;\ \ a_2^- = b_2^- = ic_1^+ = id_1^+ = \pm 1/2q_+\\
5c)\ & a_1^- = -ib_1^+ = -ic_2^- = -d_2^+ = i/2;\ \ a_2^- = -b_2^- = -ic_1^+ = id_1^+ = \pm 1/2q_+\\
5d)\ & a_1^- = -ib_1^+ = -ic_2^- = -d_2^+ = -i/2;\ \ a_2^- = -b_2^- = -ic_1^+ = id_1^+ = \pm 1/2q_+. && (5.26)
\end{aligned}$$

$$\begin{aligned}
6a)\ & a_1^- = ib_1^+ = ic_2^- = -d_2^+ = i/2;\ \ a_2^- = b_2^- = -ic_1^+ = -id_1^+ = \pm 1/2q_-\\
6b)\ & a_1^- = ib_1^+ = ic_2^- = -d_2^+ = -i/2;\ \ a_2^- = b_2^- = -ic_1^+ = -id_1^+ = \pm 1/2q_-\\
6c)\ & a_1^- = ib_1^+ = ic_2^- = -d_2^+ = i/2;\ \ a_2^- = -b_2^- = ic_1^+ = -id_1^+ = \pm 1/2q_-\\
6d)\ & a_1^- = ib_1^+ = ic_2^- = -d_2^+ = i/2;\ \ a_2^- = -b_2^- = ic_1^+ = -id_1^+ = \pm 1/2q_- && (5.27)
\end{aligned}$$

$$\begin{aligned}
7a+)\ & a_1^- = -ib_1^+ = ic_2^- = d_2^+ = 1/2,\ a_2^- = -ib_2^- = c_1^+ = -id_1^+ = \pm 1/2q_-.\\
7a-)\ & a_1^- = -ib_1^+ = ic_2^- = d_2^+ = -1/2,\ a_2^- = -ib_2^- = c_1^+ = -id_1^+ = \pm 1/2q_-.\\
7b+)\ & a_1^- = -ib_1^+ = ic_2^- = d_2^+ = 1/2,\ a_2^- = ib_2^- = -c_1^+ = -id_1^+ = \pm 1/2q_-.\\
7b-)\ & a_1^- = -ib_1^+ = ic_2^- = d_2^+ = -1/2,\ a_2^- = ib_2^- = -c_1^+ = -id_1^+ = \pm 1/2q_-.\\
7c+)\ & a_1^- = ib_1^+ = -ic_2^- = d_2^+ = 1/2,\ a_2^- = -ib_2^- = c_1^+ = -id_1^+ = \pm 1/2q_-.\\
7c-)\ & a_1^- = -ib_1^+ = ic_2^- = d_2^+ = -1/2,\ a_2^- = -ib_2^- = c_1^+ = -id_1^+ = \pm 1/2q_-.\\
7d+)\ & a_1^- = ib_1^+ = -ic_2^- = d_2^+ = 1/2,\ a_2^- = ib_2^- = -c_1^+ = -id_1^+ = \pm 1/2q_-.\\
7d-)\ & a_1^- = -ib_1^+ = ic_2^- = d_2^+ = -1/2,\ a_2^- = ib_2^- = c_1^+ = -id_1^+ = \pm 1/2q_-. && (5.28)
\end{aligned}$$

$$\begin{aligned}
8a+)\ & a_1^- = ib_1^+ = -ic_2^- = d_2^+ = 1/2,\ a_2^- = ib_2^- = c_1^+ = id_1^+ = \pm 1/2q_+\\
8a-)\ & a_1^- = ib_1^+ = -ic_2^- = d_2^+ = -1/2,\ a_2^- = ib_2^- = c_1^+ = id_1^+ = \pm 1/2q_+\\
8b+)\ & a_1^- = ib_1^+ = -ic_2^- = d_2^+ = 1/2,\ a_2^- = -ib_2^- = -c_1^+ = id_1^+ = \pm 1/2q_+\\
8b-)\ & a_1^- = ib_1^+ = -ic_2^- = d_2^+ = -1/2,\ a_2^- = -ib_2^- = -c_1^+ = id_1^+ = \pm 1/2q_+\\
8c+)\ & a_1^- = -ib_1^+ = ic_2^- = d_2^+ = 1/2,\ a_2^- = ib_2^- = c_1^+ = id_1^+ = \pm 1/2q_+\\
8c-)\ & a_1^- = -ib_1^+ = ic_2^- = d_2^+ = -1/2,\ a_2^- = ib_2^- = c_1^+ = id_1^+ = \pm 1/2q_+\\
8d+)\ & a_1^- = -ib_1^+ = ic_2^- = d_2^+ = 1/2,\ a_2^- = -ib_2^- = -c_1^+ = id_1^+ = \pm 1/2q_+\\
8d-)\ & a_1^- = -ib_1^+ = ic_2^- = d_2^+ = -1/2,\ a_2^- = -ib_2^- = -c_1^+ = id_1^+ = \pm 1/2q_+ && (5.29)
\end{aligned}$$

$$
\begin{aligned}
9a)\ & a_1^- = -b_1^+ = c_2^- = -d_2^+ = a_2^- = -d_1^+ = i/2,\ b_2^- = c_1^+ = \pm 1/2,\\
9b)\ & a_1^- = -b_1^+ = c_2^- = -d_2^+ = a_2^- = -d_1^+ = -i/2,\ b_2^- = c_1^+ = \pm 1/2,\\
9c)\ & a_1^- = -b_1^+ = c_2^- = -d_2^+ = -a_2^- = d_1^+ = i/2,\ b_2^- = c_1^+ = \pm 1/2,\\
9d)\ & a_1^- = -b_1^+ = c_2^- = -d_2^+ = -a_2^- = d_1^+ = -i/2,\ b_2^- = c_1^+ = \pm 1/2.
\end{aligned} \tag{5.30}
$$

$$
\begin{aligned}
10a)\ & a_1^- = b_1^+ = -c_2^+ = -d_2^+ = b_2^- = -c_1^+ = i/2, a_2^- = d_1^+ = \pm 1/2,\\
10b)\ & a_1^- = b_1^+ = -c_2^+ = -d_2^+ = -b_2^- = c_1^+ = i/2, a_2^- = d_1^+ = \pm 1/2,\\
10c)\ & a_1^- = b_1^+ = -c_2^+ = -d_2^+ = b_2^- = -c_1^+ = i/2, a_2^- = d_1^+ = \mp 1/2,\\
10d)\ & a_1^- = b_1^+ = -c_2^+ = -d_2^+ = -b_2^- = c_1^+ = i/2, a_2^- = d_1^+ = \mp 1/2
\end{aligned} \tag{5.31}
$$

However the algebras that are generated by these solutions are not all independent. Next we shall give only the independent ones and the substitutions in the local "tilde" frame of generators (to be introduced in what follows) which show the equivalence of some bialgebras.

- $5a$ We have for that case the following equivalences:

$$
\begin{gathered}
5a+ \equiv 5a-,\ with\ \tilde{l} \leftrightarrow \tilde{m},\ \tilde{p} \leftrightarrow \tilde{t},\\
5a+ \equiv 6b+, with\ \tilde{l} \leftrightarrow \tilde{m},\ 5a+ \equiv 5b-,\ with\ \tilde{p} \leftrightarrow \tilde{t}.
\end{gathered} \tag{5.32}
$$

- $5b$

$$
\begin{aligned}
5b+ &\equiv 5b-,\ with\ \tilde{l} \leftrightarrow \tilde{m},\ \tilde{p} \leftrightarrow \tilde{t},\\
5b+ &\equiv 6a+, with\ \tilde{l} \leftrightarrow \tilde{m},\\
5b+ &\equiv 6a-,\ with\ \tilde{p} \leftrightarrow \tilde{t}.
\end{aligned} \tag{5.33}
$$

- $5c$

$$
\begin{aligned}
5c+ &\equiv 5c-,\ with\ \tilde{l} \leftrightarrow \tilde{m},\ \tilde{p} \leftrightarrow \tilde{t},\\
5c+ &\equiv 6d+, with\ \tilde{l} \leftrightarrow \tilde{m},\\
5c+ &\equiv 6d-,\ with\ \tilde{p} \leftrightarrow \tilde{t}.
\end{aligned} \tag{5.34}
$$

- $5d$

$$
\begin{aligned}
5d+ &\equiv 5d-,\ with\ \tilde{l} \leftrightarrow \tilde{m},\ \tilde{p} \leftrightarrow \tilde{t},\\
5d+ &\equiv 6c+, with\ \tilde{l} \leftrightarrow \tilde{m},\\
5d+ &\equiv 6c-,\ with\ \tilde{p} \leftrightarrow \tilde{t}.
\end{aligned} \tag{5.35}
$$

- $7a+$

$$
\begin{aligned}
7a+ &\equiv 7b-,\ with\ \tilde{t} \to -i\tilde{t},\ \tilde{q} \to -i\tilde{q}, \tilde{l} \to -\tilde{m},\ \tilde{n} \to -\tilde{n},\\
7a+ &\equiv 8a+,\ with\ \tilde{p} \to \tilde{t},\ \tilde{t} \to i\tilde{p}, \tilde{s} \to \tilde{q},\ \tilde{q} \to i\tilde{s},\\
7a+ &\equiv 8b-,\ with\ \tilde{p} \to i\tilde{t},\ \tilde{t} \to \tilde{p},\ \tilde{s} \to i\tilde{q},\ \tilde{q} \to \tilde{s}.
\end{aligned} \tag{5.36}
$$

- $7a-$

$$
\begin{aligned}
7a- &\equiv 7b+,\ with\ \tilde{t} \to -i\tilde{t},\ \tilde{q} \to -i\tilde{q}, \tilde{l} \to -\tilde{m},\ \tilde{n} \to -\tilde{n},\\
7a- &\equiv 8a-,\ with\ \tilde{p} \to \tilde{t},\ \tilde{t} \to i\tilde{p}, \tilde{s} \to \tilde{q},\ \tilde{q} \to i\tilde{s},\\
7a+ &\equiv 8b+,\ with\ \tilde{p} \to i\tilde{t},\ \tilde{t} \to \tilde{p},\ \tilde{s} \to i\tilde{q},\ \tilde{q} \to \tilde{s}
\end{aligned} \tag{5.37}
$$

- $7c+$

$$\begin{aligned} 7c+ &\equiv 7d-, \ with\ \tilde{t}\to -i\tilde{t},\ \tilde{q}\to -i\tilde{q}, \\ 7c+ &\equiv 8c+,\ with\ \tilde{p}\to \tilde{t},\ \tilde{t}\to i\tilde{p},\ \tilde{s}\to\tilde{q},\ \tilde{q}\to i\tilde{s},\ \tilde{l}\leftrightarrow\tilde{m}, \\ 7c+ &\equiv 8d-,\ with\ \tilde{p}\leftrightarrow\tilde{t},\ \tilde{p}\leftrightarrow\tilde{t},\ \tilde{k}\leftrightarrow\tilde{n} \end{aligned} \tag{5.38}$$

- $7c-$

$$\begin{aligned} 7c- &\equiv 7d+,\ with\ \tilde{t}\to -i\tilde{t},\ \tilde{q}\to -i\tilde{q}, \\ 7c- &\equiv 8c-,\ with\ \tilde{p}\to \tilde{t},\ \tilde{t}\to i\tilde{p},\ \tilde{s}\to\tilde{q},\ \tilde{q}\to i\tilde{s},\ \tilde{l}\leftrightarrow\tilde{m}, \\ 7c- &\equiv 8d+,\ with\ \tilde{p}\leftrightarrow\tilde{t},\ \tilde{p}\leftrightarrow\tilde{t},\ \tilde{k}\leftrightarrow\tilde{n} \end{aligned} \tag{5.39}$$

- $9a+$

$$\begin{aligned} 9a+ &\equiv 9c-, \\ 9a+ &\equiv 10a+ \equiv 10c-,\ with\ \tilde{t}\to i\tilde{t},\ \tilde{q}\to i\tilde{q}, \\ &\tilde{l}\to\tilde{m},\ \tilde{m}\to -\tilde{l},\ \tilde{n}\to -\tilde{n} \\ 9a+ &\equiv 7a+,\ with\tilde{t}\to -i\tilde{t},\ \tilde{q}\to -i\tilde{q},\ \tilde{m}\to -\tilde{m},\ \tilde{n}\to -\tilde{n}. \end{aligned} \tag{5.40}$$

- $9a-$

$$\begin{aligned} 9a- &\equiv 9c+, \\ 9a- &\equiv 10a- \equiv 10c+,\ with\ \tilde{t}\to i\tilde{t},\ \tilde{q}\to i\tilde{q},\ \tilde{l}\to\tilde{m}, \\ &\tilde{m}\to -\tilde{l},\ \tilde{n}\to -\tilde{n} \\ 9a+ &\equiv 7a+,\ with\tilde{t}\to -i\tilde{t},\ \tilde{q}\to -i\tilde{q},\ \tilde{m}\to -\tilde{m},\ \tilde{n}. \end{aligned} \tag{5.41}$$

- $9b+$

$$\begin{aligned} 9b+ &=\equiv 9d-,\ 9b+ \equiv 10b- \equiv 10d+,\ with\ \tilde{l}\to\tilde{m}, \\ 9b+ &\equiv 9a+,\ with\tilde{t}\to i\tilde{t},\ \tilde{q}\to i\tilde{q},\ \tilde{m}\to -\tilde{m},\ \tilde{n}. \end{aligned} \tag{5.42}$$

- $9b-$

$$\begin{aligned} 9b- &\equiv 9d+,\ 9b- \equiv 10b+ \equiv 10d-,\ with\ \tilde{l}\to\tilde{m}, \\ 9b- &\equiv 9a-,\ with\tilde{p}\to i\tilde{p},\ \tilde{s}\to i\tilde{s},\ \tilde{m}\to -\tilde{m},\ \tilde{n}. \end{aligned} \tag{5.43}$$

So in fact we have the following set of bialgebras which we denote by:

- $$B1+ \equiv 5a+ \equiv 5a- \equiv 6b+ \equiv 6b-. \tag{5.44}$$

- $$B1- \equiv 5b+ \equiv 5b- \equiv 6a+ \equiv 6a-. \tag{5.45}$$

- $$B2+ \equiv 5c+ \equiv 5c- \equiv 6d+ \equiv 6d-. \tag{5.46}$$

- $$B2- \equiv 5d+ \equiv 5d- \equiv 6c+ \equiv 6c-. \tag{5.47}$$

- $$B3+ \equiv 7a+ \equiv 7b- \equiv 8a+ \equiv 8b- \equiv 9a+ \equiv\equiv 9b+ \equiv 10a+ \equiv 10c-\,. \tag{5.48}$$
- $$B3- \equiv 7a- \equiv 7b+ \equiv 8a- \equiv 8b+ \equiv 9a- \equiv 9b+ \equiv 10a+ \equiv 10c-\,. \tag{5.49}$$
- $$B4+ \equiv 7c+ \equiv 7d- \equiv 8c+ \equiv 8d-\,. \tag{5.50}$$
- $$B4- \equiv 7c- \equiv 7d+ \equiv 8c- \equiv 8d+\,. \tag{5.51}$$

Now we finally give the explicit forms of the independent bialgebras:

- $B1+$

$$\begin{aligned}
k^2 = n^2,\ kn = nk;\ l^2 = -m^2,\ lm = ml;\\
kl = -inm,\ nl = -ikm;\ lk = -imn, \ln = -imk;\\
p^2 = it^2,\ pt = -tp;\ s^2 = iq^2,\ sq = -qs;\\
ps = itq,\ pq = ts;\ sp = iqt,\ st = qp;\\
(k-n)r = r(k-n) = (l-im)r = r(l-im) = 0;\\
(p \mp q_+ t)r = r(p \mp q_+ t) = (s \mp q_+ q)r = r(s \mp q_+ q) = 0;\\
tm = \mp q_+ pl,\ tl = \mp q_+ pm;\ mt = \mp q_+ lp,\ lt = \mp q_+ mp;\\
pk = \pm q_+ tn,\ pn = \mp q_+ tk;\ kp = \pm q_+ nt,\ np = \mp q_+ kt;\\
qm = \mp q_+ sl,\ ql = \mp q_+ sm;\ mq = \mp q_+ ls,\ lq = \mp q_+ ms;\\
ks = \mp q_+ nq,\ ns = \pm q_+ kq; sk = \mp q_+ qn,\ sn = \pm q_+ qk.
\end{aligned} \tag{5.52}$$

As seen these formulae in fact define two algebras – the upper and lower signs of $\pm$ and $\mp$ respectively. Now we introduce the generators:

$$\begin{aligned}
k = \tilde{k} + \tilde{n},\ n = \tilde{k} - \tilde{n}; l = \tilde{l} + \tilde{m},\ m = -i(\tilde{l} - \tilde{m});\\
p = \tilde{p} + \tilde{t},\ t = q_-(\tilde{p} - \tilde{t}); s = \tilde{s} + \tilde{q},\ q = q_-(\tilde{s} - \tilde{q}).
\end{aligned} \tag{5.53}$$

So we have:

$$\begin{aligned}
\tilde{k}\tilde{n} = \tilde{n}\tilde{k} = \tilde{l}\tilde{m} = \tilde{m}\tilde{l} = 0;\ \ \tilde{k}\tilde{l} = \tilde{l}\tilde{k} = \tilde{n}\tilde{m} = \tilde{m}\tilde{n} = 0;\\
\tilde{p}^2 = \tilde{t}^2,\ \tilde{p}\tilde{t} = -\tilde{t}\tilde{p};\ \tilde{s}^2 = \tilde{q}^2,\ \tilde{s}\tilde{q} = -\tilde{q}\tilde{s};\ \tilde{p}\tilde{q} = \tilde{q}\tilde{p} = \tilde{t}\tilde{s} = \tilde{s}\tilde{t} = 0;\\
\tilde{n}r = r\tilde{n} = \tilde{m}r = r\tilde{m} = 0;\\
\tilde{t}r = r\tilde{t} = \tilde{q}r = r\tilde{q} = 0;\\
\tilde{p}\tilde{k} = \tilde{t}\tilde{n},\ \tilde{t}\tilde{k} = -\tilde{p}\tilde{n},\ \tilde{k}\tilde{p} = \tilde{n}\tilde{t},\ \tilde{k}\tilde{t} = -\tilde{n}\tilde{p};\\
\tilde{p}\tilde{l} = \tilde{t}\tilde{m},\ \tilde{t}\tilde{l} = -\tilde{p}\tilde{m},\ \tilde{l}\tilde{p} = \tilde{m}\tilde{t},\ \tilde{l}\tilde{t} = -\tilde{m}\tilde{p};\\
\tilde{s}\tilde{k} = -\tilde{q}\tilde{n},\ \tilde{q}\tilde{k} = \tilde{s}\tilde{n},\ \tilde{k}\tilde{s} = -\tilde{n}\tilde{q},\ \tilde{k}\tilde{q} = \tilde{n}\tilde{s};\\
\tilde{s}\tilde{l} = \tilde{q}\tilde{m},\ \tilde{q}\tilde{l} = -\tilde{s}\tilde{m},\ \tilde{l}\tilde{s} = \tilde{m}\tilde{q},\ \tilde{l}\tilde{q} = -\tilde{m}\tilde{s}.
\end{aligned} \tag{5.54}$$

- $B1-$

$$\begin{aligned}
k^2 = n^2,\ kn = nk;\ l^2 = -m^2,\ lm = ml;\\
kl = -inm,\ nl = -ikm;\ lk = -imn, \ln = -imk;\\
p^2 = -it^2,\ pt = tp;\ s^2 = -iq^2,\ sq = qs;\\
ps = itq,\ pq = ts;\ sp = iqt,\ st = qp;\\
(k-n)r = r(k-n) = (l-im)r = r(l-im) = 0;\\
(p \mp q_+t)r = r(p \mp q_+t) = (s \mp q_+q)r = r(s \mp q_+q) = 0;\\
tm = \pm q_+pl,\ tl = \pm q_+pm;\ mt = \pm q_+lp,\ lt = \pm q_+mp;\\
pk = \mp q_+tn,\ pn = \pm q_+tk;\ kp = \mp q_+nt,\ np = \pm q_+kt;\\
qm = \pm q_+sl,\ ql = \pm q_+sm;\ mq = \pm q_+ls,\ lq = \pm q_+ms;\\
ks = \pm q_+nq,\ ns = \mp q_+kq; sk = \pm q_+qn,\ sn = \mp q_+qk.
\end{aligned} \tag{5.55}$$

Now introducing the same generators as in the previous case we have:

$$\begin{aligned}
\tilde{k}\tilde{n} = \tilde{n}\tilde{k} = \tilde{l}\tilde{m} = \tilde{m}\tilde{l} = 0;\ \ \tilde{k}\tilde{l} = \tilde{l}\tilde{k} = \tilde{n}\tilde{m} = \tilde{m}\tilde{n} = 0;\\
\tilde{p}^2 = -\tilde{t}^2,\ \tilde{p}\tilde{t} = \tilde{t}\tilde{p};\ \tilde{s}^2 = -\tilde{q}^2,\ \tilde{s}\tilde{q} = \tilde{q}\tilde{s};\ \tilde{p}\tilde{q} = \tilde{q}\tilde{p} = \tilde{t}\tilde{s} = \tilde{s}\tilde{t} = 0;\\
\tilde{n}r = r\tilde{n} = \tilde{m}r = r\tilde{m} = 0;\\
\tilde{t}r = r\tilde{t} = \tilde{q}r = r\tilde{q} = 0;\\
\tilde{p}\tilde{k} = -\tilde{t}\tilde{n},\ \tilde{t}\tilde{k} = \tilde{p}\tilde{n},\ \tilde{k}\tilde{p} = -\tilde{n}\tilde{t},\ \tilde{k}\tilde{t} = \tilde{n}\tilde{p};\\
\tilde{p}\tilde{l} = -\tilde{t}\tilde{m},\ \tilde{t}\tilde{l} = \tilde{p}\tilde{m},\ \tilde{l}\tilde{p} = -\tilde{m}\tilde{t},\ \tilde{l}\tilde{t} = \tilde{m}\tilde{p};\\
\tilde{s}\tilde{k} = \tilde{q}\tilde{n},\ \tilde{q}\tilde{k} = -\tilde{s}\tilde{n},\ \tilde{k}\tilde{s} = \tilde{n}\tilde{q},\ \tilde{k}\tilde{q} = -\tilde{n}\tilde{s};\\
\tilde{s}\tilde{l} = -\tilde{q}\tilde{m},\ \tilde{q}\tilde{l} = \tilde{s}\tilde{m},\ \tilde{l}\tilde{s} = -\tilde{m}\tilde{q},\ \tilde{l}\tilde{q} = \tilde{m}\tilde{s}.
\end{aligned} \tag{5.56}$$

- $B2+$

$$\begin{aligned}
k^2 = n^2,\ kn = nk;\ l^2 = -m^2,\ lm = ml;\\
kl = -inm,\ nl = -ikm;\ lk = -imn, \ln = -imk;\\
p^2 = it^2,\ pt = -tp;\ s^2 = iq^2,\ sq = -qs;\\
ps = -itq,\ pq = -ts;\ sp = -iqt,\ st = -qp;\\
(k-n)r = r(k-n) = (l-im)r = r(l-im) = 0;\\
(p \mp q_+t)r = r(p \pm q_+t) = (s \mp q_+q)r = r(s \pm q_+q) = 0;\\
tm = \pm q_+pl,\ tl = \pm q_+pm;\ mt = \mp q_+lp,\ lt = \mp q_+mp;\\
pk = \mp q_+tn,\ pn = \pm q_+tk;\ kp = \pm q_+nt,\ np = \mp q_+kt;\\
qm = \pm q_+sl,\ ql = \pm q_+sm;\ mq = \mp q_+ls,\ lq = \mp q_+ms;\\
ks = \mp q_+nq,\ ns = \pm q_+kq;\ sk = \pm q_+qn,\ sn = \mp q_+qk.
\end{aligned} \tag{5.57}$$

Introducing the same generators as in the previous cases we have:

$$\begin{aligned}
\tilde{k}\tilde{n} = \tilde{n}\tilde{k} = \tilde{l}\tilde{m} = \tilde{m}\tilde{l} = 0;\;\; \tilde{k}\tilde{l} = \tilde{l}\tilde{k} = \tilde{n}\tilde{m} = \tilde{m}\tilde{n} = 0;\\
\tilde{p}^2 = \tilde{t}^2,\; \tilde{p}\tilde{t} = -\tilde{t}\tilde{p},\; \tilde{s}^2 = \tilde{q}^2,\; \tilde{s}\tilde{q} = \tilde{q}\tilde{s};\; \tilde{p}\tilde{s} = \tilde{s}\tilde{p} = \tilde{t}\tilde{q} = \tilde{q}\tilde{t} = 0\\
\tilde{n}r = r\tilde{n} = \tilde{m}r = r\tilde{m} = 0;\\
\tilde{t}r = r\tilde{p} = \tilde{q}r = r\tilde{s} = 0;\\
\tilde{p}\tilde{k} = -\tilde{t}\tilde{n},\; \tilde{t}\tilde{k} = \tilde{p}\tilde{n},\; \tilde{k}\tilde{p} = \tilde{n}\tilde{t},\; \tilde{k}\tilde{t} = -\tilde{n}\tilde{p};\\
\tilde{p}\tilde{l} = -\tilde{t}\tilde{m},\; \tilde{t}\tilde{l} = \tilde{p}\tilde{m},\; \tilde{l}\tilde{p} = \tilde{m}\tilde{t},\; \tilde{l}\tilde{t} = -\tilde{m}\tilde{p};\\
\tilde{s}\tilde{k} = \tilde{q}\tilde{n},\; \tilde{q}\tilde{k} = -\tilde{s}\tilde{n},\; \tilde{k}\tilde{s} = -\tilde{n}\tilde{q},\; \tilde{k}\tilde{q} = \tilde{n}\tilde{s};\\
\tilde{s}\tilde{l} = -\tilde{q}\tilde{m},\; \tilde{q}\tilde{l} = \tilde{s}\tilde{m},\; \tilde{l}\tilde{s} = \tilde{m}\tilde{q},\; \tilde{l}\tilde{q} = -\tilde{m}\tilde{s}.
\end{aligned} \tag{5.58}$$

- $B2-$

$$\begin{aligned}
k^2 = n^2,\; kn = nk;\; l^2 = -m^2,\; lm = ml;\\
kl = -inm,\; nl = -ikm;\; lk = -imn, \ln = -imk;\\
p^2 = -it^2,\; pt = tp;\; s^2 = -iq^2,\; sq = qs;\\
ps = -itq,\; pq = -ts;\; sp = -iqt,\; st = -qp;\\
(k-n)r = r(k-n) = (l-im)r = r(l-im) = 0;\\
(p \mp q_+ t)r = r(p \pm q_+ t) = (s \mp q_+ q)r = r(s \pm q_+ q) = 0;\\
tm = \mp q_+ pl,\; tl = \mp q_+ pm;\; mt = \pm q_+ lp,\; lt = \pm q_+ mp;\\
pk = \pm q_+ tn,\; pn = \mp q_+ tk;\; kp = \mp q_+ nt,\; np = \pm q_+ kt;\\
qm = \mp q_+ sl,\; ql = \mp q_+ sm;\; mq = \pm q_+ ls,\; lq = \pm q_+ ms;\\
ks = \pm q_+ nq,\; ns = \mp q_+ kq; sk = \mp q_+ qn,\; sn = \pm q_+ qk.
\end{aligned} \tag{5.59}$$

Introducing the same generators as in the previous cases we have:

$$\begin{aligned}
\tilde{k}\tilde{n} = \tilde{n}\tilde{k} = \tilde{l}\tilde{m} = \tilde{m}\tilde{l} = 0;\;\; \tilde{k}\tilde{l} = \tilde{l}\tilde{k} = \tilde{n}\tilde{m} = \tilde{m}\tilde{n} = 0;\\
\tilde{p}^2 = -\tilde{t}^2,\; \tilde{s}^2 = -\tilde{q}^2;\; \tilde{p}\tilde{t} = \tilde{t}\tilde{p},\; \tilde{s}\tilde{q} = \tilde{q}\tilde{s};\; \tilde{p}\tilde{s} = \tilde{s}\tilde{p} = \tilde{t}\tilde{q} = \tilde{q}\tilde{t} == 0\\
\tilde{n}r = r\tilde{n} = \tilde{m}r = r\tilde{m} = 0;\\
\tilde{t}r = r\tilde{p} = \tilde{q}r = r\tilde{s} = 0;\\
\tilde{p}\tilde{k} = \tilde{t}\tilde{n},\; \tilde{t}\tilde{k} = -\tilde{p}\tilde{n},\; \tilde{k}\tilde{p} = -\tilde{n}\tilde{t},\; \tilde{k}\tilde{t} = \tilde{n}\tilde{p};\\
\tilde{p}\tilde{l} = \tilde{t}\tilde{m},\; \tilde{t}\tilde{l} = -\tilde{p}\tilde{m},\; \tilde{l}\tilde{p} = -\tilde{m}\tilde{t},\; \tilde{l}\tilde{t} = \tilde{m}\tilde{p};\\
\tilde{s}\tilde{k} = -\tilde{q}\tilde{n},\; \tilde{q}\tilde{k} = \tilde{s}\tilde{n},\; \tilde{k}\tilde{s} = \tilde{n}\tilde{q},\; \tilde{k}\tilde{q} = -\tilde{n}\tilde{s};\\
\tilde{s}\tilde{l} = \tilde{q}\tilde{m},\; \tilde{q}\tilde{l} = -\tilde{s}\tilde{m},\; \tilde{l}\tilde{s} = -\tilde{m}\tilde{q},\; \tilde{l}\tilde{q} = \tilde{m}\tilde{s}.
\end{aligned} \tag{5.60}$$

- $B3+$

$$\begin{aligned}
k^2 = n^2,\ kn = nk;\ l^2 = m^2,\ lm = -ml;\\
kl = -inm,\ nl = -ikm;\ lk = imn, \ln = imk;\\
p^2 = t^2,\ pt = itp;\ s^2 = q^2,\ sq = iqs;\\
ps = tq,\ pq = -its;\ sp = qt,\ st = -iqp;\\
(k-n)r = r(k-n) = (l+im)r = r(l-im) = 0;\\
(p \mp q_-t)r = r(p \mp q_+t) = (s \mp q_-q)r = r(s \mp q_+q) = 0;\\
tm = \pm q_-pl,\ tl = \pm q_-pm;\ lp = \pm q_-mt,\ mp = \pm q_-lt;\\
nt = \pm q_-kp,\ kt = \mp q_-np,\ pk = \pm q_-tn,\ pn = \mp q_-tk;\\
ls = \pm q_-mq,\ ms = \pm q_-lq,\ qm = \pm q_-sl,\ ql = \pm q_-sm;\\
nq = \mp q_-ks,\ kq = \pm q_-ns;\ sk = \mp q_-qn,\ sn = \pm q_-qk.
\end{aligned} \tag{5.61}$$

Now we see that the following new generators are naturally suggested

$$\begin{aligned}
k = \tilde{k} + \tilde{n},\ n = \tilde{k} - \tilde{n}; l = \tilde{l} + \tilde{m},\ m = -i(\tilde{l} - \tilde{m});\\
p = -i\sqrt{2}(q_+\tilde{p} - q_-\tilde{t}),\ t = \mp i\sqrt{2}(\tilde{p} - \tilde{t});\\
s = -i\sqrt{2}(q_+\tilde{s} - q_-\tilde{q}),\ q = \mp i\sqrt{2}(\tilde{s} - \tilde{q}).
\end{aligned} \tag{5.62}$$

In the last two equations the upper and lower signs of t and q are irrelevant since both cases lead to one and the same set of relations. So that we have

$$\begin{aligned}
\tilde{k}\tilde{n} = \tilde{n}\tilde{k} = \tilde{l}^2 = \tilde{m}^2 = 0;\ \tilde{k}\tilde{l} = \tilde{n}\tilde{m} = \tilde{l}\tilde{n} = \tilde{m}\tilde{k} = 0;\\
\tilde{p}^2 = -i\tilde{t}^2 = 1/2(1-i)\tilde{t}\tilde{p};\ \tilde{s}^2 = -i\tilde{q}^2 = 1/2(1-i)\tilde{q}\tilde{s};\\
\tilde{p}\tilde{s} = -i\tilde{t}\tilde{q} = 1/2(1-i)\tilde{p}\tilde{q};\ \tilde{s}\tilde{p} = -i\tilde{q}\tilde{t} = 1/2(1-i)\tilde{s}\tilde{t};\\
\tilde{n}r = r\tilde{n} = \tilde{l}r = r\tilde{m} = 0;\\
\tilde{p}r = r\tilde{t} = \tilde{s}r = r\tilde{q} = 0;\\
\tilde{t}\tilde{l} = -i\tilde{t}\tilde{m},\ \tilde{p}\tilde{l} = i\tilde{p}\tilde{m} + (1-i)\tilde{t}\tilde{l};\ \tilde{l}\tilde{p} = -i\tilde{m}\tilde{p},\ \tilde{l}\tilde{t} = i\tilde{m}\tilde{t} + (1+i)\tilde{l}\tilde{p};\\
\tilde{t}\tilde{k} = -i\tilde{t}\tilde{n},\ \tilde{p}\tilde{k} = i\tilde{p}\tilde{n} + (1-i)\tilde{t}\tilde{k};\ \tilde{k}\tilde{p} = i\tilde{n}\tilde{p},\ \tilde{k}\tilde{t} = -i\tilde{n}\tilde{t} + (1+i)\tilde{k}\tilde{p};\\
\tilde{q}\tilde{l} = -i\tilde{q}\tilde{m},\ \tilde{s}\tilde{l} = i\tilde{s}\tilde{m} + (1-i)\tilde{q}\tilde{l};\ \tilde{l}\tilde{s} = -i\tilde{m}\tilde{s},\ \tilde{l}\tilde{q} = i\tilde{m}\tilde{q} + (1+i)\tilde{l}\tilde{s};\\
\tilde{q}\tilde{k} = i\tilde{q}\tilde{n},\ \tilde{s}\tilde{k} = -i\tilde{s}\tilde{n} + (1-i)\tilde{q}\tilde{k};\ \tilde{k}\tilde{s} = -i\tilde{n}\tilde{s},\ \tilde{k}\tilde{q} = i\tilde{n}\tilde{q} + (1+i)\tilde{k}\tilde{s}.
\end{aligned} \tag{5.63}$$

- $B3-$

$$\begin{aligned}
k^2 = n^2,\ kn = nk;\ l^2 = m^2,\ lm = -ml;\\
kl = -inm,\ nl = -ikm;\ lk = imn, \ln = imk;\\
p^2 = -t^2,\ pt = -itp;\ s^2 = -q^2,\ sq = -iqs;\\
ps = tq,\ pq = -its;\ sp = qt,\ st = -iqp;\\
(k-n)r = r(k-n) = (l+im)r = r(l-im) = 0;\\
(p \mp q_-t)r = r(p \mp q_+t) = (s \mp q_-q)r = r(s \mp q_+q) = 0;\\
tm = \mp q_-pl,\ tl = \mp q_-pm;\ lp = \mp q_-mt,\ mp = \mp q_-lt;\\
nt = \mp q_-kp,\ kt = \pm q_-np,\ pk = \mp q_-tn,\ pn = \pm q_-tk;\\
ls = \mp q_-mq,\ ms = \mp q_-lq,\ qm = \mp q_-sl,\ ql = \mp q_-sm;\\
nq = \pm q_-ks,\ kq = \mp q_-ns;\ sk = \pm q_-qn,\ sn = \mp q_-qk.
\end{aligned} \tag{5.64}$$

Now we see that the following new generators are naturally suggested

$$\begin{aligned} k = \tilde{k} + \tilde{n},\ n = \tilde{k} - \tilde{n}; l = \tilde{l} + \tilde{m},\ m = -i(\tilde{l} - \tilde{m}); \\ p = -i\sqrt{2}(q_+\tilde{p} - q_-\tilde{t}),\ t = \mp i\sqrt{2}(\tilde{p} - \tilde{t}); \\ s = -i\sqrt{2}(q_+\tilde{s} - q_-\tilde{q}),\ q = \mp i\sqrt{2}(\tilde{s} - \tilde{q}). \end{aligned} \quad (5.65)$$

In the last two equations the upper and lower signs of t and q are irrelevant since both cases lead to one and the same set of relations. So we have

$$\begin{aligned} \tilde{k}\tilde{n} = \tilde{n}\tilde{k} = \tilde{l}^2 = \tilde{m}^2 = 0;\ \tilde{k}\tilde{l} = \tilde{n}\tilde{m} = \tilde{l}\tilde{n} = \tilde{m}\tilde{k} = 0; \\ \tilde{t}\tilde{p} = 0,\ \tilde{p}^2 - i\tilde{t}^2 = (1-i)\tilde{p}\tilde{t};\ \tilde{q}\tilde{s} = 0,\ \tilde{s}^2 - i\tilde{q}^2 = (1-i)\tilde{s}\tilde{q}; \\ \tilde{p}\tilde{s} = -i\tilde{t}\tilde{q} = 1/2(1-i)\tilde{p}\tilde{q};\ \tilde{s}\tilde{p} = -i\tilde{q}\tilde{t} = 1/2(1-i)\tilde{s}\tilde{t}; \\ \tilde{n}r = r\tilde{n} = \tilde{l}r = r\tilde{m} = 0; \\ \tilde{p}r = r\tilde{t} = \tilde{s}r = r\tilde{q} = 0; \\ \tilde{t}\tilde{l} = i\tilde{t}\tilde{m},\ \tilde{p}\tilde{l} = -i\tilde{p}\tilde{m} + (1-i)\tilde{t}\tilde{l};\ \tilde{l}\tilde{p} = i\tilde{m}\tilde{p},\ \tilde{l}\tilde{t} = -i\tilde{m}\tilde{t} + (1+i)\tilde{l}\tilde{p}; \\ \tilde{t}\tilde{k} = i\tilde{t}\tilde{n},\ \tilde{p}\tilde{k} = -i\tilde{p}\tilde{n} + (1-i)\tilde{k}\tilde{p};\ \tilde{k}\tilde{p} = -i\tilde{n}\tilde{p},\ \tilde{k}\tilde{t} = i\tilde{n}\tilde{t} + (1+i)\tilde{k}\tilde{p}; \\ \tilde{q}\tilde{l} = i\tilde{q}\tilde{m},\ \tilde{s}\tilde{l} = -i\tilde{s}\tilde{m} + (1-i)\tilde{q}\tilde{l};\ \tilde{l}\tilde{s} = i\tilde{m}\tilde{s},\ \tilde{l}\tilde{q} = -i\tilde{m}\tilde{q} + (1+i)\tilde{l}\tilde{s}; \\ \tilde{q}\tilde{k} = -i\tilde{q}\tilde{n},\ \tilde{s}\tilde{k} = i\tilde{s}\tilde{n} + (1-i)\tilde{q}\tilde{k};\ \tilde{k}\tilde{s} = i\tilde{n}\tilde{s},\ \tilde{k}\tilde{q} = -i\tilde{n}\tilde{q} + (1+i)\tilde{k}\tilde{s}. \end{aligned} \quad (5.66)$$

- $B4+$

$$\begin{aligned} k^2 = n^2,\ kn = nk; l^2 = m^2,\ lm = -ml; \\ kl = inm,\ nl = ikm;\ lk = -imn, \ln = -imk; \\ p^2 = t^2,\ pt = -itp;\ s^2 = q^2,\ sq = -iqs; \\ ps = tq,\ pq = -its;\ sp = qt,\ st = -iqp; \\ (k-n)r = r(k-n) = (l+im)r = r(l-im) = 0; \\ (p \mp q_-t)r = r(p \mp q_+t) = (s \mp q_-q)r = r(s \mp q_+q) = 0; \\ tm = \pm q_-pl,\ tl = \mp q_-pm;\ lp = \pm q_-mt,\ mp = \mp q_-lt; \\ nt = \pm q_-kp,\ kt = \pm q_-np,\ pk = \pm q_-tn,\ pn = \pm q_-tk; \\ ls = \pm q_-mq,\ ms = \mp q_-lq,\ qm = \pm q_-sl,\ ql = \mp q_-sm; \\ nq = \pm q_-ks,\ kq = \pm q_-ns;\ sk = \pm q_-qn,\ sn = \pm q_-qk. \end{aligned} \quad (5.67)$$

Now we see that the following new generators are naturally suggested

$$\begin{aligned} k = \tilde{k} + \tilde{n},\ n = \tilde{k} - \tilde{n}; l = \tilde{l} + \tilde{m},\ m = -i(\tilde{l} - \tilde{m}); \\ p = -i\sqrt{2}(q_+\tilde{p} - q_-\tilde{t}),\ t = \mp i\sqrt{2}(\tilde{p} - \tilde{t}); \\ s = -i\sqrt{2}(q_+\tilde{s} - q_-\tilde{q},\ q = \mp i\sqrt{2}(\tilde{s} - \tilde{q}). \end{aligned} \quad (5.68)$$

In the last two equations the upper and lower signs of t and q are irrelevant since both cases lead to one and the same set of relations

$$\begin{aligned}
\tilde{k}\tilde{n} = \tilde{n}\tilde{k} = \tilde{l}^2 = \tilde{m}^2 = 0;\ \ \tilde{k}\tilde{m} = \tilde{n}\tilde{l} = \tilde{l}\tilde{k} = \tilde{m}\tilde{n} = 0;\\
\tilde{p}^2 = -i\tilde{t}^2 = 1/2(1-i)\tilde{p}\tilde{t};\ \tilde{s}^2 = -i\tilde{q}^2 = 1/2(1-i)\tilde{s}\tilde{q};\\
\tilde{p}\tilde{s} = -i\tilde{t}\tilde{q} = 1/2(1-i)\tilde{p}\tilde{q};\ \tilde{s}\tilde{p} = -i\tilde{q}\tilde{t} = 1/2(1-i)\tilde{s}\tilde{t};\\
\tilde{n}r = r\tilde{n} = \tilde{l}r = r\tilde{m} = 0;\\
\tilde{p}r = r\tilde{t} = \tilde{s}r = r\tilde{q} = 0;\\
\tilde{p}\tilde{l} = 0,\ \tilde{t}\tilde{m} = 1/2(1+i)\tilde{p}\tilde{m};\ \tilde{m}\tilde{t} = 0,\ \tilde{l}\tilde{p} = 1/2(1-i)\tilde{l}\tilde{t};\\
\tilde{p}\tilde{k} = 0,\ \tilde{t}\tilde{n} = 1/2(1+i)\tilde{p}\tilde{n};\ \tilde{k}\tilde{t} = 0,\ \tilde{n}\tilde{p} = 1/2(1-i)\tilde{n}\tilde{t};\\
\tilde{s}\tilde{l} = 0,\ \tilde{q}\tilde{m} = 1/2(1+i)\tilde{s}\tilde{m};\ \tilde{m}\tilde{q} = 0,\ \tilde{l}\tilde{s} = 1/2(1-i)\tilde{l}\tilde{q};\\
\tilde{s}\tilde{k} = 0,\ \tilde{q}\tilde{n} = 1/2(1+i)\tilde{s}\tilde{n};\ \tilde{k}\tilde{q} = 0,\ \tilde{n}\tilde{s} = 1/2(1-i)\tilde{n}\tilde{q};\ .
\end{aligned} \tag{5.69}$$

- *B*4−

$$\begin{aligned}
k^2 = n^2,\ kn = nk; l^2 = m^2,\ lm = -ml;\\
kl = inm,\ nl = ikm;\ lk = -imn, \ln = -imk;\\
p^2 = -t^2,\ pt = itp;\ s^2 = -q^2,\ sq = iqs;\\
ps = tq,\ pq = -its;\ sp = qt,\ st = -iqp;\\
(k-n)r = r(k-n) = (l+im)r = r(l-im) = 0;\\
(p \mp q_- t)r = r(p \mp q_+ t) = (s \mp q_- q)r = r(s \mp q_+ q) = 0;\\
tm = \mp q_- pl,\ tl = \pm q_- pm;\ lp = \mp q_- mt,\ mp = \pm q_- lt;\\
nt = \mp q_- kp,\ kt = \mp q_- np,\ pk = \mp q_- tn,\ pn = \mp q_- tk;\\
ls = \mp q_- mq,\ ms = \pm q_- lq,\ qm = \mp q_- sl,\ ql = \pm q_- sm;\\
nq = \mp q_- ks,\ kq = \mp q_- ns;\ sk = \mp q_- qn,\ sn = \mp q_- qk.
\end{aligned} \tag{5.70}$$

Now we see that the following new generators are naturally suggested

$$\begin{aligned}
k = \tilde{k} + \tilde{n},\ n = \tilde{k} - \tilde{n}; l = \tilde{l} + \tilde{m},\ m = -i(\tilde{l} - \tilde{m});\\
p = -i\sqrt{2}(q_+\tilde{p} - q_-\tilde{t}),\ t = \mp i\sqrt{2}(\tilde{p} - \tilde{t});\\
s = -i\sqrt{2}(q_+\tilde{s} - q_-\tilde{q}),\ q = \mp i\sqrt{2}(\tilde{s} - \tilde{q}).
\end{aligned} \tag{5.71}$$

In the last two equations the upper and lower signs of t and q are irrelevant since both cases lead to one and the same set of relations

$$\begin{aligned}
\tilde{k}\tilde{n} = \tilde{n}\tilde{k} = \tilde{l}^2 = \tilde{m}^2 = 0;\ \ \tilde{k}\tilde{m} = \tilde{n}\tilde{l} = \tilde{l}\tilde{k} = \tilde{m}\tilde{n} = 0;\\
\tilde{p}\tilde{t} = 0,\ \tilde{p}^2 - i\tilde{t}^2 = (1-i)\tilde{t}\tilde{p};\ \tilde{s}\tilde{q} = 0,\ \tilde{s}^2 - i\tilde{q}^2 = (1-i)\tilde{q}\tilde{s};\\
\tilde{p}\tilde{s} = -i\tilde{t}\tilde{q} = 1/2(1-i)\tilde{p}\tilde{q};\ \tilde{s}\tilde{p} = -i\tilde{q}\tilde{t} = 1/2(1-i)\tilde{s}\tilde{t};\\
\tilde{n}r = r\tilde{n} = \tilde{l}r = r\tilde{m} = 0;\\
\tilde{p}r = r\tilde{t} = \tilde{s}r = r\tilde{q} = 0;\\
\tilde{p}\tilde{m} = 0,\ \tilde{t}\tilde{l} = 1/2(1+i)\tilde{p}\tilde{l};\ \tilde{l}\tilde{t} = 0,\ \tilde{m}\tilde{p} = 1/2(1-i)\tilde{m}\tilde{t};\\
\tilde{p}\tilde{n} = 0,\ \tilde{t}\tilde{k} = 1/2(1+i)\tilde{p}\tilde{k};\ \tilde{n}\tilde{t} = 0,\ \tilde{k}\tilde{p} = 1/2(1-i)\tilde{k}\tilde{t};\\
\tilde{s}\tilde{m} = 0,\ \tilde{q}\tilde{l} = 1/2(1+i)\tilde{s}\tilde{l};\ \tilde{l}\tilde{q} = 0,\ \tilde{m}\tilde{s} = 1/2(1-i)\tilde{m}\tilde{q};\\
\tilde{s}\tilde{n} = 0,\ \tilde{q}\tilde{k} = 1/2(1+i)\tilde{s}\tilde{k};\ \tilde{n}\tilde{q} = 0,\ \tilde{k}\tilde{s} = 1/2(1-i)\tilde{k}\tilde{q}.
\end{aligned} \tag{5.72}$$

Journal of Physics: Conference Series **254** (2010) 012001 doi:10.1088/1742-6596/254/1/012001

6. Dual of $B1+$

In this Section we shall construct the dual of one of the exotic bialgebras that we obtained in the previous Section, namely of $B1+$.

To start with we begin with the coproducts of the elements of the T-matrix. We have:

$$\delta\begin{pmatrix} k & p & l \\ q & r & s \\ m & t & n \end{pmatrix} = \begin{pmatrix} k\otimes k+p\otimes q+l\otimes m & k\otimes p+p\otimes r+l\otimes t & k\otimes l+p\otimes s+l\otimes n \\ q\otimes k+r\otimes q+s\otimes m & q\otimes p+r\otimes r+s\otimes t & q\otimes l+r\otimes s+s\otimes n \\ m\otimes k+t\otimes q+n\otimes m & mo\times p+t\otimes r+n\otimes t & m\otimes l+t\otimes s+n\otimes n \end{pmatrix} \tag{6.1}$$

We have also

$$\epsilon\begin{pmatrix} k & p & l \\ q & r & s \\ m & t & n \end{pmatrix} = \begin{pmatrix} 1 & 0 & 0 \\ 0 & 1 & 0 \\ 0 & 0 & 1 \end{pmatrix} \tag{6.2}$$

Going to the set of "tilde" generators we see that the co-products are more complicated. Namely

$$\begin{aligned}
\delta(\tilde{k}) &= \tilde{k}\otimes\tilde{k}+\tilde{n}\otimes\tilde{n}-i\tilde{l}\otimes\tilde{l}+i\tilde{m}\otimes\tilde{m}+q_-\tilde{p}\otimes\tilde{s}-q_-\tilde{t}\otimes\tilde{q},\\
\delta(\tilde{n}) &= \tilde{k}\otimes\tilde{n}+\tilde{n}\otimes\tilde{k}+i\tilde{l}\otimes\tilde{m}-i\tilde{m}\otimes\tilde{l}-q_-\tilde{p}\otimes\tilde{q}+q_-\tilde{t}\otimes\tilde{s},\\
\delta(\tilde{l}) &= \tilde{k}\otimes\tilde{l}+\tilde{n}\otimes\tilde{m}+\tilde{l}\otimes\tilde{k}-\tilde{m}\otimes\tilde{n}+\tilde{p}\otimes\tilde{s}+\tilde{t}\otimes\tilde{q},\\
\delta(\tilde{m}) &= \tilde{k}\otimes\tilde{m}+\tilde{n}\otimes\tilde{l}-\tilde{l}\otimes\tilde{n}+\tilde{m}\otimes\tilde{k}\tilde{p}\otimes\tilde{q}+\tilde{t}\otimes\tilde{s},\\
\delta r &= r\otimes r+2q_-\tilde{s}\otimes\tilde{p}-2q_-\tilde{q}\otimes\tilde{t}\\
\delta(\tilde{p}) &= \tilde{k}\otimes\tilde{p}+\tilde{n}\otimes\tilde{t}+q_-\tilde{l}\otimes\tilde{p}-q_-\tilde{m}\otimes\tilde{t}+\tilde{p}\otimes r,\\
\delta(\tilde{t}) &= \tilde{k}\otimes\tilde{t}+\tilde{n}\otimes\tilde{p}-q_-\tilde{l}\otimes\tilde{t}+q_-\tilde{m}\otimes\tilde{p}+\tilde{t}\otimes r,\\
\delta(\tilde{q}) &= \tilde{q}\otimes\tilde{k}-q_-\tilde{q}\otimes\tilde{l}-\tilde{s}\otimes\tilde{n}+q_-\tilde{s}\otimes\tilde{m}+r\otimes\tilde{q},\\
\delta(\tilde{s}) &= \tilde{s}\otimes\tilde{k}-\tilde{q}\otimes\tilde{n}+q_-\tilde{s}\otimes\tilde{l}-q_-\tilde{q}\otimes\tilde{m}+r\otimes\tilde{s}.
\end{aligned} \tag{6.3}$$

Looking at the last four equations we see that they can be essentially simplified we introduce the new generators:

$$\tilde{k} = \hat{k}+\hat{l},\ \tilde{l} = q_+(\hat{k}-\hat{l});\quad \tilde{m} = q_+(\hat{n}-\hat{m}),\ \tilde{n} = \hat{m}+\hat{n}. \tag{6.4}$$

Thus we have:

$$\begin{aligned}
\delta(\hat{k}) &= 2\hat{k}\otimes\hat{k}+2\hat{m}\otimes\hat{n}+2q_-\tilde{p}\otimes\tilde{s},\\
\delta(\hat{l}) &= 2\hat{l}\otimes\hat{l}+2\hat{n}\otimes\hat{m}-q_-\tilde{t}\otimes\tilde{q},\\
\delta(\hat{m}) &= 2\hat{k}\otimes\hat{m}+2\hat{m}\otimes\hat{l}--q_-\tilde{p}\otimes\tilde{q},\\
\delta(\hat{n}) &= 2\hat{l}\otimes\hat{n}+2\hat{n}\otimes\hat{k}+q_-\tilde{t}\otimes\tilde{s},\\
\delta(\hat{p}) &= 2\hat{k}\otimes\hat{p}+2\hat{m}\otimes\hat{t}+\tilde{p}\otimes r,\\
\delta(\hat{q}) &= 2\hat{q}\otimes\hat{l}-2\hat{s}\otimes\hat{m}+r\otimes\tilde{q},\\
\delta(\hat{s}) &= 2\hat{s}\otimes\hat{k}-2\hat{q}\otimes\hat{n}+r\otimes\tilde{s},\\
\delta(\hat{t}) &= 2\hat{l}\otimes\hat{t}+2\hat{n}\otimes\hat{p}+\tilde{t}\otimes r,\\
\delta r &= r\otimes r+2q_-\tilde{q}\otimes\tilde{p}-2q_-\tilde{s}\otimes\tilde{t}.
\end{aligned} \tag{6.5}$$

With the "hat" generators we have

$$\begin{aligned}
\epsilon(\hat{k}) = \epsilon(\hat{l}) = 1/2,\quad \epsilon(r) = 1,\\
\epsilon(z) = 0,\ for\ z = (\hat{m},\hat{n},\hat{p},\tilde{q},\tilde{s},\tilde{t}).
\end{aligned} \tag{6.6}$$

In terms of the new generators the bialgebra is

$$
\begin{aligned}
\hat{k}^2 = \hat{l}^2,\ \hat{k}\hat{l} = \hat{l}\hat{k};\ \hat{n}^2 = \hat{m}^2,\ \hat{m}\hat{n} = \hat{n}\hat{m};\\
\hat{k}\hat{n} = -\hat{l}\hat{m},\ \hat{k}\hat{m} = -\hat{l}\hat{n};\ \hat{n}\hat{k} = -\hat{m}\tilde{l},\ \hat{m}\hat{k} = -\hat{n}\hat{l};;\\
\tilde{p}^2 = \tilde{t}^2,\ \tilde{p}\tilde{t} = -\tilde{t}\tilde{p};\ \tilde{s}^2 = \tilde{q}^2,\ \tilde{s}\tilde{q} = -\tilde{q}\tilde{s};\ \tilde{p}\tilde{q} = \tilde{q}\tilde{p} = \tilde{t}\tilde{s} = \tilde{s}\tilde{t} = 0;\\
\hat{n}r = r\hat{n} = \hat{m}r = r\hat{m} = 0; \tilde{t}r = r\tilde{t} = \tilde{q}r = r\tilde{q} = 0;\\
\tilde{p}\hat{k} = \tilde{t}\hat{n},\ \tilde{t}\hat{k} = -\tilde{p}\hat{n},\ \hat{k}\tilde{p} = \hat{n}\tilde{t},\ \hat{k}\tilde{t} = -\hat{n}\tilde{p};\\
\tilde{p}\hat{l} = \tilde{t}\hat{m},\ \tilde{t}\hat{l} = -\tilde{p}\hat{m},\ \hat{l}\tilde{p} = \hat{m}\tilde{t},\ \hat{l}\tilde{t} = -\hat{m}\tilde{p};\\
\tilde{s}\hat{k} = -\tilde{q}\hat{m},\ \tilde{q}\hat{k} = \tilde{s}\hat{m},\ \hat{k}\tilde{s} = -\hat{m}\tilde{q},\ \hat{k}\tilde{q} = \hat{m}\tilde{s};\\
\tilde{s}\hat{l} = -\tilde{q}\hat{n},\ \tilde{q}\hat{l} = \tilde{s}\hat{n},\ \hat{l}\tilde{s} = -\hat{n}\tilde{q},\ \hat{l}\tilde{q} = \hat{n}\tilde{s}.
\end{aligned}
\tag{6.7}
$$

With these new generators we have some additional relations

$$
\begin{aligned}
\tilde{p}^2\tilde{s} = \tilde{s}\tilde{p}^2 = \tilde{p}^2 r = r\tilde{p}^2 = \tilde{s}^2\tilde{p} = \tilde{p}\tilde{s}^2 = \tilde{s}^2 r = r\tilde{s}^2 = 0,\ \tilde{q}^2\tilde{t} = \tilde{t}\tilde{q}^2 = \tilde{t}^2\tilde{q} = \tilde{q}\tilde{t}^2 = 0,\\
\hat{k}^2\tilde{p} = \hat{k}^2\tilde{q} = \hat{k}^2\tilde{s} = \hat{k}^2\tilde{t} = \tilde{p}\hat{k}^2 = \tilde{q}\hat{k}^2 = \tilde{s}\hat{k}^2 = \tilde{t}\hat{k}^2 = 0,\\
\hat{l}^2\tilde{p} = \hat{l}^2\tilde{q} = \hat{l}^2\tilde{s} = \hat{l}^2\tilde{t} = \tilde{p}\hat{l}^2 = \tilde{q}\hat{l}^2 = \tilde{s}\hat{l}^2 = \tilde{t}\hat{l}^2 = 0,\\
\hat{m}^2\tilde{p} = \hat{m}^2\tilde{q} = \hat{m}^2\tilde{s} = \hat{m}^2\tilde{t} = \tilde{p}\hat{m}^2 = \tilde{q}\hat{m}^2 = \tilde{s}\hat{m}^2 = \tilde{t}\hat{m}^2 = 0,\\
\hat{n}^2\tilde{p} = \hat{n}^2\tilde{q} = \hat{n}^2\tilde{s} = \hat{n}^2\tilde{t} = \tilde{p}\hat{n}^2 = \tilde{q}\hat{n}^2 = \tilde{s}\hat{n}^2 = \tilde{t}\hat{n}^2 = 0,\\
\tilde{p}\hat{k}\hat{l} = \tilde{p}\hat{k}\hat{m} = \tilde{p}\hat{k}\hat{n} = \tilde{p}\hat{m}\hat{n} = hatk\hat{l}\tilde{p} = hatk\hat{m}\tilde{p} = hatk\hat{n}\tilde{p} = hatm\hat{n}\tilde{p} = 0,\\
\tilde{q}\hat{k}\hat{l} = \tilde{q}\hat{k}\hat{m} = \tilde{q}\hat{k}\hat{n} = \tilde{q}\hat{m}\hat{n} = hatk\hat{l}\tilde{q} = hatk\hat{m}\tilde{q} = hatk\hat{n}\tilde{q} = hatm\hat{n}\tilde{q} = 0,\\
\tilde{s}\hat{k}\hat{l} = \tilde{s}\hat{k}\hat{m} = \tilde{s}\hat{k}\hat{n} = \tilde{s}\hat{m}\hat{n} = hatk\hat{l}\tilde{s} = hatk\hat{m}\tilde{s} = hatk\hat{n}\tilde{s} = hatm\hat{n}\tilde{s} = 0,\\
\tilde{t}\hat{k}\hat{l} = \tilde{t}\hat{k}\hat{m} = \tilde{t}\hat{k}\hat{n} = \tilde{t}\hat{m}\hat{n} = hatk\hat{l}\tilde{t} = hatk\hat{m}\tilde{t} = hatk\hat{n}\tilde{t} = hatm\hat{n}\tilde{t} = 0,\\
\tilde{p}\tilde{s}\hat{k} = \tilde{p}\tilde{s}\hat{l} = \tilde{p}\tilde{s}\hat{m} = \tilde{p}\tilde{s}\hat{n} = \hat{k}\tilde{p}\tilde{s} = \hat{l}\tilde{p}\tilde{s} = \hat{m}\tilde{p}\tilde{s} = \hat{n}\tilde{p}\tilde{s} = 0,\\
\tilde{s}\tilde{p}\hat{k} = \tilde{s}\tilde{p}\hat{l} = \tilde{s}\tilde{p}\hat{m} = \tilde{s}\tilde{p}\hat{n} = \hat{k}\tilde{s}\tilde{p} = \hat{l}\tilde{s}\tilde{p} = \hat{m}\tilde{s}\tilde{p} = \hat{n}\tilde{p}\tilde{s} = 0,\\
\tilde{q}\tilde{t}\hat{k} = \tilde{q}\tilde{t}\hat{l} = \tilde{q}\tilde{t}\hat{m} = \tilde{q}\tilde{t}\hat{n} = \hat{k}\tilde{q}\tilde{t} = \hat{l}\tilde{q}\tilde{t} = \hat{m}\tilde{q}\tilde{t} = \hat{n}\tilde{q}\tilde{t} = 0,\\
\tilde{t}\tilde{q}\hat{k} = \tilde{t}\tilde{q}\hat{l} = \tilde{t}\tilde{q}\hat{m} = \tilde{t}\tilde{q}\hat{n} = \hat{k}\tilde{t}\tilde{q} = \hat{l}\tilde{t}\tilde{q} = \hat{m}\tilde{t}\tilde{q} = \hat{n}\tilde{t}\tilde{q} = 0.
\end{aligned}
\tag{6.8}
$$

Now in view of the relations between $\hat{k}$ and $\hat{l}$ and The values of ϵz we consider the following elements for the basis

$$
\begin{aligned}
\hat{k}^{2\ell}r^{\kappa},\ \hat{k}^{2\ell}\hat{l}r^{\kappa},\ \hat{k}^{2\ell+1}r^{\kappa},\ \hat{k}^{2\ell+1}\hat{l}r^{\kappa},\\
r^{\kappa}\hat{k}^{2\ell},\ r^{\kappa}\hat{k}^{2\ell}\hat{l},\ r^{\kappa}\hat{k}^{2\ell+1},\ r^{\kappa}\hat{k}^{2\ell+1}\hat{l},\\
\hat{k}^{2\ell}r^{\kappa}\hat{l},\ \hat{k}^{2\ell+1}r^{\kappa}\hat{l},\\
\hat{k}^{2\ell}r^{\kappa}\tilde{p},\ \hat{k}^{2\ell}\hat{l}r^{\kappa}\tilde{p},\ \hat{k}^{2\ell+1}r^{\kappa}\tilde{p},\ \hat{k}^{2\ell+1}\hat{l}r^{\kappa}\tilde{p},\\
\hat{k}^{2\ell}r^{\kappa}\tilde{s},\ \hat{k}^{2\ell}\hat{l}r^{\kappa}\tilde{s},\ \hat{k}^{2\ell+1}r^{\kappa}\tilde{s},\ \hat{k}^{2\ell+1}\hat{l}r^{\kappa}\tilde{s},\\
\tilde{p}r^{\kappa}\hat{k}^{2\ell},\ \tilde{p}r^{\kappa}\hat{k}^{2\ell}\hat{l},\ \tilde{p}r^{\kappa}\hat{k}^{2\ell+1},\ \tilde{p}r^{\kappa}\hat{k}^{2\ell+1}\hat{l},\\
\tilde{s}r^{\kappa}\hat{k}^{2\ell},\ \tilde{s}r^{\kappa}\hat{k}^{2\ell}\hat{l},\ \tilde{s}r^{\kappa}\hat{k}^{2\ell+1},\ \tilde{s}r^{\kappa}\hat{k}^{2\ell+1}\hat{l},\\
\hat{m},\ \hat{l},\ \tilde{q}, \tilde{t}.
\end{aligned}
\tag{6.9}
$$

Then let us denote

$$
E_1 = \exp(-\hat{k}\frac{\partial}{\partial\hat{k}}),\ E_2 = \exp(-\hat{l}\frac{\partial}{\partial\hat{l}}),\quad E = E_1E_2. \tag{6.10}
$$

Then we define the dual elements by the standard procedure of [12], namely:

$$\left\langle \hat{K}, f\right\rangle = \epsilon\left(\frac{\partial f}{\partial \hat{k}}\right), \ \left\langle \hat{L}, f\right\rangle = \epsilon\left(\frac{\partial f}{\partial \hat{l}}\right), \ \langle R, f\rangle = \epsilon\left(\frac{\partial f}{\partial r}\right),$$
$$\langle Z, f\rangle = \epsilon\left(E\frac{\partial f}{\partial z}\right), \quad z = (\hat{m}, \hat{n}, \tilde{p}, \tilde{q}, \tilde{s}, \tilde{t}). \tag{6.11}$$

In result we obtain the following dual bialgebra

$$\begin{aligned}
\left[\hat{K}, \hat{L}\right] = 0,\ \hat{K}\hat{M} = 2\hat{M},\ \hat{N}\hat{K} = 2\hat{N},\ \hat{K}\hat{N} = \hat{M}\hat{K} = 0,\\
\hat{K}\tilde{P} = 2\tilde{P},\ \tilde{S}\hat{K} = 2\tilde{S},\ \tilde{P}\hat{K} = \hat{K}\tilde{S} = \hat{K}\tilde{Q} = \tilde{Q}\hat{K} = \hat{K}\tilde{T} = \tilde{T}\hat{K} = 0,\\
\hat{M}\hat{L} = 2\hat{M},\ \hat{L}\hat{N} = 2\hat{N},\ \hat{L}\hat{M} = \hat{N}\hat{L} = 0,\\
\left[\hat{K}, R\right] = 0, \left[\hat{L}, R\right] = 0,\\
\hat{L}\tilde{P} = \tilde{P}\hat{L} = 2\tilde{P},\ \hat{L}\tilde{S} = \tilde{S}\hat{L} = 2\tilde{S},\\
\tilde{Q}\hat{L} = 2\tilde{Q},\ \hat{L}\tilde{T} = 2\tilde{T},\ \hat{L}\tilde{Q} = \tilde{T}\hat{L} = 0,\\
\hat{M}\hat{N} = 2\hat{K},\ \hat{N}\hat{M} = 2\hat{L},\ \hat{M}\tilde{P} = \tilde{P}\hat{M} = \hat{M}\tilde{Q} = \tilde{Q}\hat{M} = 0,\\
\hat{M}\tilde{T} = 2\tilde{P},\ \tilde{S}\hat{M} = -2\tilde{Q}, \hat{M}\tilde{S} = \hat{M}R = \tilde{T}\hat{M} = R\hat{M} = 0,\\
\hat{N}\tilde{P} = 2\tilde{T},\ \tilde{Q}\hat{N} = -2\tilde{S},\ \tilde{P}\hat{N} = \hat{N}\tilde{Q} = 0,\\
\hat{N}R = \hat{N}\tilde{S} = \hat{N}\tilde{T} = R\hat{N} = \tilde{S}\hat{N} = \tilde{T}\hat{N} = 0,\\
\tilde{P}\tilde{Q} = -q_-\hat{M},\ \tilde{Q}\tilde{P} = \tilde{P}\tilde{T} = \tilde{T}\tilde{P} = 0,\\
\tilde{P}\tilde{S} = q_-\hat{K},\ \tilde{S}\tilde{P} = 2q_-R,\ \left[\tilde{P}, R\right] = \tilde{P},\ \left[\tilde{S}, R\right] = -\tilde{S},\\
\tilde{T}\tilde{Q} = -q_-\hat{L},\ \tilde{Q}\tilde{T} = -2q_-R,\\
\tilde{T}\tilde{S} = q_-\hat{N},\ \tilde{T}R = \tilde{T},\ \tilde{S}\tilde{T} = R\tilde{T} = 0,\\
R\tilde{Q} = \tilde{Q}, \tilde{Q}R = \tilde{Q}\tilde{S} = \tilde{S}\tilde{Q} = 0,\\
\hat{M}^2 = \hat{N}^2 = 0,\ \tilde{P}^2 = \tilde{Q}^2 = \tilde{S}^2 = \tilde{T}^2 = 0.
\end{aligned} \tag{6.12}$$

The co-products of the duals are:

$$\begin{aligned}
\delta(\hat{K}) = \hat{K}\otimes 1_U + 1_U\otimes\hat{K},\ \delta(\hat{L}) = \hat{L}\otimes 1_U + 1_U\otimes\hat{L},\\
\delta(\hat{M}) = \hat{M}\otimes F + F\otimes\hat{M},\ \delta(\hat{N}) = \hat{N}\otimes F + F\otimes\hat{N},\\
\delta(\hat{P}) = \hat{P}\otimes F + F\otimes\hat{P},\ \delta(\hat{Q}) = \hat{Q}\otimes F + F\otimes\hat{Q},\\
\delta(\hat{S}) = \hat{S}\otimes F + F\otimes\hat{S},\ \delta(\hat{T}) = \hat{T}\otimes F + F\otimes\hat{T},\\
\delta(R) = R\otimes 1_U + 1_U\otimes R.
\end{aligned} \tag{6.13}$$

7. Summary and outlook

In the present paper we have found a multitude of exotic bialgebras and the duals of two of them. More duals should be constructed. More importantly, we plan to continue the programme fulfilled successfully for the exotic bialgebra S03, cf. summary in [4]. In particular, it is important to find the FRT duals [11] which are different from the standard duals for the exotic bialgebras. Further we should find the baxterisation of the dual algebras. Their finite-dimensional representations should be considered. Diagonalisations of the braid matrices would be used to handle the representations of the corresponding L-algebras (in the FRT formalism) and to formulate the fusion of finite-dimensional representations. Possible applications should be considered, in particular, exotic vertex models and integrable spin-chain models.

Acknowledgments

One of the authors (VKD) would like to thank for hospitality the Abdus Salam International Center for Theoretical Physics, where part of the work was done. VKD and SGM are supported in part by Bulgarian NSF grant *DO 02-257*.

References

[1] Arnaudon D, Chakrabarti A, Dobrev V K and Mihov S G 2001 Duality for exotic bialgebras *J. Phys. A: Math. Gen.* **34** 4065-4082 (*Preprint* math.QA/0101160)

[2] Arnaudon D, Chakrabarti A, Dobrev V K and Mihov S G 2002 Duality and representations for new exotic bialgebras *J. Math. Phys.* **43** 6238-6264 (*Preprint* math.QA/0206053)

[3] Arnaudon D, Chakrabarti A, Dobrev V K and Mihov S G 2003 Spectral decomposition and Baxterisation of exotic bialgebras and associated noncommutative geometries *Int. J. Mod. Phys.* **A18** 4201-4213 (*Preprint* math.QA/0209321)

[4] Arnaudon D, Chakrabarti A, Dobrev V K and Mihov S G 2006 Exotic Bialgebra S03: Representations, Baxterisation and Applications *Annales Henri Poincaré* **7** 1351-73 (*Preprint* math.QA/0601708)

[5] Chakrabarti A 2005 A nested sequence of projectors and corresponding braid matrices $\hat{R}(\theta)$ (1) odd dimensions *J. Math. Phys.* **46** 063508 (*Preprint* math.QA/0401207)

[6] Chakrabarti A 2005 Aspects of a new class of braid matrices: Roots of unity and hyper-elliptic q for triangularity, L-algebra, link-invariants, non-commutative spaces *J. Math. Phys.* **46** 063509 (*Preprint* math.QA/0412549)

[7] Abdesselam B, Chakrabarti A, Dobrev V K and Mihov S G 2007 Higher Dimensional Unitary Braid Matrices: Construction, Associated Structures and Entanglements *J. Math. Phys.* **48** 053508 (*Preprint* math.QA/0702188)

[8] Abdesselam B, Chakrabarti A, Dobrev V K and Mihov S G 2007 Higher Dimensional Multiparameter Unitary and Nonunitary Braid Matrices: Even Dimensions *J. Math. Phys.* **48** 103505 (*Preprint* arXiv:0706.2787)

[9] Abdesselam B and Chakrabarti A 2008 Multiparameter statistical models from $N^2 \times N^2$ braid matrices: Explicit eigenvalues of transfer matrices $T^{(r)}$, spin chains, factorizable scatterings for all N *Preprint* arXiv:0806.2371

[10] Abdesselam B and Chakrabarti A 2008 A new eight vertex model and higher dimensional, multiparameter generalizations *J. Math. Phys.* **49** 053301 (*Preprint* arXiv:0801.2548.)

[11] Faddeev L D, Reshetikhin N Y and Takhtajan L A 1989 *Alg. Anal.* **1** 178 (in Russian); English translation: 1990 Quantization of Lie groups and Lie algebras *Leningrad. Math. J.* **1** 193

[12] Dobrev V K 1992 Duality for the matrix quantum group $GL_{p,q}(2,\mathbb{C})$ *J. Math. Phys.* **33** 3419-3430

Quantum – a Festschrift for Tony Sudbery IOP Publishing
Journal of Physics: Conference Series **254** (2010) 012002 doi:10.1088/1742-6596/254/1/012002

Nonassociative Riemannian geometry by twisting

Edwin Beggs[1] **and Shahn Majid**[2]

[1] Department of Mathematics, Swansea University
Singleton Parc, Swansea SA2 8PP, UK
[2] School of Mathematical Sciences, Queen Mary University of London
327 Mile End Rd, London E1 4NS

E-mail: e.j.beggs@swansea.ac.uk, s.majid@qmul.ac.uk

Abstract. Many quantum groups and quantum spaces of interest can be obtained by cochain (but not cocycle) twist from their corresponding classical object. This failure of the cocycle condition implies a hidden nonassociativity in the noncommutative geometry already known to be visible at the level of differential forms. We extend the cochain twist framework to connections and Riemannian structures and provide examples including twist of the S^7 coordinate algebra to a nonassociative hyperbolic geometry in the same category as that of the octonions.

Received: 17 November 2009

1. Introduction

The idea of noncommutative geometry, of course, is to replace points in an actual space by a coordinate algebra. These are generally generated by variables x, y, z ... say enjoying algebraic properties for addition, multiplication and other relations paralleling the numbers for which they could be viewed as placeholders. Thus, solving $x^2 + y^2 + z^2 = 1$ over $\mathbb{R}$ would bring one to the actual points of a sphere, but one could also consider this equation more abstractly with x, y, z as generators of an algebra. And if in place of the usual $xy = yx$ etc we have some other non-commutation relations then, clearly, these generators could never be realised as numbers. Over the last two decades it has become obvious that the assumption of commutativity is a historical accident to do with the fact that classical mechanics was discovered before quantum mechanics and there is no particular reason, certainly from a mathematical perspective, not to allow noncommutavity as a generalisation of usual geometry more applicable to the quantum world. There are also physical reasons to think that plausibly the first quantum gravity corrections to geometry should be expressed as such noncommutativity [21]. Accordingly, a great deal of effort has been put into developing such a geometry both in Connes' 'spectral triple' approach and in a 'quantum groups' approach modelled around quantum groups such as the deformations $C_q(G)$ and their homogeneous spaces in the first instance. Both approaches are by now somewhat mature and the quantum groups approach includes specific testable predictions for Planck scale physics [4]. Meanwhile, the spectral triples approach includes specific predictions and mass relations for the standard model [11].

In this article we go one step further and consider geometry that is nonassociative in the same algebraic sense. While noncommutative geometry is motivated from quantum theory and hence somewhat familiar, this is not so for nonassociativity. If we simply drop associativity then most of what we do in algebra and geometry becomes vastly more complicated if not impossible. Therefore the first point is that we should only approach nonassociative geometry under duress. However, common to many approaches to noncommutative geometry is the idea of replacing differential structures by an algebraic formulation of 'exterior algebra of differential forms' with generators $\mathrm{d}x, \mathrm{d}y, \mathrm{d}z...$ say, formalized as an associative differential graded algebra (Ω, d) over our possibly noncommutative 'coordinate algebra'. This is pretty much a prerequisite to do any kind of physical model building on noncommutative geometry, wave equations, Maxwell's equations etc. Unfortunately, after about 20 years experience with model-building it turned out to be *not quite* possible to do this while maintaining a strict correspondence with classical differentials and classical symmetries. When quantization fails in this respect one speaks of a quantum anomaly, and in previous work we have similarly noted a fundamental anomaly or obstruction in the entire programme of noncommutative geometry.

Theorem 1.1 *[5] The standard quantum groups $C_q(G)$ of simple Lie algebras $\mathfrak{g}$ do* not *admit an associative differential calculus $\Omega(C_q(G))$ in deformation theory with left and right translation-covariance and classical dimensions.*

We showed that this obstruction, like other anomalies in physics, can be expressed at the semiclassical level as curvature, in our case of a certain Poisson preconnection, and that for $\mathfrak{g}$ semisimple there is no flat connection of the required type. The same applies to enveloping algebras $U(\mathfrak{g})$ when $\mathfrak{g}$ is semisimple, viewed as quantisation of $\mathfrak{g}^*$ with its Kirillov-Kostant Poisson bracket:

Theorem 1.2 *[6] The classical enveloping algebras $U(\mathfrak{g})$ of simple Lie algebras $\mathfrak{g}$ do* not *admit an associative differential calculus $\Omega(U(\mathfrak{g}))$ in deformation theory with* ad*-invariance under $\mathfrak{g}$ and classical dimensions.*

This was again proven using similar methods, notably Kostant's invariant theory. We note that an important first attempt at such a rigidity result was in [13] which, in particular, pointed to the role of preconnections. Moreover, attempts at such calculi at the quantum group level had been found to require extra dimensions in the cotangent bundle and this was understood now as a way to 'neutralise' the anomaly, again much as for other anomalies in physics. On the other hand, [5, 6] also provided an alternative: using cochain twisting methods it was shown that we can always keep classical dimensions and deform the classical picture, provided we allow *nonassociative geometry.* In short, if one wants a strict deformation-theoretic correspondence on quantization, we have

$$NCG \Rightarrow NAG$$

(noncommutative geometry implies nonassociative geometry). In the present paper we explore this radical alternative further. And if we want to be speculative then, once we allow nonassociative exterior algebras, we should also allow nonassociative coordinate algebras, i.e. spaces themselves not only their differential geometry could be nonassociative.

In entering this nonassociative world we also need tight controls. The key idea is to work in a 'quasiassociative' setting in which coordinate algebras are nonassociative but in a controlled way by means of a *multiplicative associator.* In mathematical terms this means that the algebra *is* associative but in some monoidal category different from that of vector spaces. A theorem of Maclane says that all constructions in such a category can be done *as if* associative i.e. without worrying about brackets. One must then insert the associator $\Phi_{X,Y,Z} : X{\otimes}(Y{\otimes}Z) \to (X{\otimes}Y){\otimes}Z$ between any three objects as needed in order to make sense of expressions and Maclane's theorem says that in a monoidal category any different ways to insert Φ as needed will give the same result. We focus in this article on a specific 'twist quantisation functor' that results in noncommutative algebras in such a monoidal category and also quantises all other covariant structures with reference to a classical symmetry.

To first explain the background, V.G. Drinfeld [12] showed that all quantum group enveloping algebras $U_q(\mathfrak{g})$ could be obtained as follows: start with the classical symmetry algebra $U(\mathfrak{g})$ *but* viewed as a quasi-Hopf algebra with respect to a certain element $\phi \in U(\mathfrak{g})^{\bar{\otimes}3}$ (a topological tensor product) obtained by solving the KZ equations. Its category of representations is a braided monoidal category with Φ given by the action of ϕ (and braiding given by the action of $\mathcal{R} = e^{ht}$ where $q = e^{\frac{h}{2}}$ and t is the split Casimir of $\mathfrak{g}$). Drinfeld showed that there exists a certain cochain $F \in U(\mathfrak{g})^{\bar{\otimes}2}$ such that conjugating the classical coproduct of $U(\mathfrak{g})$ by F gives $U_q(\mathfrak{g})$. Here ϕ also twists to some ϕ_F but F is chosen so that $\phi_F = 1$ – an ordinary Hopf algebra. In [3, 5] we developed a rather different application of the same data, namely start with completely classical data ($U(\mathfrak{g}), \phi = 1, \mathcal{R} = 1$) and twist this by Drinfeld's F (conjugate the classical coproduct). The result $U(\mathfrak{g})^F$ looks like $U_q(\mathfrak{g})$ *but* regarded as a quasi-Hopf algebra with ϕ_F closely related to Drinfeld's and $\mathcal{R}_F = F_{21}F^{-1}$, making its category of representations a (symmetric) monoidal category. There is also an important spin-off [15]: if the classical object $U(\mathfrak{g})$ acts on a classical coordinate algebra $C(M)$ then because of the functorial nature of the twist construction $C(M)$ also gets 'quantized' to an algebra $C(M)_F$, typically noncommutative and nonassociative because it lives in the category of representation of $U(\mathfrak{g})^F$. Viewed in that category it is in fact commutative and associative, but the category is no longer the usual one due to a nontrivial Φ. Similarly the classical $\Omega(M)$ gets quantised to $\Omega(M)_F$ as a calculus on $C(M)_F$. In this way any classical geometry can be systematically quantised with respect to a classical symmetry and a choice of cochain F – provided we can live with potential nonassociativity.

Note that in geometry we are interested not in enveloping algebras but in coordinate algebras. Therefore, while not essential, we will convert the above to work with classical $C(G)$ and with $C_q(G)$ viewed as a coquasiHopf algebra $C(G)^F$, and we will replace the action of $U(\mathfrak{g})$ on $C(M)$ by a *coaction* of $C(G)$. In particular G acts ($C(G)$ coacts) on $C(G)$ by left translation and induces a cochain quantisation $C(G)_F$ covariant under $C(G)^F$ [3, 6]. We recall this less familiar setting in the Preliminaries below. Depending on the choice of cochain some of the twisted algebras may remian accidentally associative – this would be the case for Drinfeld's cochain and $C(G)^F$ – and in that case the nonassociativity is hidden. But it is still present and appears typically in the differential calculus even in this case. This programme has already been carried quite far and covers the general principles and the calculus, while in the present article we now study how the next main layers of geometry, connections and curvature etc, behave under such cochain twists. The short answer is that everything works as expected if everything is covarant under the twisting classical symmetry. This also points to the limitations of the twist approach: it works *too well* – different but covariant ways to express classical constructions will all twist and only by having a general theory of noncommutative (and now nonassociative) geometry that makes sense *beyond* examples given by twisting will we be able to know which of these is most natural. We will illustrate this with the Ricci tensor in Section 7. One could consider metrics, for example, that are not invariant under the classical symmetry, although we do not do so here.

Finally, for nonassociative geometry we need concrete examples to build up our experience. About the only example that most readers may have some experience with is the octonions and these provided the first concrete example of the above 'cochain quantisation' in action [2]. The octonions are a dimension 8 algebra but with a nonassociative product, which is, however, alternative, in the sense that if two elements are repeated then they associate, so $x(yx) = (xy)x$ etc. The full picture coming out of quantum group theory amounts to the following. One can choose a basis $\{e_a\}$ labelled by $a \in \mathbb{Z}_2^3$ (a 3-vector with values in $\mathbb{Z}_2$) such that [2]

$$(e_a e_b)e_c = \phi(a, b, c)e_a(e_b e_c), \quad \phi(a, b, c) = (-1)^{a\cdot(b\times c)} = (-1)^{|abc|}.$$

This means that three elements associate if and only if their labels are linearly dependent as vectors over $\mathbb{Z}_2$, otherwise there is a -1. This was obtained in [2] precisely as a cochain quantization $C(\mathbb{Z}_2^3)_F$ of $C(\mathbb{Z}_2^3)$ under the action of $\mathbb{Z}_2^3$ by left translation and a certain choice of F. One can then proceed to differential geometry on the octonions as coordinate algebra [16]. Our reinterpretation of Drinfeld's ideas to produce $C(G)_F$ is inspired by this and exactly parallels the construction of the octonions but now for a Lie group. The octonion example also illustrates that even though twisting is 'easy' in the sense that everything is somewhat automatic by an equivalence of categories, reflected in the fact that ϕ is cohomologically trivial, this does not mean that the resulting objects are uninteresting. The octonions, for example, are a division algebra while the group algebra of $\mathbb{Z}_2^3$ is certainly not. At the same time, once we adopt our framework of geometry in monoidal categories, we are not limited to such twist or coboundary examples but can consider more general ones. It would also be interesting to apply nonassociative spectral triples to the standard model, for example based now on octonions rather than the two copies of the quaternions for the 'internal geometry' in [11].

Note added. Some of Tony Sudbery's work was towards the noncommuative geometry coming out of quantum groups and it is therefore a pleasure to contribute the present article to this Festschrift volume. There is also a degree of irony. In the late 1990s Tony left quantum groups to work on the seemingly unrelated area of octonions and other topics. Meanwhile, those who stayed behind to work purely on quantum groups and their noncommutative geometry are now learning lessons on how to do it from the octonions.

2. Preliminaries: cochain twists

We work over a field k but for physics we will often specialise this to $\mathbb{C}$ or $\mathbb{R}$. We will need to use quantum group or Hopf algebra methods and we refer to [15] for an introduction. We recall that a quantum group H is a bialgebra equipped with a notion of 'linearised inverse' or 'antipode' $S : H \to H$. Here a bialgebra means that the algebra H is also a coalgebra in a compatible way. The concept of a coalgebra is just the same as the concept of an algebra but with arrows reversed, so there is a 'coproduct' $\Delta : H \to H \otimes H$ and 'counit' $H \to k$, and we require Δ (and hence ϵ) to be algebra homomorphisms. There is a standard (Sweedler) notation $\Delta h = h_{(1)} \otimes h_{(2)}$ which we shall use for the result of Δ as an element of $H \otimes H$ (summation understood).

Suppose that H is a coalgebra. We consider its category of left comodules ${}^H\mathcal{M}$ and we write the coaction on an object V as $\lambda : V \to H \otimes V$, or as $v \mapsto v_{[-1]} \otimes v_{[0]}$ with summation understood. The concept of a comodule here is just the same as that of a module but with arrows reversed. If H is a Hopf algebra we can use its algebra product to define a tensor product on ${}^H\mathcal{M}$ by

$$\lambda(v \otimes w) = v_{[-1]}\, w_{[-1]} \otimes (v_{[0]} \otimes w_{[0]}) \ .$$

The identity in the category is just the underlying field k, with $\lambda : k \to H \otimes k$ being $1_H \otimes \mathrm{id}$. Note that at this stage, a bialgebra structure on H would suffice, but it shall be convenient to have a Hopf algebra later. An algebra in the category ${}^H\mathcal{M}$ is an object A with an associative product $\mu : A \otimes A \to A$ and a unit $1_A : k \to A$, and which are both H-comodule maps.

If G is a finite or algebraic group we are interested in the point of view where H could be its function or coordinate algebra $k(G)$ with its usual pointwise product. If similarly $A = k(X)$ is the algebra of functions on some set or algebraic variety X and G acts on X, this would be expressed in our algebraic terms as a coaction $\lambda : A \to H \otimes A$. For example, in the finite group case

$$\lambda(f)(x) = \sum_{g \in G} \delta_g\, f(gx)$$

where δ_g is the Kronecker δ-function. It may seem strange to do everything 'backwards' in terms of coactions but this will allow us to work with algebraic tensor products and algebraic groups.

Journal of Physics: Conference Series **254** (2010) 012002 doi:10.1088/1742-6596/254/1/012002

Our framework also allows A and H to be noncommutative in which case there would be no actual spaces X, G respectively.

Now we shall consider a different tensor product $\otimes^F$ for ${}^H\mathcal{M}$. We shall do this by changing the product on H while leaving the coproduct unaltered. As the coproduct is the same, we can consider the category ${}^H\mathcal{M}$ to be the same, we have only changed the tensor product. We begin with a linear map $F : H\otimes H \to k$ which is invertible, in the sense that there is another map $F^- : H\otimes H \to k$ so that

$$F(a_{(1)}\otimes b_{(1)})\,F^-(a_{(2)}\otimes b_{(2)}) \;=\; F^-(a_{(1)}\otimes b_{(1)})\,F(a_{(2)}\otimes b_{(2)}) \;=\; \epsilon(a)\,\epsilon(b)$$

and which obeys $F(a\otimes 1) = F(1\otimes a) = \epsilon(a)$. Such a map is called a 2-*cochain* on H [15]. Now we specify the tensor product $\otimes^F$ on ${}^H\mathcal{M}$ by the same vector space $V\otimes W$ as usual but with coaction

$$\lambda(v\otimes^F w) \;=\; v_{[-1]}\circ w_{[-1]}\otimes(v_{[0]}\otimes^F w_{[0]})\;,$$

where the modified product $\circ : H\otimes H \to H$ is given in terms of the original product by

$$a\circ b \;=\; F(a_{(1)}\otimes b_{(1)})\,a_{(2)}b_{(2)}\,F^-(a_{(3)}\otimes b_{(3)})\,.$$

This product may not be associative – in general $H^F = (H,\circ)$ is a coquasi-Hopf algebra [15], but this is not a problem in defining $\otimes^F$. Where it does appear as a complication is that the trivial 'identity' map from $(U\otimes^F V)\otimes^F W \to U\otimes^F(V\otimes^F W)$ is not a morphism. Instead we can use the nontrivial associator $\Phi_{U,V,W} : (U\otimes^F V)\otimes^F W \to U\otimes^F(V\otimes^F W)$ defined by

$$\begin{aligned}\Phi((u\otimes^F v)\otimes^F w) \;&=\; (\partial F)^-(u_{[-1]}\otimes v_{[-1]}\otimes w_{[-1]})\;u_{[0]}\otimes^F(v_{[0]}\otimes^F w_{[0]})\;,\\ \Phi^{-1}(u\otimes^F(v\otimes^F w)) \;&=\; (\partial F)(u_{[-1]}\otimes v_{[-1]}\otimes w_{[-1]})\;(u_{[0]}\otimes^F v_{[0]})\otimes^F w_{[0]}\;,\end{aligned}$$

where $\partial F, (\partial F)^- : H\otimes H\otimes H \to k$ is defined by

$$\begin{aligned}(\partial F)^-(a\otimes b\otimes c) \;&=\; F(a_{(1)}\otimes b_{(1)})\,F(a_{(2)}b_{(2)}\otimes c_{(1)})\,F^-(a_{(3)}\otimes b_{(3)}c_{(2)})\,F^-(b_{(4)}\otimes c_{(3)})\;,\\ (\partial F)(a\otimes b\otimes c) \;&=\; F(b_{(1)}\otimes c_{(1)})\,F(a_{(1)}\otimes b_{(2)}c_{(2)})\,F^-(a_{(2)}b_{(3)}\otimes c_{(3)})\,F^-(a_{(3)}\otimes b_{(4)})\;.\end{aligned}$$

It would be customary here to restrict to the case where F is a cocycle, i.e. $(\partial F)(a\otimes b\otimes c) = \epsilon(a)\epsilon(b)\epsilon(c)$. This means that the new tensor product would also be trivially associated. However it is not really necessary to suppose this, if the reader does not mind dealing with nontrivially associated tensor categories, and we shall not do so.

It will be useful to note that there is a natural transformation c between the two tensor product functors, $\otimes$ and $\otimes^F$, from ${}^H\mathcal{M}\times{}^H\mathcal{M}\to{}^H\mathcal{M}$, given by

$$c(v\otimes w) \;=\; F^-(v_{[-1]}\otimes w_{[-1]})\;v_{[0]}\otimes^F w_{[0]} \tag{1}$$

which, together with the identity map on objects and morphisms provids an equivalence of monoidal categories $({}^H\mathcal{M},\otimes)$ and $({}^H\mathcal{M},\otimes^F)$. The latter is the category of comodules of $H^F = (H,\circ)$.

3. Twisting algebras and modules

An algebra (A,μ) in the tensor category $({}^H\mathcal{M},\otimes)$ is twisted to an algebra $A_F = (A,\mu^F)$ in $({}^H\mathcal{M},\otimes^F)$ by

$$\mu^F(a\otimes b) \;=\; a\bullet b \;=\; F(a_{[-1]}\otimes b_{[-1]})\;a_{[0]}b_{[0]}\;. \tag{2}$$

Journal of Physics: Conference Series **254** (2010) 012002 doi:10.1088/1742-6596/254/1/012002

A quick check will show that this is associative in $({}^H\mathcal{M}, \otimes^F)$, remembering to change the order of the bracketing using Φ. Thus,

$$(a \bullet b) \bullet c = (\partial F)^-(a_{[-1]}\otimes b_{[-1]}\otimes c_{[-1]})\; a_{[0]} \bullet (b_{[0]} \bullet c_{[0]}). \tag{3}$$

We can also twist left or right A-modules V in $({}^H\mathcal{M}, \otimes)$ (meaning that the left or right actions have to preserve the H-coaction) to get left or right modules of (A, μ^F) by

$$a \triangleright^F v \;=\; F(a_{[-1]}\otimes v_{[-1]})\; a_{[0]} \triangleright v_{[0]}\;, \quad v \triangleleft^F a \;=\; F(v_{[-1]}\otimes a_{[-1]})\; v_{[0]} \triangleleft a_{[0]}\;.$$

for $v \in V$ and $a \in A$, and where we use $\triangleright$ and $\triangleleft$ for the left and right actions respectively. These obey the requirements of an action of (A, μ^F) in the category with $\otimes^F$, i.e. inserting Φ in the form ∂F^- much as in (3). When we have both left and right actions on V we can require them to commute so $(a \triangleright v) \triangleleft b = a \triangleright (v \triangleleft b)$. We say that $V \in {}_A\mathcal{M}_A$, the category of bimodules. If in addition our bimodule is left covariant under H then twisting both actions gives us a bimodule of (A, μ^F) with respect to $\otimes^F$, i.e.

$$(a \triangleright^F v) \triangleleft^F b = (\partial F)^-(a_{[-1]}\otimes v_{[-1]}\otimes b_{[-1]})\; a_{[0]} \triangleright^F (v_{[0]} \triangleleft^F b_{[0]}).$$

This all works in the same way as for the product μ^F.

For A-bimodules V and W, we can quotient the tensor product $V\otimes W$ to get the tensor product $V\otimes_A W$ over the algebra. The vector space which we quotient by is spanned by everything of the form $v \triangleleft a \otimes w - v \otimes a \triangleright w$. When our bimodules are in $({}^H\mathcal{M}, \otimes^F)$ we can analogously quotient the vector space $V\otimes^F W$ to give $V\otimes_{A^F} W$ by identifying the elements given by going the two ways round the following diagram:

$$\begin{array}{ccc} V\otimes^F(A\otimes^F W) & \xrightarrow{\quad \mathrm{id}\otimes^F \triangleright^F \quad} & V\otimes^F W \\ & {\scriptstyle \Phi^{-1}} \searrow \qquad \nearrow {\scriptstyle \triangleleft^F \otimes^F \mathrm{id}} & \\ & (V\otimes^F A)\otimes^F W & \end{array} \tag{4}$$

In fact this gives the same vector space $V\otimes_A W$ as in the untwisted case. The easiest way to see this is that the natural transformation c between the two tensor product functors, $\otimes$ and $\otimes^F$ (see equation (1)) sends elements of the form $v.a\otimes w - v\otimes a.w$ exactly to the required relations for $V\otimes_{A_F} W$, given by (4). Thus performing the quotient of vector spaces gives an isomorphism in the category.

Now $V\otimes_A W$ also has a standard A-bimodule structure, given by

$$a \triangleright (v\otimes w) \;=\; (a \triangleright v)\otimes w\;, \quad (v\otimes w) \triangleleft a \;=\; v\otimes(w \triangleleft a)\;.$$

Corresponding to this we have left and right actions of (A, μ^F) on $V\otimes_{A_F} W$ which are given at the level of $V\otimes^F W$ by the following compositions:

$$A\otimes^F(V\otimes^F W) \xrightarrow{\Phi^{-1}} (A\otimes^F V)\otimes^F W \xrightarrow{\triangleright^F\otimes \mathrm{id}} V\otimes^F W\;,$$
$$(V\otimes^F W)\otimes A \xrightarrow{\Phi} V\otimes^F(W\otimes^F A) \xrightarrow{\mathrm{id}\otimes\triangleleft^F} V\otimes^F W\;.$$

The reader can check that these quotient to well defined maps on $V\otimes_{A_F} W$. These constructions make sense for algebras and their bimodules in any k-linear monoidal category, not just $({}^H\mathcal{M}, \otimes^F)$, and work the same way as for usual algebras due to Mac Lane's coherence theorem

– we need only express usual linear algebra constructions as compositions of maps and use the same in the monoidal category with the relevant Φ inserted to make sense.

It will be useful to note that the definition of the left action on the tensor product (a similar picture holds for right) is made so that the following diagram commutes:

$$\begin{array}{ccccc} V\otimes W & \xleftarrow{\triangleright\otimes \mathrm{id}} & A\otimes V\otimes W & \xrightarrow{c_{A,V\otimes W}} & A\otimes^F(V\otimes W) \\ \downarrow c & & & & \downarrow \mathrm{id}\otimes^F c \\ V\otimes^F W & & \xleftarrow{\triangleright^F} & & A\otimes^F(V\otimes^F W) \end{array} \qquad (5)$$

This just says that $\triangleright^F$ is the image of $\triangleright$ under the functorial equivalence from the category with $\otimes$ to the category with $\otimes^F$. When all maps are 'quantized' by this equivalence then any equation expressed as a commutative diagram between maps will map over to an analogous commutative diagram on the $\otimes^F$ side. This is similarly the abstract origin of the results in this paper.

Such cochain twist ideas were applied to the octonions with $H = A = k(\mathbb{Z}_2^3)\cong k\mathbb{Z}_2^3$ and $\lambda = \Delta$ the 'left regular coaction' of any group on itself. We use the second 'group algebra' form where the algebra is spanned by basis elements e_a where $a \in \mathbb{Z}_2^3$. The cochain F and the twisted product of A_F are

$$F(e_a\otimes e_b) = (-1)^{\sum_{i\leq j} a_ib_j + a_1a_2b_3 + a_1b_2a_3 + b_1a_2a_3}, \quad e_a \bullet e_b = F(e_a\otimes e_b)e_{a+b}. \qquad (6)$$

The product $\circ$ of H is not, at it happens, changed (it remains associative) but this is largely an accident and its tensor category of comodules $\otimes^F$ is nevertheless modified. Technically, it becomes a cotriangular coquasiHopf algebra which happens to be a Hopf algebra. The category with $\otimes^F$ is a symmetric monodal one and in this category the octonions are both associative and commutative. Similarly if $C(G)$ is the algebra of functions for a finite or algebraic group and F any choice of cochain on it, we have $C(G)^F$ with product $\circ$ which can happen to remain associative (e.g. in the case of Drinfeld's F). We have a canonical coaction of $C(G)$ on itself by $\lambda = \Delta$ and hence a typically nonassociative algebra $C(G)_F$ with the induced $\bullet$ product. Note the exact parallel with the octonion construction. They are both examples of systematic quantisation by cochain twist.

4. Twisting of $\star$-structures

We now proceed to study the different layers of geometry, but allowing our 'coordinate algebras' A to be potentially nonassociative by working in a monoidal category. The first item pertains to the right concept of real form or $*$-structure. In noncommutative geometry and in physics this specifies which elements are to be represented as Hermitian or unitary in the sense that $*$ is to be respected. A categorical approch to this was introduced in [7] with the notion of a 'star object' in a bar category. We are only going to discuss it in the case relevant to cochain twists and we refer the reader to [7] for the formal category theory.

Thus we take H to be a $*$-Hopf algebra, i.e. equipped with a conjugate-linear involution $h \mapsto h^*$ on the algebra and required also to be a coalgebra map and order-reversing for the product. Then for every object V in the comodule category ${}^H\mathcal{M}$ we have another $\overline{V}$. As a set $\overline{V}$ is the same as V, but we distinguish between elements by writing $\overline{v} \in \overline{V}$ for $v \in V$. Then the left H-coaction is given by

$$\lambda(\overline{v}) = (v_{[-1]})^*\otimes\overline{v_{[0]}} \,.$$

We have natural isomorphisms $\Upsilon : \overline{V\otimes W} \to \overline{W}\otimes\overline{V}$ and $\mathrm{bb} : V \to \overline{\overline{V}}$ defined by

$$\Upsilon(\overline{v\otimes w}) = \overline{w}\otimes\overline{v}\,, \quad \mathrm{bb}(v) = \overline{\overline{v}}\,.$$

This provides the bar category associated to a Hopf $*$-algebra as a somewhat formal way of speaking about conjugate corepresentations.

We now twist this construction by altering the map Υ to $\Upsilon^F : \overline{V\otimes^F W} \to \overline{W}\otimes^F\overline{V}$ as follows:

$$\begin{aligned} \Upsilon^F(\overline{v\otimes^F w}) &= F(v_{[-2]}\otimes w_{[-2]})^*\ F^-(w_{[-1]}{}^*\otimes v_{[-1]}{}^*)\ \overline{w_{[0]}}\otimes^F\overline{v_{[0]}}\,, \\ \Upsilon^{F-1}(\overline{v}\otimes^F\overline{w}) &= F(v_{[-2]}{}^*\otimes w_{[-2]}{}^*)\ F^-(w_{[-1]}\otimes v_{[-1]})^*\ \overline{w_{[0]}\otimes^F v_{[0]}}\,. \end{aligned}$$

These formulae are made to make the following diagram commute:

$$\begin{array}{ccc} \overline{V\otimes W} & \xrightarrow{\ \Upsilon\ } & \overline{W}\otimes\overline{V} \\ {\scriptstyle \overline{c}}\downarrow & & \downarrow{\scriptstyle c} \\ \overline{V\otimes^F W} & \xrightarrow{\ \Upsilon^F\ } & \overline{W}\otimes^F\overline{V} \end{array}$$

To check that this works, we calculate

$$\begin{aligned} \Upsilon^F\left(\overline{\Upsilon^F\left(\overline{v\otimes^F w}\right)}\right) &= F(v_{[-2]}\otimes w_{[-2]})\ F^-(w_{[-1]}{}^*\otimes v_{[-1]}{}^*)^*\ \Upsilon^F\left(\overline{\overline{w_{[0]}}\otimes^F\overline{v_{[0]}}}\right) \\ &= F(v_{[-4]}\otimes w_{[-4]})\ F^-(w_{[-3]}{}^*\otimes v_{[-3]}{}^*)^*\ F(w_{[-2]}{}^*\otimes v_{[-2]}{}^*)^* \\ &\quad\ F^-(v_{[-1]}\otimes w_{[-1]})\,\overline{\overline{v_{[0]}}}\otimes^F\overline{\overline{w_{[0]}}} \\ &= \overline{\overline{v}}\otimes^F\overline{\overline{w}}\,. \end{aligned}$$

which is one of the axioms of a bar category. In this way we arrive at $({}^H\mathcal{M},\otimes^F)$ as another bar category [7]. It can be viewed as the category of comodules of H^F as a certain $*$-coquasi-Hopf algebra.

Next as in [7] we think of an H-comodule equipped with an antilinear involution categorically as a *star object*, meaning V in ${}^H\mathcal{M}$ equipped with a morphism $\star_V : V \to \overline{V}$ so that $\overline{\star_V}\,\star_V = \mathrm{bb}_V$. We shall write $\star\, v = \overline{v^*}$ to relate it to the usual $*$ as an antilinear involution on a vector space V. As $\star_V : V \to \overline{V}$ is a morphism (i.e. a right H-comodule map) we deduce that

$$v^*{}_{[-1]}\otimes\overline{v^*{}_{[0]}} = v_{[-1]}\otimes\overline{v_{[0]}{}^*}\,.$$

We can then define a star algebra A in $({}^H\mathcal{M},\otimes)$ as a star object A which has an associative product $\mu(a\otimes b) = ab$ so that $\overline{\mu}\,\Upsilon^{-1}(\star\otimes\star) = \star\mu$, which is a rather formal way to say $(ab)^* = b^*a^*$. Putting our notions this way of course now makes sense in any bar category.

Proposition 4.1 *If A is a star algebra (A,μ) in $({}^H\mathcal{M},\otimes)$ then its cochain twist (A,μ^F) obeys $\overline{\mu^F}\,\Upsilon^{F-1}(\star\otimes^F\star) = \star\mu^F$ and so is a star algebra in $({}^H\mathcal{M},\otimes^F)$. Here the $\star$ operation itself is not deformed.*

Proof: By the definitions,

$$\begin{aligned} \Upsilon^{F-1}(\star\otimes^F\star)(a\otimes^F b) &= \Upsilon^{F-1}(\overline{a^*}\otimes^F\overline{b^*}) \\ &= F(a_{[-2]}\otimes b_{[-2]})\ F^-(b_{[-1]}{}^*\otimes a_{[-1]}{}^*)^*\ \overline{b_{[0]}{}^*\otimes^F a_{[0]}{}^*}\,, \\ \overline{\mu^F}\,\Upsilon^{F-1}(\star\otimes^F\star)(a\otimes^F b) &= F(a_{[-1]}\otimes b_{[-1]})\ \overline{b_{[0]}{}^*\,a_{[0]}{}^*} \\ &= F(a_{[-1]}\otimes b_{[-1]})\ \overline{(a_{[0]}b_{[0]})^*} \end{aligned}$$

$$= \star\mu^F(a\otimes^F b)\ . \quad \square$$

Of interest in this construction is the relation between $\star$ and the tensor product. That is, given that V and W are star objects, is $V\otimes W$ a star object? To simplify this we assume that H is commutative, and then we define $(v\otimes w)^* = v^*\otimes w^*$. We define $\Gamma : V\otimes W \to W\otimes V$ to be $(\star_W^{-1}\otimes\star_V^{-1})\,\Upsilon\,\star_{V\otimes W}$ which merely recovers usual transposition in the case of $({}^H\mathcal{M},\otimes)$. In the twisted case, we define

$$\star(v\otimes^F w) = F(v_{[-2]}\otimes w_{[-2]})\ F^-(v_{[-1]}{}^*\otimes w_{[-1]}{}^*)^*\ \overline{v_{[0]}{}^*\otimes^F w_{[0]}{}^*}\ , \tag{7}$$

and this corresponds to

$$\Gamma^F(v\otimes^F w) \;=\; F(v_{[-2]}\otimes w_{[-2]})\ F^-(w_{[-1]}\otimes v_{[-1]})\ w_{[0]}\otimes^F v_{[0]}\ .$$

This is a generalised braiding associated to star objects and in some cases it obeys the braid relations [7].

The reader might want a notion of real elements that is preserved under tensor product, so that the tensor product of two real elements is real. So if we have $v = v^*$ and $w = w^*$, then we would like $(v\otimes^F w)^* = v\otimes^F w$. From (7) we can see how to ensure that this happens, we can insist on the following extra condition on the cochain:

$$F(a\otimes b)^* \;=\; F(a^*\otimes b^*)\ . \tag{8}$$

5. Twisting differential calculi and connections

Suppose that $(\Omega A, \mathrm{d}, \wedge)$ is a differential graded algebra in ${}^H\mathcal{M}$. Here the exterior algebra is a direct sum of different degrees of differential forms, with $\Omega^0 A = A$, and there is a wedge product $\wedge$ and an exterior derivative $\mathrm{d} : \Omega^n A \to \Omega^{n+1} A$ obeying $\mathrm{d}^2 = 0$ and a graded Leibniz rule. We suppose that H coacts on A and that this coaction extends to differential forms with d and $\wedge$ respecting the coaction, or put another way, that all structure maps live in ${}^H\mathcal{M}$.

The differential calculus is twisted to $(\Omega A, \mathrm{d}, \wedge^F)$, where d is unchanged, and

$$\xi \wedge^F \eta \;=\; F(\xi_{[-1]}\otimes\eta_{[-1]})\ \xi_{[0]}\wedge\eta_{[0]}\ .$$

We do not have to do any new work here, we just apply the cochain twist machinery of Section 2 to the algebra ΩA with its wedge product [3], cf. [20].

Once we have a differential calculus we can define the notion of a connection or covariant derivative. For any usual associative algebra A this is given by a map $\nabla : V \to \Omega^1 A\otimes_A V$ where V is a left A-module (more properly, a projective module for a vector bundle in noncommutative geometry), and required to obey

$$\nabla(av) = \mathrm{d}a\otimes_A v + a\nabla(v). \tag{9}$$

Its curvature is defined as a map $R_\nabla : V \to \Omega^2 A\otimes_A V$ defined by

$$R_\nabla = (\mathrm{d}\otimes\mathrm{id} - \mathrm{id}\wedge\nabla)\nabla = \nabla^{[1]}\nabla. \tag{10}$$

where the extension to higher forms is provided by

$$\nabla^{[n]} = \mathrm{d}\otimes\mathrm{id} + (-1)^n\mathrm{id}\wedge\nabla : \Omega^n A\otimes_A V \to \Omega^{n+1}\otimes_A V. \tag{11}$$

Proposition 5.1 *Given a covariant derivative ∇ on a left A-module V which is also a left H-comodule map,*

$$\nabla^F(v) \;=\; F^-(\xi_{[-1]}\otimes w_{[-1]})\;\xi_{[0]}\otimes^F w_{[0]}$$

where $\nabla(v) = \xi\otimes w$ (or a sum of such terms), provides a well-defined covariant derivative $\nabla^F : V \to \Omega^1 A_F\otimes_{A_F} V$ with V twisted to an A_F-module as in Section 2.

Proof: We can say loosely that $\nabla^F = c\circ\nabla$ or that ∇^F is defined by commutativity of

$$\begin{array}{ccc} V & \xrightarrow{\ \nabla^F\ } & \Omega^1 A\otimes^F V \\ \downarrow{\scriptstyle\nabla} & \nearrow{\scriptstyle c} & \\ \Omega^1 A\otimes V & & \end{array} \qquad (12)$$

but what we mean more precisely is a choice of representatives for the result of ∇ allowing us to work in $\Omega^1 A\otimes V$ followed by a check that ∇^F then descends to $\otimes_{A_F}$ if ∇ descends to $\otimes_A$. A brief check will further show that this is a morphism in ${}^H\mathcal{M}$. To check the left Liebniz rule, we begin by writing, for $v\in V$ and $b\in A$,

$$\begin{aligned}
\nabla v &= \xi\otimes w\ ,\\
v_{[-1]}\otimes\nabla v_{[0]} &= \xi_{[-1]}w_{[-1]}\otimes\xi_{[0]}\otimes w_{[0]}\ ,\\
v_{[-1]}\otimes\nabla(b.v_{[0]}) &= \xi_{[-1]}w_{[-1]}\otimes b.\xi_{[0]}\otimes w_{[0]} + v_{[-1]}\otimes \mathrm{d}b\otimes v_{[0]}\ ,\\
v_{[-1]}\otimes\nabla^F(b.v_{[0]}) &= F^-(b_{[-1]}\xi_{[-1]}\otimes w_{[-1]})\;\xi_{[-2]}w_{[-2]}\otimes(b_{[0]}{\cdot}\xi_{[0]}\otimes^F w_{[0]})\\
&\quad + F^-(b_{[-1]}\otimes v_{[-1]})\;v_{[-2]}\otimes(\mathrm{d}b_{[0]}\otimes^F v_{[0]})\ .
\end{aligned}$$

Now we calculate, using the last equation, and writing $\bullet = \triangleright^F$ for clarity,

$$\begin{aligned}
\nabla^F(a\bullet v) &= F(a_{[-1]}\otimes v_{[-1]})\,\nabla^F(a_{[0]}v_{[0]})\\
&= F(a_{[-2]}\otimes\xi_{[-2]}w_{[-2]})\,F^-(a_{[-1]}\xi_{[-1]}\otimes w_{[-1]})\;(a_{[0]}{\cdot}\xi_{[0]}\otimes^F w_{[0]})\\
&\quad + F(a_{[-2]}\otimes v_{[-2]})\,F^-(a_{[-1]}\otimes v_{[-1]})\;(\mathrm{d}a_{[0]}\otimes^F v_{[0]})\\
&= F(a_{[-2]}\otimes\xi_{[-2]}w_{[-2]})\,F^-(a_{[-1]}\xi_{[-1]}\otimes w_{[-1]})\;(a_{[0]}{\cdot}\xi_{[0]}\otimes^F w_{[0]}) + \mathrm{d}a\otimes^F v\ .
\end{aligned}$$

We now recognise the first term as

$$\begin{aligned}
a\bullet\nabla^F v &= F^-(\xi_{[-1]}\otimes w_{[-1]})\;a\bullet(\xi_{[0]}\otimes^F w_{[0]})\\
&= F^-(\xi_{[-2]}\otimes w_{[-2]})\;(\partial F)(a_{[-1]}\otimes\xi_{[-1]}\otimes w_{[-1]})\;(a_{[0]}\bullet\xi_{[0]})\otimes^F w_{[0]}\\
&= F^-(\xi_{[-3]}\otimes w_{[-2]})\;(\partial F)(a_{[-2]}\otimes\xi_{[-2]}\otimes w_{[-1]})\;F(a_{[-1]}\otimes\xi_{[-1]})\;(a_{[0]}{\cdot}\xi_{[0]})\otimes^F w_{[0]}\\
&= F(a_{[-2]}\otimes\xi_{[-2]}w_{[-2]})\,F^-(a_{[-1]}\xi_{[-1]}\otimes w_{[-1]})\;(a_{[0]}{\cdot}\xi_{[0]}\otimes^F w_{[0]})\ .
\end{aligned}$$

□

We next check what happens to maps which preserve the covariant derivatives between left modules. That is, given left modules and covariant derivatives (V,∇_V) and (U,∇_U), we have a left module map $\theta : V\to U$ for which $(\mathrm{id}\otimes\theta)\nabla_V = \nabla_U\theta : V\to\Omega^1 A\otimes_A U$. If we also suppose that $\theta : V\to U$ is an H-comodule map, then it is immediate from Proposition 5.1 that $(\mathrm{id}\otimes^F\theta)\nabla_V^F = \nabla_U^F\theta : V\to\Omega^1 A\otimes_{A_F} U$. We likewise extend the covariant derivative to maps

$$\nabla^{F[n]} = \mathrm{d}\otimes^F\mathrm{id} + (-1)^n\mathrm{id}\wedge\nabla^F : \Omega^n A_F\otimes_{A_F} V\to\Omega^{n+1}A_F\otimes_{A_F} V\ ,$$

where we remember that we have to use the associator to calculate $\mathrm{id}\wedge\nabla^F$. Then the curvature is defined analogously to before, as

$$R_{\nabla^F} = \nabla^{F[1]}\,\nabla^F : V\to\Omega^2 A_F\otimes_{A_F} V\ .$$

Proposition 5.2 *If the curvature of the covariant derivative (V,∇) is given by $R_\nabla(v)=\omega\otimes w\in\Omega^2A\otimes_A V$ (or a sum of such terms), then the curvature of the corresponding connection (V,∇^F) is given by*

$$R_{\nabla^F}(v) \;=\; F^-(\omega_{[-1]}\otimes w_{[-1]})\;\omega_{[0]}\otimes^F w_{[0]}\;.$$

Proof: We compute the curvature, taking $\nabla v=\xi\otimes w$, as

$$\begin{aligned}\nabla^{F[1]}\nabla^F v &= F^-(\xi_{[-1]}\otimes w_{[-1]})\;\nabla^{F[1]}(\xi_{[0]}\otimes^F w_{[0]})\\ &= F^-(\xi_{[-1]}\otimes w_{[-1]})\;\mathrm{d}\xi_{[0]}\otimes^F w_{[0]} - F^-(\xi_{[-1]}\otimes w_{[-1]})\;\xi_{[0]}\wedge^F\nabla^F w_{[0]}\;.\end{aligned}$$

Now we set $\nabla w=\eta\otimes u$, and

$$w_{[-1]}\otimes\nabla^F(w_{[0]}) \;=\; F^-(\eta_{[-1]}\otimes u_{[-1]})\;\eta_{[-2]}u_{[-2]}\otimes(\eta_{[0]}\otimes^F u_{[0]})$$

Now we calculate

$$\begin{aligned}\nabla^{F[1]}\nabla^F v &= F^-(\xi_{[-1]}\otimes w_{[-1]})\;\mathrm{d}\xi_{[0]}\otimes^F w_{[0]}\\ &\quad - F^-(\xi_{[-1]}\otimes\eta_{[-2]}u_{[-2]})\;F^-(\eta_{[-1]}\otimes u_{[-1]})\;\;\xi_{[0]}\wedge^F(\eta_{[0]}\otimes^F u_{[0]})\\ &= F^-(\xi_{[-1]}\otimes w_{[-1]})\;\mathrm{d}\xi_{[0]}\otimes^F w_{[0]}\\ &\quad - F^-(\xi_{[-2]}\otimes\eta_{[-3]}u_{[-3]})\;F^-(\eta_{[-2]}\otimes u_{[-2]})\;(\partial F)(\xi_{[-1]}\otimes\eta_{[-1]}\otimes u_{[-1]})(\xi_{[0]}\wedge^F\eta_{[0]})\otimes^F u_{[0]}\\ &= F^-(\xi_{[-1]}\otimes w_{[-1]})\;\mathrm{d}\xi_{[0]}\otimes^F w_{[0]} - F^-(\xi_{[-1]}\eta_{[-1]}\otimes u_{[-1]})\;(\xi_{[0]}\wedge\eta_{[0]})\otimes^F u_{[0]}\;.\end{aligned}$$

We compare this with the straight computation of $R_\nabla(v)$ and change notations to the form in the statement. We can say loosely that $R_{\nabla^F}=c\circ R_\nabla$. □

6. Bimodule covariant derivatives

In applying these ideas to Riemannan geometry we will be in the case of $V=\Omega^1A$, which is a bimodule. When V is a bimodule we may have a well-defined generalised braiding $\sigma: V\otimes_A\Omega^1A\to\Omega^1A\otimes_A V$ such that

$$\nabla(v\triangleleft a)=(\nabla v)\triangleleft a+\sigma(v\otimes_A\mathrm{d}a).$$

A covariant derivative on a bimodule equipped with such an object, if it exists, is called a *bimodule covariant derivative.* We already saw in Section 2 how to twist an H-covariant bimodule to a bimodule of (A,μ^F). We suppose that ∇ and σ are also H-covariant i.e. expressed in the category ${}^H\mathcal{M}$.

Proposition 6.1 *If (V,∇,σ) is an H-covariant bimodule covariant derivative then (V,∇^F,σ^F) is a bimodule covariant derivative for (A,μ^F) where (writing $\sigma(v_{[0]}\otimes\xi_{[0]})=\eta\otimes w$ or a sum of such representatives)*

$$\sigma^F(v\otimes^F\xi) \;=\; F(v_{[-1]}\otimes\xi_{[-1]})\;F^-(\eta_{[-1]}\otimes w_{[-1]})\eta_{[0]}\otimes^F w_{[0]}$$

descends to the quotients as required.

Proof: One should either work with $\otimes_A$ and $\otimes_{A_F}$ or representatives with $\otimes$ and $\otimes^F$ and an eventual quotient understood. By definition,

$$\begin{aligned}\sigma^F(v\otimes^F\mathrm{d}a) &= \nabla^F(v\triangleleft^F a)-\nabla^F(v)\triangleleft^F a\\ &= F(v_{[-1]}\otimes a_{[-1]})\;\nabla^F(v_{[0]}\,a_{[0]})-\nabla^F(v)\triangleleft^F a\;.\end{aligned}$$

If we write $\nabla v = \xi\otimes w$, then $v_{[-1]}\otimes\nabla v_{[0]} = \xi_{[-1]}\, w_{[-1]}\otimes\xi_{[0]}\otimes w_{[0]}$, so

$$\sigma^F(v\otimes^F \mathrm{d}a) \quad = \quad F(v_{[-1]}\otimes a_{[-1]})\ \nabla^F(v_{[0]}\,a_{[0]}) - F^-(\xi_{[-1]}\otimes w_{[-1]})\ (\xi_{[0]}\otimes^F w_{[0]})\triangleleft^F a\ .$$

Now we use

$$\begin{aligned} v_{[-1]}\otimes a_{[-1]}\otimes\nabla(v_{[0]}a_{[0]}) &= v_{[-1]}\otimes a_{[-1]}\otimes\nabla(v_{[0]})a_{[0]} + v_{[-1]}\otimes a_{[-1]}\otimes\sigma(v_{[0]}\otimes \mathrm{d}a_{[0]}) \\ &= \xi_{[-1]}\, w_{[-1]}\otimes a_{[-1]}\otimes\xi_{[0]}\otimes w_{[0]}a_{[0]} + v_{[-1]}\otimes a_{[-1]}\otimes\sigma(v_{[0]}\otimes \mathrm{d}a_{[0]})\ , \\ v_{[-1]}\otimes a_{[-1]}\otimes\nabla^F(v_{[0]}a_{[0]}) &= \xi_{[-2]}\, w_{[-2]}\otimes a_{[-2]}\otimes F^-(\xi_{[-1]}\otimes w_{[-1]}a_{[-1]})\ (\xi_{[0]}\otimes^F w_{[0]}a_{[0]}) \\ &\quad + v_{[-1]}\otimes a_{[-1]}\otimes F^-(\sigma^{(1)}(v_{[0]}\otimes \mathrm{d}a_{[0]})_{[-1]}\otimes\sigma^{(2)}(v_{[0]}\otimes \mathrm{d}a_{[0]})_{[-1]}) \\ &\quad \sigma^{(1)}(v_{[0]}\otimes \mathrm{d}a_{[0]})_{[0]}\otimes\sigma^{(2)}(v_{[0]}\otimes \mathrm{d}a_{[0]})_{[0]}\ . \end{aligned}$$

The first term in this equation cancels with $\nabla^F(v)\triangleleft^F a$, giving the answer. □

The main purpose of the bimodule covariant derivative construction is to allow us to tensor product covariant derivatives. Given A-bimodule covariant derivatives (V,∇_V,σ_V) and (W,∇_W,σ_W), we have the following covariant derivative on $V\otimes_A W$ defined by

$$\nabla_{V\otimes_A W}(v\otimes w) \quad = \quad \nabla_V(v)\otimes w + (\sigma_V\otimes_A \mathrm{id})(v\otimes\nabla_W w) \tag{13}$$

quotiented to $V\otimes_A W$. By Proposition 6.1 we have a bimodule connection ∇^F on $V\otimes_A W$ as an A_F-bimodule $V\otimes_{A_F} W$.

Proposition 6.2 *Given A-bimodule covariant derivatives (V,∇_V,σ_V) and (W,∇_W,σ_W), the twisted ∇^F on $V\otimes_{A_F} W$ obeys*

$$\nabla^F_{V\otimes_{A_F} W} \quad = \quad \Phi(\nabla^F_V\otimes^F \mathrm{id}) + (\sigma^F_V\otimes^F \mathrm{id})\Phi^{-1}(\mathrm{id}\otimes^F\nabla^F_W))$$

where the composition on the right descends to the required quotient over A_F.

Proof: Unwinding the definitions here, ∇^F is characterised by

$$\begin{array}{ccccc} V\otimes W & & \xrightarrow{\quad c \quad} & & V\otimes^F W \\ \Big\downarrow{\scriptstyle\nabla} & & & & \Big\downarrow{\scriptstyle\nabla^F} \\ \Omega^1 A\otimes V\otimes W & \xrightarrow{c} & \Omega^1 A\otimes^F(V\otimes W) & \xrightarrow{\mathrm{id}\otimes^F c} & \Omega^1 A\otimes^F(V\otimes^F W) \end{array} \tag{14}$$

where we again abuse notations as in previous proofs. To check that the map is indeed well defined over $V\otimes_{A_F} W$, we check that the paths from the top left to bottom right corners of the following diagrams are identical (up to the relations on $V\otimes_{A_F} W$ in the bottom right corner), which can be done easily just by considering the left and bottom paths of the diagrams:

$$\begin{array}{ccccc} V\otimes A\otimes W & \xrightarrow{c} & (V\otimes A)\otimes^F W & \xrightarrow{c\otimes^F \mathrm{id}} & (V\otimes^F A)\otimes^F W \\ \Big\downarrow{\scriptstyle\triangleleft\otimes \mathrm{id}} & & & & \Big\downarrow{\scriptstyle\triangleleft^F\otimes^F \mathrm{id}} \\ V\otimes W & & \xrightarrow{\quad c \quad} & & V\otimes^F W \\ \Big\downarrow{\scriptstyle\nabla} & & & & \Big\downarrow{\scriptstyle\nabla^F} \\ \Omega^1 A\otimes V\otimes W & \xrightarrow{c} & \Omega^1 A\otimes^F(V\otimes W) & \xrightarrow{\mathrm{id}\otimes^F c} & \Omega^1 A\otimes^F(V\otimes^F W) \end{array}$$

Journal of Physics: Conference Series **254** (2010) 012002 doi:10.1088/1742-6596/254/1/012002

$$\begin{array}{ccccc}
V\otimes A\otimes W & \xrightarrow[c]{} & V\otimes^F(A\otimes W) & \xrightarrow[\mathrm{id}\otimes^F c]{} & V\otimes^F(A\otimes^F W)\\
\downarrow{\scriptstyle \mathrm{id}\otimes\triangleright} & & & & \downarrow{\scriptstyle \mathrm{id}\otimes^F\triangleright^F}\\
V\otimes W & & \xrightarrow{\quad c\quad} & & V\otimes^F W\\
\downarrow{\scriptstyle \nabla} & & & & \downarrow{\scriptstyle \nabla^F}\\
\Omega^1A\otimes V\otimes W & \xrightarrow{c} & \Omega^1A\otimes^F(V\otimes W) & \xrightarrow{\mathrm{id}\otimes^F c} & \Omega^1A\otimes^F(V\otimes^F W)
\end{array}$$

We do not actually need to check this nor that we have a bimodule connection, in view of Proposition 6.1. However, it is instructive to do so for good measure. For the left Liebniz rule, we use

$$\begin{array}{ccccc}
A\otimes V\otimes W & \xrightarrow[c]{} & A\otimes^F(V\otimes W) & \xrightarrow[\mathrm{id}\otimes^F c]{} & A\otimes^F(V\otimes^F W)\\
\downarrow{\scriptstyle \triangleright\otimes\mathrm{id}} & & & & \downarrow{\scriptstyle \triangleright^F}\\
V\otimes W & & \xrightarrow{\quad c\quad} & & V\otimes^F W\\
\downarrow{\scriptstyle \nabla} & & & & \downarrow{\scriptstyle \nabla^F}\\
\Omega^1A\otimes V\otimes W & \xrightarrow{c} & \Omega^1A\otimes^F(V\otimes W) & \xrightarrow{\mathrm{id}\otimes^F c} & \Omega^1A\otimes^F(V\otimes^F W)
\end{array} \tag{15}$$

Summarising the right side of diagram (15) as 'right', we get two parts to (15), the first being

$$\begin{array}{ccccc}
A\otimes V\otimes W & \xrightarrow[c]{} & A\otimes^F(V\otimes W) & \xrightarrow[\mathrm{id}\otimes^F c]{} & A\otimes^F(V\otimes^F W)\\
\downarrow{\scriptstyle \mathrm{d}\otimes\mathrm{id}\otimes\mathrm{id}} & & & & \downarrow{\scriptstyle \text{first right}}\\
\Omega^1A\otimes V\otimes W & \xrightarrow{c} & \Omega^1A\otimes^F(V\otimes W) & \xrightarrow{\mathrm{id}\otimes^F c} & \Omega^1A\otimes^F(V\otimes^F W)
\end{array}$$

which gives the right hand side $\mathrm{d}\otimes^F\mathrm{id}$ by functoriality. The second term is rather more difficult:

$$\begin{array}{ccccc}
A\otimes(V\otimes W) & \xrightarrow[c]{} & A\otimes^F(V\otimes W) & \xrightarrow[\mathrm{id}\otimes^F c]{} & A\otimes^F(V\otimes^F W)\\
\downarrow{\scriptstyle \mathrm{id}\otimes\nabla} & & & & \\
A\otimes\Omega^1A\otimes V\otimes W & & & & \downarrow{\scriptstyle \text{second right}}\\
\downarrow{\scriptstyle \triangleright\otimes\mathrm{id}\otimes\mathrm{id}} & & & & \\
\Omega^1A\otimes V\otimes W & \xrightarrow{c} & \Omega^1A\otimes^F(V\otimes W) & \xrightarrow{\mathrm{id}\otimes^F c} & \Omega^1A\otimes^F(V\otimes^F W)
\end{array}$$

Functoriality of c applied to the top and bottom sides gives

$$\begin{array}{ccccc}
A\otimes(V\otimes W) & \xrightarrow[\mathrm{id}\otimes c]{} & A\otimes(V\otimes^F W) & \xrightarrow[c]{} & A\otimes^F(V\otimes^F W)\\
\downarrow{\scriptstyle \mathrm{id}\otimes\nabla} & & & & \\
A\otimes\Omega^1A\otimes V\otimes W & & & & \downarrow{\scriptstyle \text{second right}}\\
\downarrow{\scriptstyle \triangleright\otimes\mathrm{id}\otimes\mathrm{id}} & & & & \\
\Omega^1A\otimes V\otimes W & \xrightarrow{\mathrm{id}\otimes c} & \Omega^1A\otimes(V\otimes^F W) & \xrightarrow{c} & \Omega^1A\otimes^F(V\otimes^F W)
\end{array}$$

and since the bottom left two operators commute, we get

$$\begin{array}{ccccc}
A\otimes(V\otimes W) & \xrightarrow{\mathrm{id}\otimes c} & A\otimes(V\otimes^F W) & \xrightarrow{c} & A\otimes^F(V\otimes^F W) \\
\downarrow{\scriptstyle \mathrm{id}\otimes\nabla} & & & & \\
A\otimes\Omega^1 A\otimes V\otimes W & & & & \downarrow{\scriptstyle \text{second right}} \\
\downarrow{\scriptstyle \mathrm{id}\otimes\mathrm{id}\otimes c} & & & & \\
A\otimes\Omega^1 A\otimes(V\otimes^F W) & \xrightarrow{\triangleright\otimes\mathrm{id}} & \Omega^1 A\otimes(V\otimes^F W) & \xrightarrow{c} & \Omega^1 A\otimes^F(V\otimes^F W)
\end{array}$$

Now use diagram (14) to get

$$\begin{array}{ccccc}
A\otimes(\Omega^1 A\otimes^F(V\otimes^F W)) & \xleftarrow{\mathrm{id}\otimes\nabla^F} & A\otimes(V\otimes^F W) & \xrightarrow{c} & A\otimes^F(V\otimes^F W) \\
\downarrow{\scriptstyle \mathrm{id}\otimes c^{-1}} & & & & \downarrow{\scriptstyle \text{second right}} \\
A\otimes\Omega^1 A\otimes(V\otimes^F W) & \xrightarrow{\triangleright\otimes\mathrm{id}} & \Omega^1 A\otimes(V\otimes^F W) & \xrightarrow{c} & \Omega^1 A\otimes^F(V\otimes^F W)
\end{array}$$

More functoriality of c gives

$$\begin{array}{ccccc}
A\otimes^F(\Omega^1 A\otimes(V\otimes^F W)) & \xleftarrow{\mathrm{id}\otimes c^{-1}} & A\otimes^F(\Omega^1 A\otimes^F(V\otimes^F W)) & \xleftarrow{\mathrm{id}\otimes^F\nabla^F} & A\otimes^F(V\otimes^F W) \\
\downarrow{\scriptstyle c^{-1}} & & & & \downarrow{\scriptstyle \text{second right}} \\
A\otimes\Omega^1 A\otimes(V\otimes^F W) & \xrightarrow{\triangleright\otimes\mathrm{id}} & \Omega^1 A\otimes(V\otimes^F W) & \xrightarrow{c} & \Omega^1 A\otimes^F(V\otimes^F W)
\end{array}$$

Now a direct calculation will show that this is the required $\triangleright^F(\mathrm{id}\otimes^F\nabla^F : A\otimes^F(V\otimes^F W) \to \Omega^1 A\otimes^F(V\otimes^F W)$.

For the bimodule part of the bimodule covariant derivative, consider

$$\begin{array}{ccccc}
V\otimes W\otimes A & \xrightarrow{c} & (V\otimes W)\otimes^F A & \xrightarrow{c\otimes^F\mathrm{id}} & (V\otimes^F W)\otimes^F A \\
\downarrow{\scriptstyle \mathrm{id}\otimes\triangleleft} & & & & \downarrow{\scriptstyle \triangleleft^F} \\
V\otimes W & & \xrightarrow{c} & & V\otimes^F W \\
\downarrow{\scriptstyle \nabla} & & & & \downarrow{\scriptstyle \nabla^F} \\
\Omega^1 A\otimes V\otimes W & \xrightarrow{c} & \Omega^1 A\otimes^F(V\otimes W) & \xrightarrow{\mathrm{id}\otimes^F c} & \Omega^1 A\otimes^F(V\otimes^F W)
\end{array}$$

This splits into two terms, the first of which is

$$\begin{array}{ccccc}
V\otimes W\otimes A & \xrightarrow{c} & (V\otimes W)\otimes^F A & \xrightarrow{c\otimes^F\mathrm{id}} & (V\otimes^F W)\otimes^F A \\
\downarrow{\scriptstyle \mathrm{id}\otimes\mathrm{id}\otimes\mathrm{d}} & & & & \\
V\otimes W\otimes\Omega^1 A & & & & \downarrow{\scriptstyle \text{first term}} \\
\downarrow{\scriptstyle \sigma_{V\otimes W}} & & & & \\
\Omega^1 A\otimes V\otimes W & \xrightarrow{c} & \Omega^1 A\otimes^F(V\otimes W) & \xrightarrow{\mathrm{id}\otimes^F c} & \Omega^1 A\otimes^F(V\otimes^F W)
\end{array}$$

By functoriality of d and the formula for $\sigma_{V\otimes W}$, this is

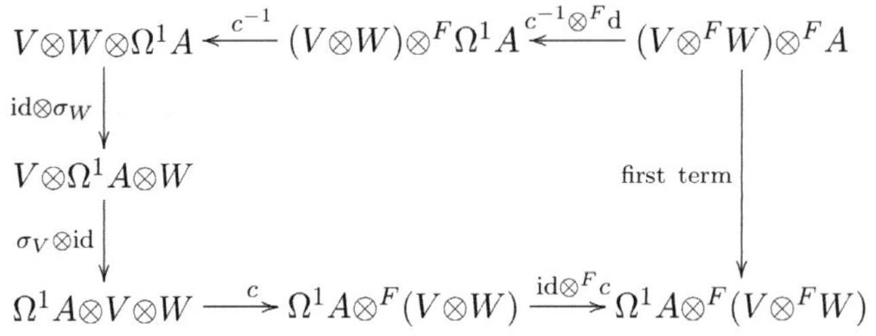

The second term is

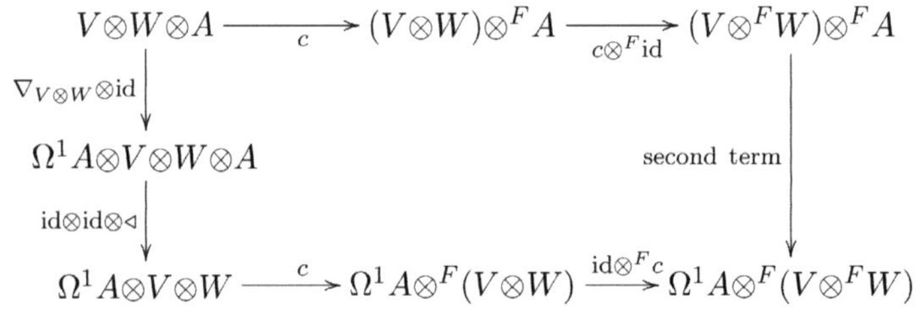

By using the functoriality of c, this is

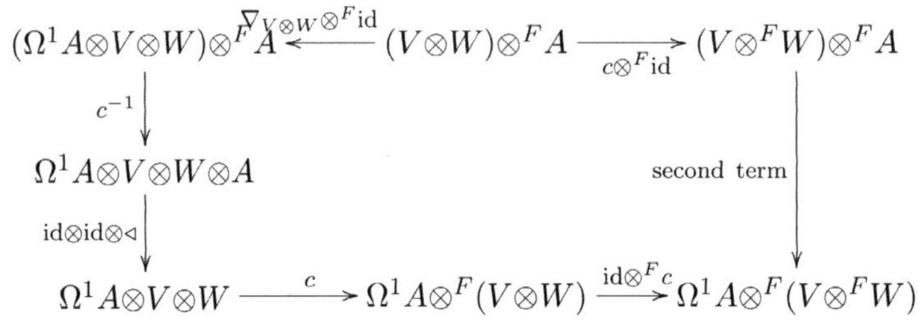

The first term of the equation in the statement is given by differentialting the v, and can be read off the diagram as the following, where we set $\nabla v=\xi\otimes u$:

$$F(\xi_{[-2]}u_{[-3]}\otimes w_{[-3]})\,F^-(\xi_{[-1]}u_{[-2]}\otimes w_{[-2]})\,F^-(u_{[-1]}\otimes w_{[-1]})\,\xi_{[0]}\otimes^F(u_{[0]}\otimes^F w_{[0]})\ ,$$

which gives $\Phi(\nabla^F_V\otimes^F\mathrm{id})$. To find the second term, we differentiate the w and substitute into diagram (14) to get

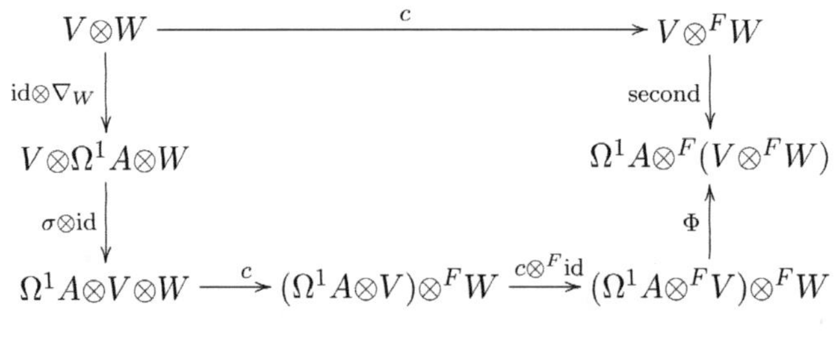

Journal of Physics: Conference Series **254** (2010) 012002 doi:10.1088/1742-6596/254/1/012002

By the functoriality of c we get

$$
\begin{array}{ccc}
V\otimes W & \xrightarrow{\quad\quad\quad\quad c \quad\quad\quad\quad} & V\otimes^F W \\
\Big\downarrow{\scriptstyle \mathrm{id}\otimes\nabla_W} & & \Big\downarrow{\scriptstyle \mathrm{second}} \\
V\otimes\Omega^1 A\otimes W & & \Omega^1 A\otimes^F(V\otimes^F W) \\
\Big\downarrow{\scriptstyle c} & & \Big\uparrow{\scriptstyle \Phi} \\
(V\otimes\Omega^1 A)\otimes^F W \xrightarrow{\sigma\otimes^F\mathrm{id}} & (\Omega^1 A\otimes V)\otimes^F W & \xrightarrow{c\otimes^F\mathrm{id}} (\Omega^1 A\otimes^F V)\otimes^F W
\end{array}
$$

and by the definition of σ^F,

$$
\begin{array}{ccc}
V\otimes W & \xrightarrow{\quad\quad\quad\quad c \quad\quad\quad\quad} & V\otimes^F W \\
\Big\downarrow{\scriptstyle \mathrm{id}\otimes\nabla_W} & & \Big\downarrow{\scriptstyle \mathrm{second}} \\
V\otimes\Omega^1 A\otimes W & & \Omega^1 A\otimes^F(V\otimes^F W) \\
\Big\downarrow{\scriptstyle c} & & \Big\uparrow{\scriptstyle \Phi} \\
(V\otimes\Omega^1 A)\otimes^F W \xrightarrow{c\otimes^F\mathrm{id}} & (V\otimes^F\Omega^1 A)\otimes^F W & \xrightarrow{\sigma^F\otimes^F\mathrm{id}} (\Omega^1 A\otimes^F V)\otimes^F W
\end{array}
$$

Finally we use the definition of Φ^{-1} and ∇^F_W. $\square$

Finally, we need to see how the notion of $\star$-compatibilty behaves. This is an important extra condition in noncommutative Riemannian geometry [8] and corresponds in our application to ensuring reality constraints. To keep things simple we again assume that the Hopf algebra H is now commutative (it then has $S^2=\mathrm{id}$), and that (A,Ω^*A,d) is a classical differential geometry. So we assume that A is commutative, for bimodule A-covariant derivatives the map σ is transposition, and that $\wedge$ is given by antisymmetrisation.

Proposition 6.3 *Suppose for simplicity that we start in the classical situation with A commutative. If ∇ is compatible with $\star$, then ∇^F is compatible with $\star$.*

Proof: In classical differential geometry, the condition that ∇ is compatible with $\star$ is

$$\overline{\sigma}\,\Upsilon^{-1}(\star\otimes\star)\nabla(v) \;=\; \overline{\nabla(v^*)}\;,$$

and this reduces to $\nabla(v^*)=\xi^*\otimes w^*$ where $\nabla(v)=\xi\otimes w$. Now

$$
\begin{aligned}
\overline{\sigma^F}\,\Upsilon^{F-1}(\star\otimes^F\star)\nabla^F v &= F^-(\xi_{[-1]}\otimes w_{[-1]})\,\overline{\sigma^F}\,\Upsilon^{F-1}(\star\otimes^F\star)(\xi_{[0]}\otimes^F w_{[0]}) \\
&= F^-(\xi_{[-1]}\otimes w_{[-1]})\,\overline{\sigma^F}\,\Upsilon^{F-1}(\overline{\xi_{[0]}{}^*}\otimes^F\overline{w_{[0]}{}^*}) \\
&= F^-(w^*_{[-1]}\otimes\xi^*_{[-1]})^*\,\overline{\sigma^F}\,(\overline{w_{[0]}{}^*\otimes^F\xi_{[0]}{}^*}) \\
&= F^-(w^*_{[-3]}\otimes\xi^*_{[-3]})^*\,F(w_{[-2]}{}^*\otimes\xi_{[-2]}{}^*)^*\,F^-(\xi_{[-1]}{}^*\otimes w_{[-1]}{}^*)^*(\overline{\xi_{[0]}{}^*\otimes^F w_{[0]}{}^*}) \\
&= F^-(\xi_{[-1]}{}^*\otimes w_{[-1]}{}^*)^*(\overline{\xi_{[0]}{}^*\otimes^F w_{[0]}{}^*}) \\
&= \overline{\nabla^F(v^*)}\;.\quad\square
\end{aligned}
$$

7. Twisting of metrics and Riemannian structures

We are now ready to put much of this together into an algebraic framework for Riemannian geometry. In this section we will start with differential calculus over an (associative) algebra A, but the example to bear in mind is of the smooth functions on a manifold. As previously, we take a differential calculus ΩA in the category ${}^H\mathcal{M}$, where (because of our previous simplifications) we assume that H is commutative. We again assume that the connection ∇ to be deformed is an H-comodule map. To fit the twisting procedure, we assume that the Riemannian metric is H-invariant. Within the standard framework of real differential geometry, star preservation is automatic (as star is the identity), so we assume that the connection preserves star.

We look at connections $\nabla : \Omega^1 A \to \Omega^1 A \otimes_A \Omega^1 A$ where $\Omega A = \oplus_n \Omega^n A$ is the exterior algebra on A. We use the formalism invented for (associative) noncommutative geometry but we are mainly interested in using it in the classical commutative case and then twisting it to generate nonassociative examples. Thus, the curvature R_∇ now has the meaning of Riemann curvature and ∇ will be some kind of Levi-Civita connection for a metric. Torsion for any connection on $\Omega^1 A$ is defined as

$$\mathrm{Tor}_\nabla = \mathrm{d} - \wedge\nabla : \Omega^1 A \to \Omega^2 A \tag{16}$$

so torsion-free makes sense.

To formulate the metric and metric compatibility there are currently two approaches. The original one [18] is

$$g \in \Omega^1 A \otimes_A \Omega^1 A, \quad \wedge(g) = 0 \tag{17}$$

where the second expresses symmetry with respect to the notion of skew-symmetrization used in the exterior algebra. We also require some form of nondegeneracy best expressed in terms of projective modules or in a frame bundle approach. Finally, this setting has a remarkable symmetry between tangent and cotangent bundles and dual to torsion (i.e. torsion on the dual bundle regarded as cotangent bundle) is a notion of *cotorsion*. If we assume that the torsion vanishes then cotorsion reduces to

$$\mathrm{coTor} \;=\; (\wedge\nabla\otimes\mathrm{id} - (\wedge\otimes\mathrm{id})(\mathrm{id}\otimes\nabla))g \in \Omega^2 A \otimes_A \Omega^1 A \tag{18}$$

and we see that its vanishing is a weaker (skew-symmetrized) version of usual metric compatibility. A generalised Levi-Civita connection is then a torsion free cotorsion free (or skew-compatible) connection. This weakening of the usual notion of metric compatibility seems to be necessary in several examples where the generalised Levi-Civita connection then exists and is unique for the chosen metric. If A is commutative we can of course impose $\nabla g = 0$ in the usual way extending it as a derivation to the two tensor factors but keeping the 'left output' of ∇ to the far left where it can be evaluated against a vector field.

A second, more recent, approach [8] makes sense for bimodule connections. First of all, there is a useful weaker notion to torsion free, namely that the torsion Tor_∇ be a bimodule map (we call this torsion-compatible). We also make use of bar categories and work with Hermitian metrics $g \in \Omega^1 A \otimes_A \overline{\Omega^1 A}$ and in this case full metric compatibility $\nabla g = 0$ makes perfect sense. There is now a slightly different notion of cotorsion in place of (18),

$$\mathrm{coTor} \;=\; (\wedge\otimes\mathrm{id})\nabla g \in \Omega^2 A \otimes_A \overline{\Omega^1 A}\,. \tag{19}$$

In [8] we showed that these notions lead, for example, to a unique torsion free metric and $*$-compatible connection with classical limit on a class of metrics on $C_q(SU_2)$ with its left covariant 3D calculus.

When all of these structures are H-covariant, we can twist them. We have already seen how $\wedge, \nabla$ twist and checking our formulae for $\wedge^F$ and ∇^F we see that

$$\mathrm{Tor}_{\nabla^F} = \mathrm{d} - \wedge^F \nabla^F = \mathrm{Tor}_\nabla$$

is unchanged as a linear map $\Omega^1 A \to \Omega^2 A$. For $g \in \Omega^1 A \otimes_A \Omega^1 A$, we take

$$g^F = c(g) \in \Omega^1 A_F \otimes_{A_F} \Omega^1 A_F$$

and the cotorsion or skew-metric compatibility computed in the category with $\otimes^F$ takes the form

$$\mathrm{coTor}^F \;=\; (\wedge^F \nabla^F \otimes^F \mathrm{id} - (\wedge^F \otimes^F \mathrm{id})\Phi^{-1}(\mathrm{id}\otimes^F \nabla^F))\, g^F \in \Omega^2 A_F \otimes_{A_F} \Omega^1 A_F$$

where an associator is inserted. The reader can easily check the following commutative diagram,

$$\begin{array}{ccc} \Omega^1 A\otimes\Omega^1 A & \xrightarrow{\;c\;} & \Omega^1 A\otimes^F\Omega^1 A \\ \big\downarrow{\scriptstyle X} & & \big\downarrow{\scriptstyle X^F} \\ \Omega^2 A\otimes\Omega^1 A & \xrightarrow{\;c\;} & \Omega^2 A\otimes^F\Omega^1 A \end{array}$$

where $X = (\wedge\otimes\mathrm{id})(\mathrm{id}\otimes\nabla)$ and $X^F = (\wedge^F\otimes^F\mathrm{id})\Phi^{-1}(\mathrm{id}\otimes^F\nabla^F)$. It follows that the twisted cotorsion vanishes if and only if the original cotorsion vanishes. Similarly in the hermitian metric framework. The following result follows directly from this discussion:

Proposition 7.1 *Let $g \in \Omega^1 A\otimes_A\Omega^1 A$ be a Riemannian metric for A, invariant under H, and assume for simplicity that we are in the classical situation. Then a torsion free, cotorsion free connection ∇ on A which is an H-comodule map is twisted to a torsion free, cotorsion free connection ∇^F on A_F.*

In particular, the classical Levi-Civita connection will be covariant and hence twist as required. We can also consider a Riemannian metric as an inner product $\langle,\rangle : \Omega^1 A\otimes_A\overline{\Omega^1 A} \to A$ (using nondegeneracy), which we assume is an H-comodule map. We can use a little more generality to get the following result for bimodule connections:

Lemma 7.2 *Begin with A-bimodule connections (V,∇_V) and (W,∇_W) in ${}^H\mathcal{M}$. Suppose that we have a map $\kappa : V\otimes_A W \to A$ in ${}^H\mathcal{M}$ which preserves the connections (using (A,d) as a connection on A), i.e.*

$$\mathrm{d}\kappa \;=\; (\mathrm{id}\otimes_A\kappa)\nabla_{V\otimes_A W} : V\otimes W \to \Omega^1 A\ .$$

Then for the twisted connections we have

$$\mathrm{d}\kappa^F \;=\; (\mathrm{id}\otimes_{A_F}\kappa^F)\nabla^F_{V\otimes_{A_F}W} : V\otimes_{A_F} W \to \Omega^1 A_F\ ,$$

where

$$\kappa^F(v\otimes^F w) \;=\; F(v_{[-1]}\otimes w_{[-1]})\,\kappa(v_{[0]}\otimes w_{[0]})$$

descends to $\kappa^F : V\otimes_{A_F} W \to A_F$.

Proof: From (14) we have (where μ is multiplication)

$$\begin{array}{ccccc} & V\otimes W & \xrightarrow{\qquad\qquad c \qquad\qquad} & & V\otimes^F W \\ {\scriptstyle \kappa}\swarrow & \big\downarrow{\scriptstyle \nabla} & & & \big\downarrow{\scriptstyle \nabla^F} \\ A & \Omega^1 A\otimes V\otimes W & \xrightarrow{\;c\;} \Omega^1 A\otimes^F(V\otimes W) & \xrightarrow{\mathrm{id}\otimes^F c} & \Omega^1 A\otimes^F(V\otimes^F W) \\ \big\downarrow{\scriptstyle \mathrm{d}} & \big\downarrow{\scriptstyle \mathrm{id}\otimes\kappa} & \big\downarrow{\scriptstyle \mathrm{id}\otimes^F\kappa} & \swarrow{\scriptstyle \mathrm{id}\otimes^F\kappa^F} & \\ \Omega^1 A & \xleftarrow{\;\mu\;} \Omega^1 A\otimes A & \xrightarrow{\;c\;} \Omega^1 A\otimes^F A & & \end{array}$$

The result can be read off from the diagram. □

Journal of Physics: Conference Series **254** (2010) 012002 doi:10.1088/1742-6596/254/1/012002

Corollary 7.3 *Let $g\in\Omega^1A\otimes_A\overline{\Omega^1A}$ be a hermitian Riemannian metric on A invariant under H and assume for simplicity that we are in the classical situation and that F obeys (8). Then a torsion free bimodule connection ∇ on A which is metric and star-preserving and an H-comodule map is twisted to a star preserving, torsion free bimodule connection ∇^F on A_F which preserves the twisted Riemannian metric.*

Again, we can take a classical connection which is trivially a bimodule connection with the flip map for σ and we will obtain a twisted one. Clearly one can go on in this line. One can use a metric to define an interior product of 1-forms on n-forms and use this to define all relevant quantities in physics including candidates for the Ricci tensor and the stress energy tensor of scalar and other fields. The problem at this point is not that we can't do this, but that there may be several different covariant formalisations all with the same classical limit and they will all twist. For this one really needs a theory that works beyond the twist examples. Until then, twisting nevertheless provides a source of examples.

This is illustrated by the Ricci tensor where the classical 'formula' of contracting Riemann could be expressed in different ways depending on where we contract. While not unique, probably the simplest formulation is as follows, as a version similar to [18, 19]. First, we need a splitting map such that

$$i:\Omega^2A\to\Omega^1A\otimes_A\Omega^1A,\quad \wedge\circ i=\mathrm{id}. \tag{20}$$

In the classical case this is $i(\xi\wedge\eta)=(\xi\otimes\eta-\eta\otimes\xi)/2$. Then we can define Ricci in our first formulation of metric as

$$\mathrm{Ricci}=((\ ,\ )\otimes\mathrm{id})(\mathrm{id}\otimes i\otimes\mathrm{id})(\mathrm{id}\otimes R_\nabla)g\in\Omega^1A\otimes_A\Omega^1A \tag{21}$$

where $g\in\Omega^1A\otimes_A\Omega^1A$ and where we write its inverse as $\langle,\rangle:\Omega^1A\otimes_A\Omega^1A\to A$. The use of the metric here is for convenience and is not visible classically as the metric and inverse metric are contracted. It would be better to use that Ω^1A is a projective module and assume the existence of a suitable trace, but this would involve more formalism. We will check shortly that in the classical case (21) is $-1/2$ of the usual Ricci tensor, i.e. the natural normalisation from a structural point of view is not quite the usual one. The Ricci scalar $S=\langle\ ,\ \rangle(\mathrm{Ricci})$ is likewise $-1/2$ the usual one.

We now twist this. In view of the way that $\wedge^F$ is defined, clearly we can set

$$i^F(\xi\wedge\eta)\ =\ c\circ i(\xi\wedge\eta)=F^-(v_{[-1]}\otimes w_{[-1]})v_{[0]}\otimes^F w_{[0]}$$

if $i(\xi\wedge\eta)=v\otimes w$ (or rather a sum of such terms) and descend to quotients. For example, if we start with classical geometry then clearly

$$i^F(\xi\wedge\eta)=\frac{1}{2}(F^-(\xi_{[-1]}\otimes\eta_{[-1]})\,\xi_{[0]}\otimes^F\eta_{[0]}-F^-(\eta_{[-1]}\otimes\xi_{[-1]})\,\eta_{[0]}\otimes^F\xi_{[0]}).$$

We then compute the Ricci tensor defined analogously in the $\otimes^F$ category,

$$\mathrm{Ricci}^F=\Big((\langle,\rangle^F\otimes^F\mathrm{id})\,\Phi^{-1}\,(\mathrm{id}\otimes^F i^F)\otimes^F\mathrm{id}\Big)\,\Phi^{-1}\,(\mathrm{id}\otimes^F R_{\nabla^F})\,g^F\ \in\ \Omega^1A_F\otimes_{A_F}\Omega^1A_F\ .$$

Proposition 7.4 $\mathrm{Ricci}^F=c(\mathrm{Ricci})$. *The Ricci scalar $S^F=S$ is unchanged under twist.*

Proof: To compare the result with classical geometry, we use the following commutative diagram, applying the top line to $g \to g^F$:

$$
\begin{array}{ccc}
\Omega^1 A \otimes_A \Omega^1 A & \xrightarrow{\quad c \quad} & \Omega^1 A \otimes^F_A \Omega^1 A \\
\downarrow{\scriptstyle \mathrm{id}\otimes R_\nabla} & & \downarrow{\scriptstyle \mathrm{id}\otimes^F R_{\nabla^F}} \\
\Omega^1 A \otimes_A (\Omega^2 A \otimes_A \Omega^1 A) & \xrightarrow{(\mathrm{id}\otimes c)c} & \Omega^1 A \otimes^F_A (\Omega^2 A \otimes^F_A \Omega^1 A) \\
\downarrow{\scriptstyle =} & & \downarrow{\scriptstyle \Phi^{-1}} \\
(\Omega^1 A \otimes_A \Omega^2 A) \otimes_A \Omega^1 A & \xrightarrow{(c\otimes \mathrm{id})c} & (\Omega^1 A \otimes^F_A \Omega^2 A) \otimes^F_A \Omega^1 A \\
\downarrow{\scriptstyle (\mathrm{id}\otimes i)\otimes \mathrm{id}} & & \downarrow{\scriptstyle (\mathrm{id}\otimes^F i^F)\otimes^F \mathrm{id}} \\
(\Omega^1 A \otimes_A (\Omega^1 A \otimes_A \Omega^1 A)) \otimes_A \Omega^1 A & \xrightarrow{((\mathrm{id}\otimes c)c\otimes \mathrm{id})c} & (\Omega^1 A \otimes^F_A (\Omega^1 A \otimes^F_A \Omega^1 A)) \otimes^F_A \Omega^1 A \\
\downarrow{\scriptstyle =} & & \downarrow{\scriptstyle \Phi^{-1}\otimes^F \mathrm{id}} \\
((\Omega^1 A \otimes_A \Omega^1 A) \otimes_A \Omega^1 A) \otimes_A \Omega^1 A & \xrightarrow{((c\otimes \mathrm{id})c\otimes \mathrm{id})c} & ((\Omega^1 A \otimes^F_A \Omega^1 A) \otimes^F_A \Omega^1 A) \otimes^F_A \Omega^1 A \\
\downarrow{\scriptstyle (\langle,\rangle\otimes \mathrm{id})\otimes \mathrm{id}} & & \downarrow{\scriptstyle (\langle,\rangle^F\otimes^F \mathrm{id})\otimes^F \mathrm{id}} \\
\Omega^1 A \otimes_A \Omega^1 A & \xrightarrow{\quad c \quad} & \Omega^1 A \otimes^F_A \Omega^1 A
\end{array}
\qquad (22)
$$

and filling in the cells as commutative in view of the definitions of the various twisted objects in relation to the originals. The method here is the same as for other results above and therefore this point given with greater brevity. □

This kind of argument is an explicit elaboration of the statement that the twist functor is an equivalence of categories. Nevertheless, there may be readers who would like an even more explicit treatment with tensors as familiar in Riemannian geometry. There are good reasons *not* to do this in noncommutative (and nonassociative) geometry, namely we would need first to elaborate a notion of *local coordinates*. We shall, however, check the definition of Ricci in (21) in the classical case. We use standard formulae, for vector fields U, V and W and Christoffel symbols:

$$
\begin{aligned}
R(U,V)(W)^l &= R^l{}_{ijk}\, W^i\, U^j\, V^k \,, \\
R(U,V)(W) &= \nabla_U \nabla_V W - \nabla_V \nabla_U W - \nabla_{[U,V]} W \,, \\
(\nabla_V W)^l &= V^k\,(W^l{}_{,k} + \Gamma^l_{ki}\, W^i) \,, \\
R^l{}_{ijk} &= \frac{\partial \Gamma^l_{ki}}{\partial x^j} - \frac{\partial \Gamma^l_{ji}}{\partial x^k} + \Gamma^m_{ki}\,\Gamma^l_{jm} - \Gamma^m_{ji}\,\Gamma^l_{km}
\end{aligned}
$$

We use ξ^a for the differential form $\mathrm{d}x^a$ in terms of coordinates on the manifold M. Then $\nabla \xi^a = -\Gamma^a_{bc}\,\xi^b \otimes \xi^c$ and

$$
\begin{aligned}
R_\nabla(\xi^a) &= (\mathrm{d}\otimes \mathrm{id} - \mathrm{id}\wedge \nabla)(-\Gamma^a_{bc}\,\xi^b\otimes\xi^c) \\
&= -\Gamma^a_{bc,d}\,\xi^d \wedge \xi^b \otimes \xi^c \;-\; \Gamma^a_{bc}\,\Gamma^c_{pq}\,\xi^b \wedge \xi^p \otimes \xi^q \\
&= -(\Gamma^a_{bc,d} + \Gamma^a_{de}\,\Gamma^e_{bc})\,\xi^d \wedge \xi^b \otimes \xi^c \\
&= -\frac{1}{2} R^a{}_{cdb}\,\xi^d \wedge \xi^b \otimes \xi^c \,.
\end{aligned}
$$

Using this to continue our calculation,

$$
\begin{aligned}
g &= g_{fa}\,\xi^f \otimes \xi^a \,, \\
(\mathrm{id}\otimes R_\nabla)\, g &= -\frac{1}{2} R^a{}_{cdb}\, g_{fa}\,\xi^f \otimes (\xi^d \wedge \xi^b \otimes \xi^c) \,.
\end{aligned}
$$

Journal of Physics: Conference Series **254** (2010) 012002 doi:10.1088/1742-6596/254/1/012002

When we apply i, we use the fact that we already have antisymmetry due to the symmetries of $R^a{}_{cdb}$, and so do not have to explicitly antisymmetrise. Referring to (21) we have

$$\begin{aligned}
\mathrm{Ricci} &= -R^a{}_{cdb}\, g_{fa}\, (\langle\, ,\rangle\otimes\mathrm{id})\, (\mathrm{id}\otimes i)(\xi^f\otimes\xi^d\wedge\xi^b)\otimes\xi^c/2 \\
&= -R^a{}_{cdb}\, g_{fa}\, (\langle\, ,\rangle\otimes\mathrm{id})\, (\xi^f\otimes(\xi^d\otimes\xi^b))\otimes\xi^c/2 \\
&= -R^a{}_{cdb}\, g_{fa}\, g^{fd}\, \xi^b\otimes\xi^c/2 \\
&= -R^a{}_{cab}\, \xi^b\otimes\xi^c/2 \\
&= -\frac{1}{2}R_{cb}\xi^b\otimes\xi^c
\end{aligned}$$

in our conventions (with $\otimes_A$ understood throughout). The $-$ sign has its roots in the fact that we work with Riemann as an operator on forms not on vector fields. The factor of $1/2$ has its roots in the fact that R_∇ has values in $\Omega^2 A$ rather than in antisymmetric tensors. Doing the same computations using the twisted objects similarly gives $\mathrm{Ricci}^F = -\frac{1}{2}R_{cb}F^-(\xi^b_{[-1]}\otimes\xi^c_{[-1]})\xi^b_{[0]}\otimes^F\xi^c_{[0]}$ as it should.

In view of the last proposition, we see that the Einstein tensor $\mathrm{Ricci}-\frac{2}{n}gS$ (the trace-reversal of Ricci) also twists as $\mathrm{Einstein}^F = c(\mathrm{Einstein})$. We can similarly extend the metric and i to interior products, a Hodge *-operation, etc, in particular to define the stress-energy tensor T of matter fields of various types as a composition of maps, all of which twist. Then (in suitable normalisation)

$$\mathrm{Einstein} = T \quad\Leftrightarrow\quad \mathrm{Einstein}^F = T^F.$$

Details will be elaborated elsewhere but the same remarks apply that this approach of itself works *too* well: we need experience and insight from examples that go beyond twisting in order to know which of various ways to do these constructions is the most natural.

8. Some examples

Examples of interest fall into two groups. The first consists of examples where the algebra of coordinates is associative and of independent interest. In this case knowing that such examples are twists opens the door to associated constructions for these algebras. Quite often in the literature, however, it is not realised that even when the algebra happens to be associative, this does not mean that the nonassociativity has gone away. Rather, as we explained in [6], it is merely *hidden* and typically resurfaces elsewhere, for example in the differential geometry. The second class of examples are 'manufactured' in the sense that the coordinate algebra, typically nonassociative, is defined for the first time, by twisting, and then studied. Since twisting works automatically, the content here is to have interesting classical data consisting of a geometry with a symmetry, which could be a discrete one, and an interesting cochain F. We shall look in detail at examples of both types. Let us note that, although it is not our focus, we can also apply our theory to the case where F is a cocycle and we stay within associative but noncommutative geometry, for example our results imply noncommutative conformal geometry on θ-deformed twistor spaces in the approach of [9].

8.1. Drinfeld twist examples

We have already covered the first class of examples in [5, 6], namely the coordinate Hopf algebras $C_q(G)$ associated to semisimple Lie algebras $\mathfrak{g}$, using a different point of view on Drinfeld's twist element F that obtains this as $C(G)^F$. We have explained the situation in the introduction. Unfortunately, the form of F is not really known other than to lowest order. It indeed takes the form

$$F(a\otimes b) = \epsilon(a)\epsilon(b) + \lambda f(a\otimes b) + O(\lambda^2)$$

where $f \in \mathfrak{g}\otimes\mathfrak{g}$ and λ is the deformation parameter. For matrix groups a Lie algebra element ξ is evaluated against matrix element functions $t^i{}_j \in C(G)$ by

$$\xi(t^i{}_j) = \rho^i{}_j(\xi)$$

where ρ is a representation, and then extended to products via the additive coproduct on ξ. In Drinfeld's theory F is given as a formal power-series of enveloping algebra elements but it can be similarly evaluated on elements to give formal power-series in λ. This was sufficient for the deformation-theory analysis in [5] but it is expected that Drinfeld's F defined in this way is actually finite and hence defined for numerical values of λ. The reason is that F is quite similar in character to the 'universal R-matrix' which makes $U_q(\mathfrak{g})$ formally quasitriangular, and this does dualise to a well-defined linear functional making $C_q(G)$ coquasitriangular over the field. The underling reason is that the relevant elements in the power-series are nilpotent and hence some power of them vanish against any finite product of representative functions. Moreover, from Drinfeld's theory one knows that $f_{21} - f = r_-$ the antisymmetric part of the Drinfeld-Sklyanin solution r of the modified classical Yang-Baxter equations (which is known explicitly). In this context it was shown in [5] that while $C(G)^F$ happens to be associative, this necessarily does not apply to $\Omega(G)^F$.

We note that although the H^F construction means rather more, its algebra can be viewed as an example of the $(\ )_F$ algebra cochain twist. To do this, take $A = C(G)$ and $H = C(G)\otimes C(G)^{cop}$ where left and right translation are viewed as a left coaction of H. The standard $\Omega(G)$ is left-and-right translation invariant (bicovariant) and hence indeed twists by

$$F(a\otimes b\otimes c\otimes d) = F(a\otimes c)F^{-}(b\otimes d)$$

where F on the right is Drinfeld's F viewed as a linear functional. We therefore have a twisted calculus $\Omega(G)_F$ for $C_q(G)$ as another point of view on [5].

In order to apply our further twisting theory we take the classical metric defined by the Killing form K,

$$g = K(\omega\otimes_A\omega)$$

where $\omega \in \Omega^1(G)\otimes\mathfrak{g}$ is the left Maurer-Cartan form and $K : \mathfrak{g}\otimes\mathfrak{g} \to \mathbb{C}$ is supposed to be nondegenerate and ad-invariant. This is both left and right invariant and hence invariant under H. The Levi-Civita connection determined by this is

$$\nabla\omega = -\frac{1}{2}[\omega\otimes_A\omega]$$

where we take the Lie bracket of the values of ω. The latter obeys the Maurer-Cartan equations

$$\mathrm{d}\omega + \frac{1}{2}[\omega\wedge\omega] = 0$$

and hence $\mathrm{Tor}_\nabla = 0$. That ∇ is metric compatible follows easily from ad-invariance of K and a short computation yields R_∇ and Ricci $\propto g$. Now because K is ad-invariant, the metric defined as above in terms of the left-invariant Maurer-Cartan form is biinvariant, i.e. both left and right translations on the group are isometries. Hence $H = C(G)\otimes C(G)^{cop}$ coacts and leaves the metric invariant and the results in Section 7 apply. Therefore we now also have a metric g^F and Levi-Civita connection ∇^F etc.

The above is an example of hidden nonassociativity in that the algebra $A_F\cong C_q(G)$ is associative. We can also do the one-sided twist using only left-covariance of all the same structures and working directly with the dual version of Drinfeld's F on $H = C(G)$ (not doubled up). Then as in [3] we have $A_F = C(G)_F$ a nonassociative algebra, now with $\Omega(A_F) = \Omega(G)_F$, g^F and ∇^F all defined by twisting. It remains, however, the case that working with Drinfeld's cochain is difficult due to lack of formulae.

8.2. Podleś sphere and nonassociative black holes

In this section we look at $\mathfrak{g} = sl_2$ and cochain F of the form

$$F = 1 + \sum_{n=1}^{\infty} c_n(\lambda)(x \otimes y)^n = f(x \otimes y)$$

for some function f. Here $[x, y] = h$, $[h, x] = 2x$, $[h, y] = -2y$ are the sl_2 relations. Although for practical purposes one can think of $F \in U(su_2)\bar{\otimes}U(su_2)$ we actually think of F in the convolution algebra of linear functionals $C(SU_2)\otimes C(SU_2) \to \mathbb{C}$, or even better as an element of the smaller space $(C(SU_2)^*)^{\otimes 2}$ i.e. as a tensor product of linear functionals. Here x (say) is regarded as such a functional and note that

$$x^n(t^{i_1}{}_{j_1} \cdots t^{i_m}{}_{j_m}) = 0$$

for all $n > m \geq 1$ due to $x^2(t^i{}_j) = 0$ in the defining representation. With these remarks we are going to work formally with F viewed in $U(su_2)\bar{\otimes}U(su_2)$ but everything can be made precise in the above way. Clearly we have $(\epsilon\otimes\mathrm{id})F = (\mathrm{id}\otimes\epsilon)F = 1$ as needed for a cochain of this form, and we require F to be invertible. For this we require that the power-series f has a non-zero radius of convergence about the origin.

According to our theory any such cochain induces a noncommutative and nonassociative geometry on any classical manifold which has an SU_2 symmetry. For Riemannian geometry it means any manifold M with an su_2-triple of Killing vectors fields ξ_i obeying,

$$[\xi_i, \xi_j] = -\epsilon_{ijk}\xi_k$$

and we set $h\triangleright = 2\imath\xi_3$, $x\triangleright = -\xi_+$ and $y\triangleright = -\xi_-$ where $\xi_\pm = \xi_1 \pm \imath\xi_2$ to match our above conventions. Note that since we based our twists on left coactions this corresponds to a right action of $U(su_2)$ now. However, this Hopf algebra is isomorphic to its opposite so we can work with familiar left actions as stated. For our global theory we should suppose that these vector fields integrate to a right action of SU_2 and that everything is algebraic (so that we have a coaction). However, none of these is really necessary in deformation theory where we work only with power-series.

Proposition 8.1 *We let $M = S^2$ defined as usual with coordinate relation $\sum_i x_i^2 = 1$ and $\xi_i = \epsilon_{ijk}x_j\frac{\partial}{\partial x_k}$. A nontrivial F of the form above can* never *induce an associative product. However, if $c_2 = c_1^2$ the product of $C(S^2)_F$ coincides on the generators with that of the Podleś sphere defined by generators K, z, z^* with relations*

$$zK = q^2Kz, \quad z^*z = 1 - K^2, \quad zz^* = 1 - q^2K^2,$$

where $q^2 = 1 - 2c_1$. We assume that c_1 is real and $\neq 1, \frac{1}{2}$.

Proof: We work in complex coordinates $x_\pm = x_1 \pm \imath x_2$ and begin by computing the action as

$$x\triangleright\begin{pmatrix} x_+ \\ x_- \\ x_3 \end{pmatrix} = \imath\begin{pmatrix} 0 \\ -2x_3 \\ x_+ \end{pmatrix}, \quad y\triangleright\begin{pmatrix} x_+ \\ x_- \\ x_3 \end{pmatrix} = \imath\begin{pmatrix} 2x_3 \\ 0 \\ -x_- \end{pmatrix}.$$

We compute A_F from $a \bullet b = \cdot(F\triangleright(a\otimes b)) = \cdot f(x\otimes y)(a\otimes b)$ to obtain

$$x_3 \bullet x_3 = x_3^2 - c_1x_+(-x_-), \quad x_+ \bullet x_+ = x_+^2, \quad x_- \bullet x_- = x_-^2$$

$$x_+ \bullet x_- = x_+x_-, \quad x_- \bullet x_+ = x_-x_+ - c_1(-2x_3)2x_3 - 4c_2x_+(-x_-)$$
$$x_+ \bullet x_3 = x_+x_3, \quad x_3 \bullet x_+ = x_3x_+ - c_1x_+2x_3$$
$$x_- \bullet x_3 = x_-x_3 - c_1(-2x_3)(-x_-), \quad x_3 \bullet x_- = x_3x_-.$$

Now, from the last four relations we see that $x_\pm$ indeed q^2-commute with x_3 as required for $K \propto x_3$ and $z \propto x_-$. From the first relation and $x_+x_- = r^2 - x_3^2$ for a sphere of radius r we see that

$$x_3 \bullet x_3 = x_3^2 + c_1(r^2 - x_3^2) = (1 - c_1)x_3^2 + r^2c_1$$

and hence $x_3^2(1 - c_1) = x_3 \bullet x_3 - r^2c_1$. We also see that

$$x_+ \bullet x_- = x_+x_- = r^2 - x_3^2 = \frac{(r^2 - x_3 \bullet x_3)}{(1 - c_1)}.$$

Accordingly we set $z = x_-/\sqrt{1 - c_1}$ and $K = x_3/r$ and have the 2nd relation of the Podleś sphere. Note now that *if we assume associativity* among the generators,

$$x_- \bullet (x_+ \bullet x_-) = x_- \bullet \frac{(r^2 - x_3 \bullet x_3)}{1 - c_1} = \frac{x_-r^2}{1 - c_1} - \frac{(1 - 2c_1)^2}{1 - c_1}(x_3 \bullet x_3) \bullet x_-$$
$$= (x_- \bullet x_+) \bullet x_- = \frac{(A - Bx_3 \bullet x_3)}{(1 - c_1)} \bullet x_-$$

forces $A = r^2$ and $B = (1 - 2c_1)^2$ (in other words, the 3rd of the Podleś sphere relations is dictated by the first two and associativity). Comparing with what we computed for $x_- \bullet x_+$ we need $c_1^2 = c_2$ and in this case we obtain the 3rd relation as well. A look at $\bullet$ of some higher degree monomials, however, confirms that in this case $\bullet$ is nevertheless not associative. □

We have included this example because it is striking that, while nonassociative, the nonassociativity is almost hidden and on the generators gives us a Podleś sphere, i.e. a 'mildly nonassociative' Podleś sphere. Notice that the true Podleś sphere is $C_q(SU_2)$-covariant which suggests that our form of F has features similar to the Drinfeld cochain for sl_2. Also, the true Podleś sphere has recently been interpreted as a q-fuzzy sphere or more geometrically as a slice in the q-hyperboloid of unit Lorentzian distance in Minkowski space [17] and indeed arises in an effective description of 3D quantum gravity *with* cosmological constant [21].

We can apply this construction to spheres of any radius and to manifolds nested by spheres such as $\mathbb{R}^3$, the entire hyperboloid in Minkowski space, and the Schwarzschild black-hole. Clearly, it will induce a 'mildly nonassociative' Podleś or q-fuzzy sphere in place of each classical sphere. In particular, as the rotational symmetry is an isometry of the Schwarzschild solution, the metric is invariant and we have g^F, ∇^F on these spaces by our construction.

It is worth noting that these results are in the spirit of but rather different from recent attempts to apply *projective* Drinfeld twists to quantize black holes [22] based on a certain different $F = f(x{\otimes}y)$ in [1] where c_n are generated by a beta-function. As we have seen, using such a cochain in a standard way cannot give an associative algebra. Rather, the point of view is quite different; F acts on $S^2 = SU_2/U(1)$ via *right-translation* to induce a left-translation SU_2-invariant bidifferential operator. Here $U(su_2)$ itself does not act on $C(SU_2/U(1))$ in such a way at all, so this does not fit our framework. However, the failure of F to be a cocycle is contained ($\phi = \partial F$ has an h to the right in its middle factor, and this acts trivially) and as a result the quantizaiton is accidentally associative, giving a fuzzy-sphere and a residual SU_2 symmetry. Although the framework is different from our more conventional use of Drinfeld twists, the lesson is the same that even though the algebra quantizing each sphere is associative, there remains a hidden nonassociativity in the model as F is not exactly a cocycle.

8.3. *Octonionic-S^7 as second quantisation*

In this section we give a first concrete example of nontrivial nonassociative Riemannian geometry, with curvature, of interest because it lies in the same category as the one in which the octonions $\mathbb{O}$ live in the approach of [2]. This is the category of $\mathbb{Z}_2^3$-graded spaces and we use the same cochain F that 'quantizes' $k(\mathbb{Z}_2^3)$ to obtain $\mathbb{O} = k(\mathbb{Z}_2^3)_F$ as explained in Section 3. We again work algebraically over a field k, and in the sequel we will assume this to be of characteristic zero (the reader can keep in mind $k = \mathbb{R}$). In physics when the result of first quantisation is (in some form) used as the basis of another (infinite dimensional) quantization, one speaks of *second quantization.* In the same spirit, we regard the result of the first (finite) quantization as a space in its own right and look at the submanifold of unit length octonions. This submanifold is S^7 and we 'requantize' this via a twist, albeit quite mildly by an action of $\mathbb{Z}_2^3$. These methods could be used to similarly twist any geometry admitting a finite group symmetry.

As usual, the effort is all in the preparation. We recall that the description of octonions we use has basis $\{e_a\}$ labelled by $a \in \mathbb{Z}_2^3$. The norm on the octonions in this basis is the usual euclidean one. Hence our natural description of S^7 is with coordinates $\{x^a\}$ again labelled by $a \in \mathbb{Z}_2^3$ and the relation $\sum_a x_a^2 = 1$. Following [14] we also have a coproduct, counit and antipode,

$$\Delta x_c = \sum_{a+b=c} x_a \otimes x_b F(a,b), \quad \epsilon x_a = \delta_{a,0}, \quad S x_a = F(a,a) x_a$$

making $k(S^7)$ into a Hopf coquasigroup. It fails to be an ordinary Hopf algebra just because S^7 is not a group. Notice that all these maps and the algebra relations themselves respect a $\mathbb{Z}_2^3$ grading defined by $|x_a| = a$, i.e. $H = k\mathbb{Z}_2^3 \cong k(\mathbb{Z}_2^3)$ coacts on everything here.

Next, in [14] Hopf-algebra-like methods are used to define the differential geometry of S^7 and we continue this programme now to compute its Riemannian geometry. The starting point is that a natural differential structure defined by the ideal $(k(S^7)^+)^2$ as for algebraic groups, is parallelizable with the basis of 'left-invariant' 1-forms $\{\omega^i\}$ where $\omega^i = S x^i{}_{(1)} \mathrm{d} x^i{}_{(2)} = \sum_a F(a,i) x_a \mathrm{d} x_{a+i}$ is defined as for Hopf algebras and computes as stated. Our convention is that middling indices i, j, k, l, m, n run in the range $1, \cdots, 7$. This induces 'left-invariant vector fields' ∂_i and 'Lie-like structure constants' $c^i{}_{jk} \in k(S^7)$ by

$$\mathrm{d} f = \sum_i (\partial_i f) \omega^i, \quad \mathrm{d}\omega^i + c^i{}_{jk} \omega^j \omega^k = 0$$

which can be computed as

$$\partial_i = -\sum_a F(a,i) x^{a+i} \frac{\partial}{\partial x^a}$$

$$c^i{}_{jk} = -F(i,j) F(i+j,k) \sum_a \phi(a,i,j) \phi(a,i+j,k) F(a,i+j+k) x^a x^{a+i+j+k}.$$

Notice that the former do not close under commutator and the latter are not constants but functions; both effects are due to the nonassociativity of the product on S^7. Moreover,

Proposition 8.2 *The 'Killing form' is $c^m{}_{in} c^n{}_{jm} = -6\delta_{ij}$ (summation of repeated indices understood). Using δ_{ij} from now on to lower indices, c_{ijk} is totally antisymmetric.*

We refer to [14] for the methods used to obtain these results. However, F and ϕ are given by particular formulae hence all our assertions can be checked by direct computation using Mathematica. Being antisymmetric in the outer two indices then means that the metric

$$g = \delta_{ij} \omega^i \otimes \omega^j$$

is preserved by

$$\nabla\omega^i = -c^i{}_{jk}\omega^j\otimes\omega^k$$

as

$$\nabla g = -c^i{}_{mn}\delta_{ij}\omega^m\otimes\omega^n\otimes\omega^j - c^j{}_{mn}\delta_{ij}\omega^i\otimes\omega^m\otimes\omega^n = 0.$$

In calculations we leave $\otimes_A$ understood. Here ∇ acts as a derivation and its left output (which might be evaluated against a vector field) is kept to the far left and all $\otimes$ products are over $k(S^7)$ so we can bring coefficients to the left as well. This connection is also manifestly torsion free and is therefore the Levi-Civita connection for the natural metric constructed in this way. These steps are exactly as for the Lie group case in Section 8.1 but now $c^i{}_{jk}$ are functions and do not define a Lie algebra (they define a Mal'tsev algebra at the identity of S^7), and we have given the computations in a basis rather than more abstractly.

It remains to compute the Riemann and Ricci curvatures to make sure we have the usual constant curvature metric. From (10) we compute

$$\begin{aligned}
R_\nabla\omega^i &= (\mathrm{d}\otimes\mathrm{id} - \mathrm{id}\wedge\nabla)(-c^i{}_{jk}\omega^j\otimes\omega^k)\\
&= -\mathrm{d}c^i{}_{jk}\omega^j\otimes\omega^k + c^i{}_{jk}c^j{}_{mn}\omega^m\omega^n\otimes\omega^k - c^i{}_{jk}\omega^j c^k{}_{mn}\omega^m\otimes\omega^n\\
&= -\left(c^i{}_{mj}c^j{}_{nk} - c^i{}_{jk}c^j{}_{mn} + \partial_m c^i{}_{nk}\right)\omega^m\omega^n\otimes\omega^k\ .
\end{aligned}$$

To obtain the Ricci curvature we apply a map i that lifts the 2-form-valued part of R_∇ to an antisymmetrised element of $\Omega^1\otimes\Omega^1$ (this is one way to be able to define an interior product). We then contract the first leg of the metric and the first leg of the lifting with the inverse metric $\langle\omega^i,\omega^j\rangle = \delta_{ij}$ as in (21),

$$\begin{aligned}
\mathrm{Ricci} &= \delta_{pi}\langle\omega^p,(i\otimes\mathrm{id})R_\nabla\omega^i\rangle\\
&= -\delta_{pi}\langle\omega^p,\frac{1}{2}\left(c^i{}_{mj}c^j{}_{nk} - c^i{}_{nj}c^j{}_{mk} - 2c^i{}_{jk}c^j{}_{mn} + \partial_m c^i{}_{nk} - \partial_n c^i{}_{mk}\right)\omega^m\rangle\omega^n\otimes\omega^k\\
&= -\frac{1}{2}\left(c^i{}_{ij}c^j_{nk} - c^i{}_{nj}c^j{}_{ik} - 2c^i{}_{jk}c^j{}_{in} + \partial_i c^i{}_{nk} - \partial_n c^i{}_{ik}\right)\omega^n\otimes\omega^k\\
&= \frac{1}{2}\left(c^i{}_{nj}c^j{}_{ki} - \partial_i c^i{}_{nk}\right)\omega^n\otimes\omega^k = -3g
\end{aligned}$$

using the symmetries of c and the Killiing form identity. Here the term $\partial_i c^i{}_{nk}$ is antisymmetric in n, k hence it must vanish as the Ricci tensor is necessarily symmetric. This turns out to be the case as one can compute using Mathematica from the formulae already given. Hence this S^7 is an Einstein space as it should be (the usual answer here for a unit 7-sphere would be $6g$; the minus sign and factor of 1/2 is due to our conventions as explained at the end of Section 7). In short, the main difference that the $c^i{}_{jk}$ are functions does show up in the full Riemann tensor but not in Ricci.

We have given an algebraic derivation of the usual geometry of S^7 using Hopf coquasigroup methods. This is actually a uniform approach that applies to all the parallellizable spheres. Now notice that $|\omega^i| = i$ (here d does not change the $\mathbb{Z}_2^3$-degree) and $|c^i{}_{jk}| = i+j+k$. Hence g has zero degree and ∇ does change the degree, i.e. everything is covariant under the grading Hopf algebra $k\mathbb{Z}_2^3$ and we can apply our cochain twist. We obtain $k(S^7)_F$ and a calculus $\Omega(S^7)_F$ on it, with everything nonassociative. The product $x^a \bullet x^b = F(a,b)x^a x^b$ obeys

$$(x^a\bullet x^b)\bullet x^c = \phi(a,b,c)x^a\bullet(x^b\bullet x^c),\quad x^b\bullet x^a = \mathcal{R}(a,b)x^a\bullet x^b$$

where $\phi(a,b,c) = -1$ if and only if a,b,c are linearly independent, as before, and $\mathcal{R}(a,b) = -1$ if and only if a,b are linearly independent, and otherwise 1, again as for the octonions. The second feature means that the algebra is *altercommutative* in the sense that elements anticommute if and only if they have distinct non-zero degrees. Otherwise they commute. This is not, however, the octonion algebra. For products of generators we have degree (and the same for their $\bullet$ products)

$$|(x^0)^m(x^1)^{n_1}\cdots(x^7)^{n_7}| = d(N) = (n_4+n_5+n_6+n_7, n_2+n_3+n_6+n_7, n_1+n_3+n_5+n_7)$$

where $N = (n_1,\cdots,n_7)$ and the degree is computed modulo 2 to give an element of $\mathbb{Z}_2^3$. We have enumerated the $\{x^a\}$ in binary order $x^{000}, x^{001},\cdots,x^{111}$ for brevity. Then

$$(\prod(x^i)^{n_i})\bullet(\prod(x^j)^{m_j}) = F(d(N),d(M))\prod(x^i)^{n_i+m_i}$$

and associativity remains via ϕ on the relevant degrees (so three monomials fail to associate if and only if their $\mathbb{Z}_2^3$-degrees are linearly independent) and commutativity remains via $\mathcal{R}$ (so two monomials fail commute if and only if they have different non-zero degrees). The generator x^0 has zero degree, has unchanged product with everything and any power of it associates with everything. Finally, we have the sphere relation which now takes the form

$$1 = x^0\bullet x^0 - \sum_i x^i\bullet x^i.$$

The even powers $x^i\bullet x^i = -(x^i)^2$ and $1, x^0$ generate a commutative associative subalgebra of degree 0, i.e. which commutes and associates with everything else. Similarly, we can compute the differential calculus obtained by twisting the classical one. The products

$$x^a\bullet \mathrm{d}x^b = F(a,b)x^a\mathrm{d}x^b,\quad \mathrm{d}x^a\bullet\mathrm{d}x^b = F(a,b)\mathrm{d}x^a\mathrm{d}x^b$$

obey parallel formulae controlled by $\phi,\mathcal{R}$. Explicitly,

$$0 = x^0\bullet\mathrm{d}x^0 - \sum_i x^i\bullet\mathrm{d}x^i$$

with relations involving x^0 or $\mathrm{d}x^0$ unchanged, as is $\mathrm{d}x^a\bullet\mathrm{d}x^a = 0$ for all a. The changed relations for distinct i,j,k are

$$x^i\bullet\mathrm{d}x^j = -\mathrm{d}x^j\bullet x^i,\quad \mathrm{d}x^i\bullet\mathrm{d}x^j = \mathrm{d}x^j\bullet\mathrm{d}x^i,\quad (x^i\bullet\mathrm{d}x^j)\bullet x^k = \phi(i,j,k)x^i\bullet(\mathrm{d}x^j\bullet\mathrm{d}x^k)$$

and similar. In degree three we have

$$(\mathrm{d}x^i\bullet\mathrm{d}x^j)\bullet\mathrm{d}x^k = -\mathrm{d}x^i\bullet(\mathrm{d}x^j\bullet\mathrm{d}x^k)$$

if and only if $i+j+k\neq 0$.

By our theory the metric and connection also twist:

$$g^F = \omega^0\otimes\omega^0 - \sum_i \omega^i\otimes\omega^i,\quad \nabla^F\omega^i = \sum_{j,k} F(i,k)c^i{}_{jk}\omega^i\otimes\omega^k$$

where we made use of $F(i+k,k) = F(i,k)F(k,k) = -F(i,k)$ to simplify. We write the summations explicitly for clarity. The inverse metric $(\ ,\ )^F$ also acquires -1 values on the $i=1,\cdots,7$ diagonal entries and

$$i^F(\omega^i\bullet\omega^j) = \frac{1}{2}(\omega^i\otimes\omega^j + \omega^j\otimes\omega^i),\quad i\neq j$$

because $F(i,j) = -F(j,i)$ when we compute $i^F(\omega^i\omega^j)$, which we can then cancel as the same factor occurs in $\omega^i \bullet \omega^j$. Following through the calculation of Ricci_F in terms of these quantities and remembering the associator, one can verify

$$\mathrm{Ricci}^F = -3g^F.$$

as implied by Propostion 7.4 (we again have an Einstein space, with constant curvature). This provides a useful check of our nonassociative theory.

Further insight into this 'octonionic' S^7 can be gained by looking at the same construction for second-quantized quaternions and complex numbers. The quaternions are constructed in the same way but with $\mathbb{Z}_2^2$-grading. F has the same form but without the 'cubic' term. As a result, F is a cocycle and $\phi = 1$. The above geometry gives the classical Riemannian geometry of S^3 viewed as the unit quaternions, and $k(S^3)_F$ is associative and altercommutative. It therefore has bosonic central generators $1, x^0$, and noncommutative generators x^1, x^2, x^3 with relations

$$\{x^1, x^2\} = \{x^2, x^3\} = \{x^3, x^1\} = 0, \quad 1 = (x^0)^2 - (x^1)^2 - (x^2)^2 - (x^3)^2$$

where the bullet product is understood. The noncommutative generators are 'Grassman-like' but do not anticommute with themselves, indeed the algebra is commutative in the category of $\mathbb{Z}_2^2$-graded spaces with a specific symmetric braiding different from that of the category of supervector spaces. If we go down one level more to $\mathbb{Z}_2$-gradings we have one degree 1 generator x^1 and $1, x^0$ bosonic. As a result everything is commutative and associative and $k(S^1)_F = k[x^0, x^1]$ modulo the relation $1 = (x^0)^2 - (x^1)^2$, i.e. a hyperbola. Again one has metric g^F and connection ∇^F with constant curvature. So all of these 'twisted spheres' are actually some kind of hyperbolae. One could view them as baby versions of the q-deformation twist in Section 8.2 which also turned spheres into mildly nonassociative q-hyperboloids.

References

[1] Alekseev A and Lachowska A 2005 Invariant ∗-products on coadjoint orbits and the Shapovalov pairing *Comment. Math. Helv.* **80** 795-810
[2] Albuquerque H and Majid S 1999 Quasialgebra structure of the octonions *J. Algebra* **220** 188-224
[3] Akrami S E and Majid S 2004 Braided cyclic cocycles and nonassociative geometry *J. Math. Phys.* **45** 3883-3911
[4] Amelino-Camelia G and Majid S 2000 Waves on noncommutative spacetime and gamma-ray bursts *Int. J. Mod. Phys. A* **15** 4301-4324
[5] Beggs E J and Majid S 2006 Semiclassical differential structures, *Pac. J. Math.* **224** 1-44
[6] Beggs E J and Majid S (in press) Quantization by cochain twists and nonassociative differentials *J. Math. Phys.*
[7] Beggs E J and Majid S 2009 Bar categories and star operations *Alg. and Representation Theory* **12** 103-152
[8] Beggs E J and Majid S 2009 *-Compatible connections in noncommutative Riemannian geometry *Preprint* arXiv:0904.0539
[9] Brain S J and Majid S 2008 Quantisation of twistor theory by cocycle twist *Commun. Math. Phys.* **284** 713-774
[10] Connes A 1994 *Noncommutative Geometry* (London and San Diego: Academic Press)
[11] Connes A, Chamseddine A and Marcolli M 2007 Gravity and the standard model with neutrino mixing *Adv. Theor. Math. Phys.* **11** 991
[12] Drinfeld V G 1990 Quasi-Hopf algebras *Leningrad Math. J.* **1** 1419-1457
[13] Hawkins E 2004 Noncommutative rigidity *Commun. Math. Phys.* **246** 211-235
[14] Klim J and Majid S (in press) Hopf quasigroups and the algebraic 7-sphere *J. Algebra*
[15] Majid S 2000 *Foundations of Quantum Group Theory* (Cambridge: Cambridge University Press 1995)
[16] Majid S 2005 Gauge theory on nonassociative spaces *J. Math. Phys.* **46** 103519
[17] Majid S 2008 q-Fuzzy spheres and quantum differentials on $B_q[SU_2]$ and $U_q(su_2)$ *Preprint* arXiv:0812.4942

[18] Majid S 2002 Riemannian geometry of quantum groups and finite groups with nonuniversal differentials *Commun. Math. Phys.* **225** 131-170

[19] Majid S 2005 Noncommutative Riemannian and spin geometry of the standard q-sphere *Commun. Math. Phys.* **256** 255-285

[20] Majid S and Oeckl R 1999 Twisting of quantum differentials and the Planck scale Hopf algebra *Commun. Math. Phys.* **205** 617-655

[21] Majid S and Schroers B 2009 q-Deformation and semidualisation in 3D quantum gravity *J. Phys A* **42** 425402

[22] Schupp P and Solodukhin S N 2009 Exact black hole solutions in noncommutative gravity *Preprint* arXiv:0906.2724

Journal of Physics: Conference Series **254** (2010) 012003 doi:10.1088/1742-6596/254/1/012003

The infinite Dirac operator

Thomas Bieske[1] **and John Ryan**[2]

[1]Department of Mathematics and Statistics, University of South Florida, Tampa, FL 33620-5700, USA

[2]Department of Mathematical Sciences, University of Arkansas, Fayetteville, AR 72701, USA

E-mail: tbieske@mail.usf.edu, jryan@uark.edu

Abstract. In this article, we define the infinite Dirac operator and explore some key properties, particularly its conformal invariance. En route, we also establish the conformal invariance of the p-Dirac equation. We also introduce the infinite Dirac operator on the sphere S^n and establish the relationship between the two infinite Dirac operators via the Cayley transformation. Also we introduce an infinite Laplace operator on S^n.

Received: 15 March 2010

1. Introduction

Associated to most Laplace operators is a first order Dirac-type operator. For the Laplacian in Euclidean space, there is the Dirac operator arising in Clifford analysis in euclidean space. For the Laplace-Beltrami operator on a Riemannian manifold, there is the Hodge-Dirac operator $d + *d*$. Furthermore, on S^n, for the conformal Laplacian, or Yamabe operator, there is the spherical Dirac operator, and for a spin manifold associated to the Atiyah-Singer-Dirac operator, there is the spinorial Laplacian.

In [6] it is shown that associated to the second-order non-linear p-Laplace operator $(1 < p < \infty)$ there is also a first order non-linear p-Dirac operator, and the conformal structure associated with the p-Laplace operator carries over naturally to the p-Dirac operator. In this paper we extend that work by introducing a first order non-linear infinite Dirac operator associated to the non-linear infinite Laplacian in $\mathbb{R}^n$. We introduce its "fundamental solution". We establish its conformal invariance and the conformal covariance of its solution space. En route we need to establish the conformal invariance of the p-Dirac operator and the conformal covariance of the solution set of the p-Dirac equation. This differs from the p-Laplace case where conformal invariance of the operator only exists for $p = n$. For all other choices of p the p-Laplace operator transforms to an (A, p)-Laplace operator where $A(x)$ is a weight function dependant on the choice of Möbius transformation. We conclude by using the Cayley transformation to extend some of our results to the sphere, and introducing a p-Laplace operator, and infinite Laplace operator on S^n together with their fundamental solutions.

2. Background

One way to derive the Laplace equation in $\mathbb{R}^n$ is from the Dirichlet integral $\int_U \|\nabla f\|^2 dx^n$, where $f : U \to \mathbb{R}$ and U is a domain in $\mathbb{R}^n$. Assuming U is bounded and also assuming appropriate conditions on the boundary of U, minimizing this integral leads to the vanishing of the first moment integral. So we have

$$\int_U \langle \nabla f, \nabla \psi \rangle dx^n = 0,$$

where $\psi \in C_0^\infty(U)$. Applying Stokes' Theorem to this integral leads to Laplace's equation.

To get to the p-Laplace equation we replace the 2 in the Dirichlet integral by p to get $\int_U \|\nabla f\|^p dx^n$ where $1 < p < \infty$. Again minimizing leads to the integral equation

$$\int_U \langle \|\nabla f\|^{p-2} \nabla f, \nabla \psi \rangle dx^n = 0.$$

Applying Stokes' Theorem gives rise to the equation

$$-\operatorname{div}(\|\nabla f\|^{p-2} \nabla f) = 0.$$

This is the p-Laplace equation, and the term $-\operatorname{div}(\|\nabla f\|^{p-2}\nabla f)$ is denoted by $\Delta_p f$. For $p \neq 2$ the p-Laplace equation is nonlinear. Letting $g : \partial U \to \mathbb{R}$ be a continuous function $g : \partial U \to \mathbb{R}$, there exist unique solutions to the Dirichlet problem

$$\begin{cases} \Delta_p f = 0 & \text{in} \quad U \\ f = g & \text{on} \quad \partial U. \end{cases}$$

In addition, the fundamental solution to the p-Laplace equation is given by

$$h_p(x) = \begin{cases} \|x\|^{\frac{n-p}{1-p}} & \text{for} \quad p \neq n \\ \log \|x\| & \text{for} \quad p = n. \end{cases}$$

That is, for appropriate constants $C_{p,n} = C(p,n)$ we have

$$\Delta_p h_p(x) = C_{p,n} \delta_0$$

in the sense of distributions. For further discussion on the p-Laplace equation in the Euclidean environment, the interested reader is directed to [3, 4] and the references therein.

Taking the limit as $p \to \infty$ we obtain the infinite Laplace operator.

3. Preliminaries

We wish to extend the nonlinear infinite Laplace equation by an infinite Dirac equation. To do this we need some background on Clifford algebras and some results from [6] on p-Dirac operators.

We shall consider Euclidean n space, $\mathbb{R}^n$, as embedded in the real 2^n dimensional Clifford algebra Cl_n where for each $x \in \mathbb{R}^n$ we have $x^2 = -\|x\|^2$. In terms of an orthonormal basis $e_1, \ldots, e_n$ this gives the anticommutation relationship

$$e_i e_j + e_j e_i = -2\delta_{ij}$$

and a basis

$$1, e_1, \ldots, e_n, \ldots, e_{j1} \ldots e_{j_r}, \ldots, e_1 \ldots e_n$$

for Cl_n, where $j_1 < \ldots < j_r$ and $1 \leq r \leq n$.

We shall need the antiautomorphism $\sim: Cl_n \to Cl_n :\sim (e_{j1} \ldots e_{j_r}) = e_{j_r} \ldots e_{j1}$. For $A \in Cl_n$ we write $\tilde{A}$ for $\sim A$.

We also need the antiautomorphism $- : Cl_n \to Cl_n : -(e_{j1} \ldots e_{j_r}) = (-1)^r e_{j_r} \ldots e_{j1}$. We write $\overline{A}$ for $-(A)$. The scalar part of $A\overline{A}$ gives the square of the norm of A. More precisely, if $A = a_0 + \ldots + a_{1\ldots n} e_1 \ldots e_n$ then the scalr part of $A\overline{A}$ is $a_0^2 + \ldots + a_{1\ldots n}^2$.

In [1] and elsewhere it is shown that if $y = M(x)$ is a Möbius transformation over $\mathbb{R}^n \cup \{\infty\}$ then $y = (ax+b)(cx+d)^{-1}$ where a, b, c and $d \in Cl_n$ with the following properties:

Journal of Physics: Conference Series **254** (2010) 012003 doi:10.1088/1742-6596/254/1/012003

(i) a, b, c and d are products of vectors in $\mathbb{R}^n$

(ii) $a\tilde{c}$, $c\tilde{d}$, $d\tilde{b}$ and $b\tilde{a} \in \mathbb{R}^n$

(iii) $a\tilde{d} - b\tilde{c} = \pm 1$.

The Dirac operator in $\mathbb{R}^n$ is the first order differential operator $D := \Sigma_{j=1}^n e_j \frac{\partial}{\partial x_j}$. Note that $D^2 = -\Delta$, the Laplacian in $\mathbb{R}^n$. The fundamental solution to the Dirac equation, $Df = 0$ is, up to a constant, $\frac{x}{\|x\|^n}$. Here f is defined on a domain U in $\mathbb{R}^n$ and f takes its values in Cl_n.

In [2] it is shown that

$$D_y = J_{-1}(M,x)^{-1} D_x J_1(M,x)$$

where D_y is the Dirac operator with respect to $y = M(x)$, D_x is the Dirac operator with respect to x with $J_{-1}(M,x) = \frac{\widetilde{cx+d}}{\|cx+d\|^{n+2}}$ and $J_1(M,x) = \frac{\widetilde{cx+d}}{\|cx+d\|^n}$.

In [6] the p-Dirac equation is defined to be $\mathcal{D}_p f = D(\|f\|^{p-2} f) = 0$. When $p \neq 2$, this equation is nonlinear. When $p = n$ a solution is given by

$$u_n(x) = x^{-1} = -\frac{x}{\|x\|^2}.$$

For other choices of p a solution is

$$u_p(x) = \frac{x}{\|x\|^{\frac{n+p-2}{p-1}}}.$$

If $f = Dg$ then the p-Dirac equation becomes $D(\|Dg\|^{p-2} Dg) = 0$, and when g is a scalar valued function then the scalar part of this equation becomes the p-Laplace equation.

Consider the integral

$$\int_U \|f(x)\|^{p-2} \overline{f}(x) D\psi(x) dx^n$$

where $f : U \to Cl_n$ and $\psi \in C_0^\infty(U, Cl_n)$, where $C_0^\infty(U, Cl_n)$ is the space of Cl_n-valued test functions with support in U. If this integral vanishes for all $\psi \in C_0^\infty(U, Cl_n)$ then f is called a weak solution to the p-Dirac equation. Note that

$$\frac{1}{\omega_n} \int_U \|u_p(x-y)\|^{p-2} \overline{u}_p(x-y) D\psi(x) dx^n = \psi(y)$$

for all $y \in U$, where ω_n is the surface area of the unit sphere S^{n-1}. This identity follows from the fact that

$$\|u_p(x-y)\|^{p-2} u_p(x-y) = \frac{x-y}{\|x-y\|^n},$$

and this, up to a scalar, is the fundamental solution of the Dirac operator D. In this sense u_p can be regarded as the fundamental solution to the p-Dirac equation. Alternatively, as ψ is smooth on $\mathbb{R}^n$ and $u_p(x-y)$ is smooth on $\mathbb{R}^n \backslash \{y\}$, applying Stokes' Theorem to the previous integral we obtain the identity

$$\int_U \mathcal{D}_p u_p(x-y) \psi(x) dx^n = \psi(y).$$

This last equation is to be interpreted in the distributional sense. Furthermore

$$D \frac{1}{\omega_n} \int_U \|u_p(x-y)\|^{p-2} u_p(x-y) \psi(x) dx^n = \psi(y).$$

4. The infinite Dirac operator in $\mathbb{R}^n$

We now define the infinite Dirac operator acting on a function f as $\mathcal{D}_\infty f(x)$ by taking the formal limit of $\mathcal{D}_p f(x)$ as $p \to \infty$. Taking the limit of the fundamental solutions $u_p(x)$, as $\lim_{p\to\infty} \frac{n+p-2}{p-1} = 1$ we obtain

$$u_\infty(x) = \lim_{p\to\infty} u_p(x) = \frac{x}{\|x\|}$$

which leads to the following theorem.

Theorem 4.1 *Let u_∞ be defined as above. Then for all $x \in \mathbb{R}^n \setminus \{0\}$, we have*

$$\mathcal{D}_\infty u_\infty(x) = 0.$$

Note that as $\frac{1}{\omega_n}\int_U \|u_p(x-y)\|^{p-2} u_p(x-y) D\psi(x) dx^n = \psi(y)$ for all $y \in U$ and $\psi \in C_0^\infty(U, Cl_n)$ then applying Stokes' Theorem to this expression and taking limits in p we have

$$\frac{1}{\omega_n}\int_U \mathcal{D}_\infty u_\infty(x-y)\psi(x)dx^n = \psi(y).$$

This equation is valid in the distributional sense.

As previously noted, if $f = Dg$ then the p-Dirac operator becomes $D(\|Dg(x)\|^{p-2}Dg(x))$ and the scalar part of this is the usual p-Laplace operator whenever g is scalar valued. If we take the limit of this operator as p tends to infinity then we obtain a Clifford valued operator acting on $g(x)$ which we denote by $\Delta'_\infty g(x)$. So $\Delta'_\infty g(x) = \mathcal{D}_\infty Dg(x)$. Note that when g is scalar-valued we retrieve the usual infinite Laplacian acting on g, $\Delta_\infty g(x)$. Further when $g(x) = \|x\|$ then $Dg(x) = n\frac{x}{\|x\|}$, and in this case $\Delta'_\infty g(x) = \Delta_\infty g(x) = n\mathcal{D}_\infty \frac{x}{\|x\|}$. These provide some links between Δ_∞ and $\mathcal{D}_\infty$. In fact if we have a Cl_n valued differentiable function, $g(x)$, defined on U and we place $Dg(x) = f(x)$ we see that $\Delta'_p g(x) = \mathcal{D}_p f(x)$. So if $\|Dg(x)\|^{p-2}Dg(x)$ converges in the Cl_n valued Sobolev space $W_0^{1,1}(U, Cl_n)$ as p tends to infinity then $\Delta'_\infty g(x) = \mathcal{D}_\infty f(x)$ in the weak sense. In greater generality suppose that for each $p \in (1, \infty)$ the function $g_p : U \to Cl_n$ is differentiable almost everywhere and converges pointwise almost everywhere to the function $g_\infty(x)$. Suppose further that f_p is defined to be Dg_p and that f_p converges almost everywhere to f_∞ and $\|f_p(x)\|^{p-2}f_p(x)$ converges as p tends to infinity in the Sobolev space $W_0^{1,1}(U, Cl_n)$ then $\Delta'_\infty g_\infty = \mathcal{D}_\infty f_\infty$ in the weak sense.

When g is scalar-valued, we have

$$\Delta'_\infty g(x) = \Delta_\infty g(x) + \Delta''_\infty g(x)$$

where $\Delta''_p g(x)$ is the part of $D(\|Dg(x)\|^{p-2}Dg(x))$ taking its values in the subspace $< e_1e_2, \ldots, e_{n-1}e_n >$ of Cl_n and $\Delta''_\infty g(x) = \lim_{p\to\infty} \Delta''_p g(x)$. When $n = 3$ applying the Hodge star map reveals that when g is scalar-valued then

$$\Delta''_\infty g(x) \equiv \lim_{p\to\infty} \nabla \times \|\nabla g(x)\|^{p-2}\nabla g(x).$$

Now let us return to the integral $\int_U \|f(y)\|^{p-2}\overline{f}(y)D_y\psi(y)dy^n$. If $y = M(x) = (ax+b)(cx+d)^{-1}$ then using the intertwining operators for D_x and D_y and the fact that the Jacobian for $M(x)$ is $\frac{1}{\|cx+d\|^{2n}}$, we obtain

$$\int_U \|f(y)\|^{p-2}\overline{f}(y)D_y\psi(y)dy^n =$$

$$\int_{M^{-1}(U)} \|f(M(x))\|^{p-2}\overline{\frac{\widetilde{cx+d}}{\|cx+d\|^n}} f(M(x))D_xJ_1(M,x)\psi(M(x))dx^n.$$

Redistributing terms in $\|cx+d\|$ this integral becomes

$$\int_{M^{-1}(U)} \|k(M,p,x)f(M(x))\|^{p-2}\overline{k(M,p,x)f}(M(x))D_x J_1(M,x)\psi(M(x))dx^n$$

where

$$k(M,p,x) = \frac{\widetilde{cx+d}}{\|cx+d\|^{\frac{n+p-2}{p-1}}}.$$

This establishes the conformal invariance of the operator $\mathcal{D}_p$ and the conformal covariance of its solution set.

As ψ is arbitrary in $C_0^\infty(U, Cl_n)$, we then have:

Theorem 4.2 *The function $f : U \to Cl_n$ is a weak solution to the p-Dirac equation if and only if*

$$k(M,p,x)f(M(x))$$

is a weak solution to the p-Dirac equation on $M^{-1}(U)$ where $y = M(x) = (ax+b)(cx+d)^{-1}$.

In particular the fundamental solution

$$u_p(y) = \frac{y}{\|y\|^{\frac{n+p-2}{p-1}}}$$

is transformed to

$$u_p = \frac{\widetilde{cx+d}}{\|cx+d\|^{\frac{n+p-2}{p-1}}} \frac{(ax+b)(cx+d)^{-1}}{\|(ax+b)(cx+d)^{-1}\|^{\frac{n+p-2}{p-1}}}.$$

Now, $(ax+b)(cx+d)^{-1}$ is a vector in $\mathbb{R}^n$ and so $(ax+b)(cx+d)^{-1} = (\widetilde{cx+d})^{-1}(\widetilde{ax+b})$. We then have

$$u_p = \frac{\widetilde{ax+b}}{\|ax+b\|^{\frac{n+p-2}{p-1}}}$$

and we obtain a solution to the p-Dirac equation.

Furthermore, for each $\psi \in C_0^\infty(U, Cl_n)$ and for each Möbius transformation $y = M(x) = (ax+b)(cx+d)^{-1}$, we have that

$$\lim_{p\to\infty} \int_U \|f(y)\|^{p-2}\overline{f}(y)D_y\psi(y)dy^n$$

can be expressed as

$$\lim_{p\to\infty} \int_{M^{-1}(U)} \|k(M,p,x)f(M(x))\|^{p-2}\overline{k(M,p,x)f}(M(x))D_x J_1(M,x)\psi(M(x))dx^n.$$

Suppose $\lim_{p\to\infty} \int_U \|f(y)\|^{p-2}f(y)D_y\psi(y)dy^n = 0$ then we have established:

Theorem 4.3 *A function $f : U \to Cl_n$ satisfies $D_\infty f(y) = 0$ if and only if $D_\infty \frac{\widetilde{cx+d}}{\|cx+d\|} f(M(x)) = 0$ where $y = M(x) = (ax+b)(cx+d)^{-1}$.*

This establishes the conformal invariance of the operator D_∞ and the conformal covariance of its solution set.

In the particular case where $f(y) = u_\infty(y) = \frac{y}{\|y\|}$ we get that the function $\frac{cx+d}{\|cx+d\|}(ax+b)(cx+d)^{-1}\|(ax+b)(cx+d)^{-1}\|^{-1}$ is a solution to the infinite Dirac equation. By the same arguments presented here for the p-Dirac equation this simplifies to $\frac{\widetilde{ax+b}}{\|ax+b\|}$. So $\frac{\widetilde{ax+b}}{\|ax+b\|}$ is a solution to the infinite Dirac equation.

5. The infinite Dirac operator on the sphere

As done in [6], we may define the p-Dirac operator on the sphere S^n. Recall that the Dirac operator on the sphere is given by

$$D_s = x(\Gamma_x + \frac{n}{2})$$

where $x \in S^n$ and

$$\Gamma_x = \sum_{i<j}^{n+1} e_i e_j \left(x_i \frac{\partial}{\partial x_j} - x_j \frac{\partial}{\partial x_i} \right).$$

The p-Dirac operator acting on a function $f : V \to Cl_{n+1}$, with V a domain on S^n, is defined on the sphere for $1 < p < \infty$ by

$$\mathcal{D}_{s,p} f = D_s(\|f\|^{p-2} f).$$

See [6]. Further f is called a weak solution on V to the operator $D_{s,p}$ if for each $\psi \in C_0^\infty(V, Cl_{n+1})$

$$\int_V \|f(x)\|^{p-2} \overline{f}(x) D\psi(x) dV = 0$$

where dV is the volume measure of the domain V on S^n.

Given a point $y \in S^n$, the fundamental solution to the equation $D_{s,p} h_p = 0$ with singularity y is given by

$$h_p(x) = \frac{x - y}{\|x - y\|^{\frac{n+p-2}{p-1}}}.$$

Similar to the previous section, we can define the infinite Dirac operator on the sphere, $\mathcal{D}_{s,\infty}$ as the limit of the operators $\mathcal{D}_{s,p}$.

The Cayley transformation from $\mathbb{R}^n$ to S^n is given by $y = C(x) = (e_{n+1}x + 1)(x + e_{n+1})^{-1}$ and the relationship between D and D_s is given by $D_s = (x + e_{n+1})^{-1}\|x + e_{n+1}\|^{n+2} D \frac{x+e_{n+1}}{\|x+e_{n+1}\|^n}$. See [5]. Adapting arguments from the previous section we can now see that under the Cayley transformation the p-Dirac operator $\mathcal{D}_{s,p} f = D(\|f(y)\|^{p-2} f(y))$ on a domain V on S^n transforms to the p-Dirac operator

$$D \left\| \frac{x + e_{n+1}}{\|x + e_{n+1}\|^{\frac{n+p-2}{p-1}}} \right\|^{p-2} \frac{x + e_{n+1}}{\|x + e_{n+1}\|^{\frac{n+p-2}{p-1}}} f(M(x))$$

on the domain $C^{-1}(V)$ in $\mathbb{R}^n$. In particular if $\mathcal{D}_{s,p} f(y) = 0$ then

$$\mathcal{D}_p \frac{x + e_{n+1}}{\|x + e_{n+1}\|^{\frac{n+p-2}{p-1}}} f(M(x)) = 0.$$

By taking limits as p tends to infinity we see that $\mathcal{D}_{s,\infty} f(y) = 0$ if and only if

$$\mathcal{D}_\infty \frac{x + e_{n+1}}{\|x + e_{n+1}\|} f(M(x)) = 0.$$

In [5] it is shown for x and $y \in S^n$ that $\|x - y\|^2 = 2(1- < x, y >)$ and that $\Gamma_x \|x - y\|^\alpha = 2^\alpha \Gamma_x (1- < x, y >) = 2^{\frac{\alpha}{2}} \frac{\alpha}{2} (1- < x, y >)^{\alpha - 1} x \wedge y$. It follows from [5] that $(x\Gamma_x + \frac{x\alpha}{2}) \|x - y\|^\alpha = 2^{\frac{\alpha}{2}} (x\Gamma_x + x\frac{x\alpha}{2})(1- < x, y >)^{\frac{\alpha}{2}} = 2^{\frac{\alpha}{2}} \frac{\alpha}{2} x (1- < x, >)^{\frac{\alpha}{2} - 1} (x \wedge y + 1- < x, y >)$. But $x \wedge y = xy + < x, y >$ and $x^2 = -1$. So this term simplifies to give the equation

$$(x\Gamma_x + \frac{x\alpha}{2}) \|x - y\|^\alpha = \frac{\alpha}{2} (x - y) \|x - y\|^{\alpha - 1}.$$

Now consider the case of $\|x-y\|^{\frac{p-n}{p-1}}$. In this case

$$x(\Gamma_x + \frac{p-n}{2p-2})\|x-y\|^{\frac{p-n}{p-1}} = \frac{\alpha}{2}\frac{x-y}{\|x-y\|^{\frac{n+p-2}{p-1}}}.$$

As $D_s = x(\Gamma_x + \frac{n}{2})$ this previous equation can be rewritten as

$$(D_s + x\frac{p(n-1)}{p-1})\|x-y\|^{\frac{p-n}{p-1}} = \frac{\alpha}{2}\frac{x-y}{\|x-y\|^{\frac{n+p-2}{p-1}}}.$$

We shall denote the differential operator $D_s + x\frac{p(n-1)}{p-1}$ by $D'_{s,p}$.

Note that $\|D'_{s,p}\|x-y\|^{\frac{p-n}{p-1}}\|^{p-2}D'_{s,p}\|x-y\|^{\frac{p-n}{p-1}}$ is up to a constant $\frac{x-y}{\|x-y\|^n}$. In turn $\frac{1}{\omega_n}\frac{x-y}{\|x-y\|^n}$ is the fundamental solution to the operator D_s. See [5].

It follows that up to a constant, $\alpha(p)$, dependant on p

$$\alpha(p)\int_{S^n}\|D'_{s,p}\|x-y\|^{\frac{p-n}{p-1}}\|^{p-2}D'_{s,p}\|x-y\|^{\frac{p-n}{p-1}}D_{s,p}\psi(x)dS^n = \psi(y)$$

for any $\psi \in C^\infty(S^n)$.

It follows that the p-Laplace operator on S^n acting on $g(x)$is

$$\Delta_{s,p}g(x) := D_s\|D'_{s,p}g(x)\|^{p-2}D'_{s,p}g(x).$$

This gives a correction to [6]. We have also shown that $\|x-y\|^{\frac{p-n}{p-1}}$ is, up to a constant, the fundamental solution for the operator $\Delta_{s,p}$.

We can also say that a function $g(x)$ on a domain $V \subset S^n$ is a weak solution to the operator $\Delta_{s,p}$ if for each $\psi \in C_0^\infty(V, Cl_{n+1})$

$$\int_V \|D'_{s,p}g(x)\|^{p-2}\overline{D'_{s,p}}g(x)D_s\psi(x)dS^n = 0.$$

Further we can formally define the infinite Laplace operator on S^n acting on $g(x)$ to be $\Delta_{s,\infty}g(x)$ as $\lim_{p\to\infty}\Delta_{s,p}g(x)$. It follows that $\|x-y\|$ is a solution to the infinite Laplace operator on S^n. We see that $\lim_{p\to\infty}\frac{2p-2}{p-n}D'_{s,p}\|x-y\|^{\frac{p-n}{p-1}} = \frac{x-y}{\|x-y\|}$, the fundamental solution of $\mathcal{D}_{s,\infty}$. So we have a link between the operators $\Delta_{s,\infty}$ and $\mathcal{D}_{s,\infty}$. In fact if for a differentiable Cl_{n+1} valued function $g(x)$ defined on V we place $D'_{s,p}g(x) = f_p(x)$ then $\Delta_{s,p}g(x) = D'_{s,p}f_p(x)$. So if $D'_{s,p}g(x)\|^{p-2}D'_{s,p}g(x)$ converges as p tends to infinity in the Cl_{n+1} valued Sobolev space $W^{1,1}(S^n, Cl_{n+1})$ then $\Delta_{s,\infty}g(x) = \mathcal{D}_{s,\infty}f_\infty(x)$ in the weak sense.

Acknowledgements
This paper is dedicated to Tony Sudbery.

References

[1] Ahlfors L 1986 Möbius transformations in $\mathbb{R}^n$ expressed through 2×2 matrices of Clifford numbers *Complex Variables: Theory and Application* **5** 215-224
[2] Bojarski B 1989 Conformally covariant differential operators (Tehran: Proc. XXth Iranian Math Congress)
[3] Heinonen J, Kilpeläinen T and Martio O 1993 *Nonlinear Potential Theory of Degenerate Elliptic Equations* (New York: Oxford University Press)
[4] Lindqvist P 2006 *Notes on the p-Laplace equation*, Report by the Department of Mathematics and Statistics (Univ. of Jyvskyl)
[5] Liu H and Ryan J 2002 Clifford analysis techniques for spherical pde *J. Fourier Analysis Appl.* **8** 235-264
[6] Nolder C and Ryan J 2009 p-Dirac equations *Adv. Appl. Clifford Alg.* **19** 391-402

On Morley's theorem, determinantal identities and the QR algorithm

A J Macfarlane

Centre for Mathematical Sciences, D.A.M.T.P., Wilberforce Road, Cambridge CB3 0WA, UK

E-mail: a.j.macfarlane@damtp.cam.ac.uk

Abstract. Three results are shown in this contribution. First, the diagram used in the proof of Morley's theorem is shown to provide a template for the creation of a closed solid figure. Second, it is shown how a proposition governing a class of determinants can be used to produce a rich supply of identities involving, *e.g.*, Fibonacci and related families of numbers. Finally, it is shown in the case of an arbitrary 2×2 symmetric matrix A, how to obtain explicit expressions for all the elements of all the matrices involved in implementing the determination of the eigenvalues of A by means of the QR algorithm.

Received: 18 October 2009

1. Introduction

This article is based on material presented at the Symposium *Quantum Groups, Foundations, Information* held in honour of Tony Sudbery on the occasion of his retirement. Tony, one of my first PhD students in Cambridge, became and is a valued lifelong friend. It is therefore my pleasure to offer this contribution to his Festschrift.

The article consists of three sections. The first shows how the diagram used in the proof of Morley's theorem provides a template for the creation of a closed solid figure. The second shows how a proposition governing a class of determinants can be used to produce a rich supply of identities involving, *e.g.*, Fibonacci and related families of numbers. The third section shows, in the case of an arbitrary 2×2 symmetric matrix A, how to obtain explicit expressions for all the elements of all the matrices involved in implementing the determination of the eigenvalues of A by means of the QR algorithm.

2. The Morley tricorn

Morley's theorem states that the points of intersection of adjacent trisectors of the angles of any triangle are the vertices of an equilateral triangle, known as Morley's triangle. The content of the theorem is illustrated in Fig. 1. In Fig. 1 additional points P, Q, R and dotted lines DP, EQ, FR have been included for use in the construction of the Morley tricorn, to be described. The points P, Q, R points are uniquely determined so that

$$AQ = AR = s - a, \quad BR = BP = s - b, \quad CP = CQ = s - c, \tag{1}$$

where $2s = a + b + c$ in standard notation.

Concise proofs of Morley's theorem are given in [1, 2], and a good picture of other studies on the subject is to be found at P1199 of [3].

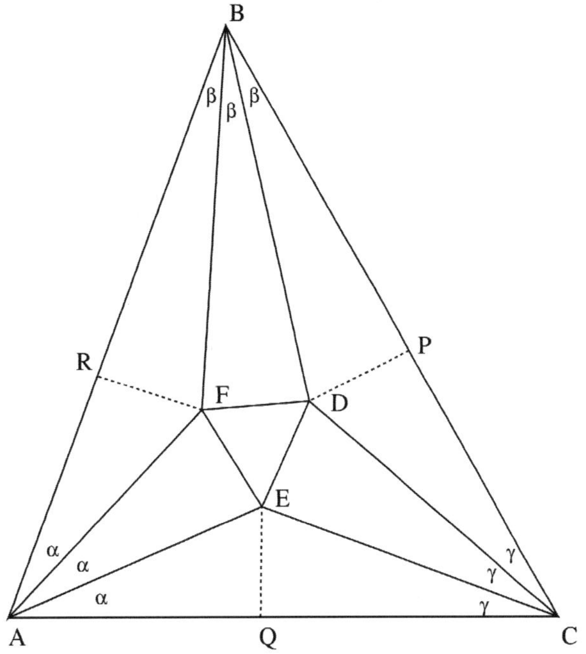

FIGURE 1

Fig. 1 was considered at the symposium as the high point of a sequence of demonstrations of how the diagrams used for proofs of theorems in the geometry of plane triangles can be folded (origami-wise: valley folds on dotted lines and mountain folds on solid ones) so as to realise three-dimensional entities. It was shown there how Fig. 1 can be regarded as a template for the creation of a unique rigid exactly closed solid, called [4] the Morley tricorn, with ten undistorted plane faces.

To understand the process, suppose cuts along the dotted lines FR, EQ, DP have been made. Then fold the triangles ΔAQE and ΔARF up over AE and AF respectively to bring AQ and AR into coincidence. Let W_A denote the point of coincidence of Q and R. This step yields a pyramid, or horn, with ΔEFW_A providing the outline of its base, and vertex A. Similarly pyramids, on ΔFDW_B and ΔDEW_C as base and vertices B and C respectively, can be formed. The points W_B and W_C are the points at which the respective pairs of points R and P, and P and Q, are thereby brought into coincidence. It is then easy to see that the three pyramids can be folded up over EF, FD, DE to bring W_A, W_B, W_C into coincidence at a point W, with exact closure that produces a 'three horned' three-dimensional figure, or tricorn. Full geometric justification of this statement can be given.

In fact, it is not necessary to cut along any lines at all. The foldings of the three pyramids just described, can be performed simultaneously, and in fact easily, in a process in which inward (valley) folds along the three dotted lines are made. This produces the Morley tricorn, a figure with central pyramid $DEFW$, which has the Morley triangle as base and vertex W, and pyramidal horns with vertices A, B, C, standing on the outlined slant faces of the central pyramid.

3. Difference equations and determinantal identities

Let $\mathcal{S}$ denote a sequence $\{y_0, y_1, y_2, \ldots\}$ of real quantities y_n governed by a difference equation

$$y_n = ay_{n-1} + by_{n-2}, \tag{2}$$

where a, b, y_0, y_1 are given real numbers.
Let $\mathcal{T}$ denote a sequence of $n \times n$ determinants

$$T_n(x, y_0, y_1, a, b) = \begin{vmatrix} y_1 & y_2 & y_3 & \cdots & y_{n-2} & y_{n-1} & y_n \\ -x & y_1 & y_2 & \cdots & y_{n-3} & y_{n-2} & y_{n-1} \\ 0 & -x & y_1 & \cdots & y_{n-4} & y_{n-3} & y_{n-2} \\ \vdots & \vdots & \vdots & \ddots & \vdots & \vdots & \vdots \\ 0 & 0 & 0 & \cdots & y_1 & y_2 & y_3 \\ 0 & 0 & 0 & \cdots & -x & y_1 & y_2 \\ 0 & 0 & 0 & \ldots & 0 & -x & y_1 \end{vmatrix}, \tag{3}$$

where x is an indeterminate. We now quote a result stated and proved in [5].

Proposition 1: If the $y_n \in \mathcal{S}$ are subject to (2), with x, a, b, y_0, y_1 all fixed real numbers, then the determinants $T_n = T_n(x, y_0, y_1, a, b) \in \mathcal{T}$ are related by the difference equation

$$T_n = (y_1 + ax)T_{n-1} + bx(y_0 + x)T_{n-2}. \tag{4}$$

At the symposium, the proposition was used to present a miscellany of determinantal identities involving the Fibonacci (and related families of) numbersin either the role of $\mathcal{S}$ or $\mathcal{T}$ or both. A paper covering such material [5] has now appeared. Here, one illustrative example is given, and the origin of the ideas in the theory of one-dimensional cellular automata [6] is outlined.

3.1. An example involving Fibonacci numbers

The Fibonacci numbers F_n are such that

$$F_{n+2} = F_{n+1} + F_n, \quad F_0 = F_1 = 1, \tag{5}$$

and hence

$$F_n = \frac{\alpha^n - \beta^n}{\alpha - \beta}, \quad \alpha, \beta = \frac{1 \pm \sqrt{5}}{2}. \tag{6}$$

For the example, take $\mathcal{S} = \{y_0 = -1, y_1 = 1, -1, 1, \ldots\}$ so that $y_{n+2} = y_n, a = 0, b = 1$. Then, by Proposition 1, $T_n = T_n(\beta, -1, 1, 0, 1) \in \mathcal{T}$ satisfies

$$T_{n+2} = T_{n+1} + T_n, \tag{7}$$

in the case in which $x = \beta$. It follows easily that

$$T_n = F_n - \beta F_{n-1} = \alpha^{n-1}, \tag{8}$$

as can be checked. The choice $x = \alpha$ leads similarly to $T_n = \beta^{n-1}$.

3.2. The origin of Proposition 1

A one dimensional cellular automaton involves a linear array of cells, usually displayed as squares, with a cell variable $x_n(t)$ attached to the n-th cell at time t. Each cell variable takes on the value 1 or 0, usually displayed as dark or light coloured square. Consider the situation in which the cellular automaton CA150, governed by Wolfram rule 150 [6], evolves from the initial state of 'single seed', for which $x_0(t = 0) = 1$ and $x_n(t = 0) = 0$ for $n \neq 0$. Time advances in discrete steps through values $t = 0, 1, 2, \ldots$, and the state at time $(t+1)$ of CA150 is determined by the state at time t by rule 150, *i.e.*

$$x_n(t + 1) = x_{n+1}(t) + x_n(t), \tag{9}$$

by which all cells are simultaneously updated using modulo two arithmetic. The weight $\Omega(t)$ of the state of CA150 at time t is defined to be the sum of all its cell variables at time t, or the number of dark squares in a display of the state at time t. Wolfram [6] gave a nice algorithm for writing down an explicit value for $\Omega(t)$.

Let $t = t_B$ denote the binary representation t_B of t. Then, if t_B involves strings of ones of length a_k, $k = 1, 2, \ldots, p$, separated by strings of zeros of lengths greater than zero, these lengths being irrelevant for the algorithm, then the algorithm states that

$$\Omega(t) = \prod_{k=1}^{p} \chi_{a_k}. \tag{10}$$

The numbers χ_n are shifted Jacobsthal numbers, [3], page 951, governed by

$$\chi_{n+2} = \chi_{n+1} + 2\chi_n, \quad \chi_0 = 1, \ \chi_1 = 3. \tag{11}$$

For $t = 5$, that is $t_B = 101$ in binary, the algorithm gives $\Omega(5) = (\chi_1)2 = 9$, while for $t = 13$ or $t_B = 1101$ in binary, $\Omega(13) = \chi_1\chi_2 = 15 = \Omega(11)$, since 11 has binary representation 1011. Such results can be checked against a display of the evolution of CA150 from single seed.

In [7] a proof of the algorithm is given. Also given there is a result for the total weight $W(n)$ of the states of CA150 at the first 2^n time steps of its evolution from a state of single seed at $t = 0$

$$W(n) = 2^n F_{n+2}, \tag{12}$$

involving the Fibonacci numbers. To cast (12) into a suitable form, evaluate the $\Omega(t)$ from the Wolfram algorithm, and write the answers in terms of variables $y_r = \chi_{r-1}$, noting $y_1 = 1$

and $y_2 = 3$. This leads to

$$\begin{aligned} W(1) &= \Omega(0)+\Omega(1) = 1+\chi_1 = 1+y_2 = y_12+y_2 = 4 = 2F_3 \\ W(2) &= \Omega(0)+\Omega(1)+\Omega(2)+\Omega(3) = 1+2y_2+y_3 \\ &= y_13+2y_1y_2+y_3 = 12 = 4F_4 \\ W(3) &= 1+3y_2+2y_3+y_4+y_22 \\ &= y_14+3y_12y_2+2y_1y_3+y_22+y_4 = 40 = 8F_5, \end{aligned} \tag{13}$$

$W(4) = 16F_6$, and so on. The key step here is to insert factors $y_1 = 1$ into the output from the Wolfram algorithm to put the terms in each $W(n)$ into one-to-one correspondence with partitions of n. The results in (13) for the W_n can be seen thereby to coincide with the expansions of the determinants $T_{n+1}(1,1,1,1,2) \in \mathcal{T}$. Use of Proposition 1 now allows proof for all n of the result

$$F_{n+1} = \frac{2}{2^n} T_n(1,1,1,1,2), \tag{14}$$

a determinantal identity expressing Fibonacci in terms of Jacobsthal numbers.

4. The QR algorithm

Confine attention to real symmetric $n \times n$ matrices with eigenvalues

$$\lambda_1 > \lambda_2 \,\dots\, \lambda_n > 0. \tag{15}$$

The QR factorisation of such a matrix A expresses it uniquely in the form

$$\mathrm{A} = \mathrm{QR}, \tag{16}$$

where Q is orthogonal, and R is upper triangular with positive diagonal elements. The QR algorithm, based, unsurprisingly, on QR factorisation, is of major importance in numerical linear algebra as a means, with widespread applications, of obtaining good approximations to all the eigenvalues of A. Detailed expositions can be found *e.g.* in [8–10]. We confine attention here to the simplest version of the algorithm, with a view entirely to exposing algebraic structure.

The QR algorithm

For $r = 0, 1, 2, \ldots$, the algorithm for A defines a sequence of matrices A_r. It does so by repetition of the following two steps

(i) factorisation of A_r

$$\mathrm{A}_r = \mathrm{Q}_r\,\mathrm{R}_r, \tag{17}$$

(ii) definition of A_{r+1}

$$\mathrm{R}_r\,\mathrm{Q}_r = \mathrm{A}_{r+1}. \tag{18}$$

The matrices A_r of the sequence share with $\mathrm{A} = \mathrm{A}_0$ many of its key properties – symmetry, trace, determinant, eigenvalues.

The importance of the algorithm depends on its

Convergence property

Journal of Physics: Conference Series **254** (2010) 012004 doi:10.1088/1742-6596/254/1/012004

$$\lim_{r\to\infty} \mathrm{A}_r = \Lambda = \mathrm{diag}\,(\lambda_1, \lambda_2, \ldots, \lambda_n). \tag{19}$$

It is proved, in contexts much more general than those to which we restrict ourselves here, in the cited references [8–10].

By consideration of the $n = 2$ case, this article illustrates a study which has already been completed [11] for real symmetric tridiagonal $n \times n$ matrices. The purpose is to indicate the nice algebraic structure that underlies the QR algorithm by giving closed expressions for all the elements of all the matrices $\mathrm{A}_r, \mathrm{Q}_r, \mathrm{R}_r$: see (24–26) below. Eq. (24) yields a simple proof of (19) in the limited context of the paper.

4.1. 2 × 2 *matrices*

Consider the matrix

$$\mathrm{A} = \mathrm{A}_0 = \begin{pmatrix} a & f \\ f & b \end{pmatrix}, \tag{20}$$

with eigenvalues $\lambda_+ > \lambda_- > 0$, so that

$$\det\mathrm{A} = \Delta = ab - f2 = \lambda_+\lambda_-, \quad \mathrm{tr}\mathrm{A} = T = a + b = \lambda_+ + \lambda_-. \tag{21}$$

Let the sequence $\mathcal{S} = \{x_r, \quad r \in \mathbb{Z}\}$ be governed by the difference equation

$$x_{r+2} = Tx_{r+1} - \Delta x_r, \tag{22}$$

and initial conditions $x_0 = 1, \quad x_1 = a$. This gives $x_2 = a2 + f2$, $\Delta x_{-1} = b$, and the general solution in the equivalent forms

$$\begin{aligned} x_r &= \alpha\lambda_+{}^r + \beta\lambda_-{}^r, \quad \alpha\beta = \frac{f2}{(\lambda_+ - \lambda_-)2} \\ &= ay_{r-1} - \Delta y_{r-2}, \quad \text{where} \quad y_r = \frac{\lambda_+{}^{r+1} - \lambda_-{}^{r+1}}{\lambda_+ - \lambda_-}. \end{aligned} \tag{23}$$

Parametrisations of all the elements of all the matrices A_r, Q_r, R_r for all $r = 0, 1, 2, \ldots$, can now be presented in the form:

$$\mathrm{A}_r = \frac{1}{x_{2r}} \begin{pmatrix} x_{2r+1} & f\Delta^r \\ f\Delta^r & x_{2r-1}\Delta \end{pmatrix} \tag{24}$$

$$\mathrm{Q}_r = \frac{1}{\sqrt{x_{2r}\, x_{2r+2}}} \begin{pmatrix} x_{2r+1} & -f\Delta^r \\ f\Delta^r & x_{2r+1} \end{pmatrix} \tag{25}$$

$$\mathrm{R}_r = \frac{1}{\sqrt{x_{2r}\, x_{2r+2}}} \begin{pmatrix} x_{2r+2} & fT\Delta^r \\ 0 & x_{2r}\Delta \end{pmatrix}. \tag{26}$$

By performing explicitly the factorisation $\mathrm{A}_0 = \mathrm{Q}_0\ \mathrm{R}_0$, it is easy to see that the expressions (24–26) for $r = 0$ are correct for the given matrix $\mathrm{A} = \mathrm{A}_0$ of (20). To show that they satisfy (17) and (18) for all $r = 0, 1, 2, \ldots$, requires some identities

$$x_{r+1}\,x_{r-1} - x_r 2 = f2\Delta^{r-1} \tag{27}$$
$$x_{r+2}\,x_{r-1} - x_{r+1}\,x_r = f2T\Delta^{r-1} \tag{28}$$
$$x_{2r+1}\,x_r - x_{2r}\,x_{r+1} = f2\Delta^r y_{r-1} \tag{29}$$
$$x_{2r}\,y_r - x_{2r+1}\,y_{r-1} = \Delta^r x_r \tag{30}$$
$$x_r 2 + f2\,y_{r-1}2 = x_{2r}. \tag{31}$$

An easy way to prove these identities involves insertion of explicit expressions from (23) into their left-hand sides.

Turning now to the proof that the parametrisations (24–26) satisfy (17), it is found that the $(1,1)$ and $(2,1)$ elements of the product on the right give $(\mathrm{A}_r)_{11}$ and $(\mathrm{A}_r)_{21}$ immediately. The $(1,2)$ element of the product must equal the $(2,1)$ element because all the A_r symmetric, and this is seen to hold upon use of (22). For the $(1,1)$ elements to agree, completing the proof of (17), requires the identity (28), with r replaced by $2r$.

Similarly, the $(2,1)$ and $(2,2)$ elements of the product on the right side of (18) give $(\mathrm{A}_{r+1})_{21}$ and $(\mathrm{A}_{r+1})_{22}$ immediately. The $(1,2)$ element of the product must equal the $(2,1)$ element, and is found to do so, again upon use of (22). For the $(2,2)$ elements to agree, completing the proof of (18), requires the identity (28), this time with r replaced by $(2r+1)$.

Since $\lambda_+ > \lambda_-$, it is found that

$$\lim_{r\to\infty} x_r = \alpha {\lambda_+}^r. \tag{32}$$

Recalling $\Delta = \lambda_+\lambda_-$, proof of (19) now follows, element by element.

4.2. Diagonalisation

Begin, following standard practice, by writing

$$\Lambda = \begin{pmatrix} \lambda_+ & 0 \\ 0 & \lambda_- \end{pmatrix} = \mathrm{P}^T\mathrm{A}\mathrm{P}, \tag{33}$$

where

$$\mathrm{P} = \begin{pmatrix} \cos\theta & -\sin\theta \\ \sin\theta & \cos\theta \end{pmatrix} = \frac{1}{\sqrt{\lambda_+ - \lambda_-}} \begin{pmatrix} \sqrt{a-\lambda_-} & -\sqrt{\lambda_+ - a} \\ \sqrt{\lambda_+ - a} & \sqrt{a-\lambda_-} \end{pmatrix}. \tag{34}$$

The results $(\lambda_+ - a) > 0$, $\quad (a - \lambda_-) > 0$ and

$$f\cot\theta = a - \lambda_-, \tag{35}$$

may all be proved.

Next, note that (17) and (18) imply

$$\mathrm{A}_{r+1} = {\mathrm{Q}_r}^T \mathrm{A}_r \mathrm{Q}_r, \tag{36}$$

and more generally

$$\mathrm{A}_r = {\Omega_r}^T \mathrm{A}_0 \Omega_r, \tag{37}$$

where

$$\Omega_r = \mathrm{Q}_0\mathrm{Q}_1 \ldots \mathrm{Q}_{r-1}. \tag{38}$$

We have already proved the key result (19). We now wish to prove that

$$\lim_{r\to\infty} \Omega_r = \mathrm{P}. \tag{39}$$

For this purpose, it is necessary justify use of the representation

$$\Omega_r = \frac{1}{\sqrt{x_{2r}}} \begin{pmatrix} x_r & -fy_{r-1} \\ fy_{r-1} & x_r \end{pmatrix}. \tag{40}$$

Known, or easily available, results for $\Omega_1 = \mathrm{Q}_0$ and $\Omega_2 = \mathrm{Q}_0\mathrm{Q}_1$ correctly provide the cases $r = 1, 2$ of (40). It remains to prove that the inductive step

$$\Omega_{r+1} = \Omega_r \mathrm{Q}_r \tag{41}$$

is true. For the $(1,1)$ and $(2,2)$ elements of (41) use identity (29). For the $(1,2)$ and $(2,1)$ elements use identity (30). To verify that (40) defines an orthogonal matrix requires (31).

Writing Ω_r as

$$\Omega_r = \begin{pmatrix} \cos\theta_r & -\sin\theta_r \\ \sin\theta_r & \cos\theta_r \end{pmatrix}, \tag{42}$$

leads to

$$f\cot\theta_r = \frac{x_r}{y_{r-1}} = a - \Delta\,\frac{y_{r-2}}{y_{r-1}}. \tag{43}$$

Since P and Ω_r define rotations of angles θ and θ_r respectively, it is sufficient, to complete the proof of (39), to show that $\cot\theta_r$ approaches $\cot\theta$ as r approaches infinity. This follows directly from the result

$$\lim_{r\to\infty} \frac{y_{r-2}}{y_{r-1}} = \frac{1}{\lambda_+}. \tag{44}$$

References

[1] Barbara R 1997 Two short proofs of Morley's theorem *Mathematical Gazette* **81** 447-450
[2] Scott J A 2002 Morley's theorem once again *Mathematical Gazette* **86** 80-81
[3] Weisstein E W 1999 *CRC Concise Encyclopedia of Mathematics* (Boca Raton: Chapman & Hall)
[4] Macfarlane A J (in print) The Morley tricorn: a solid figure constructed from the diagram for Morley's theorem *Mathematical Gazette*
[5] Macfarlane A J 2010 Use of determinants to present identities involving Fibonacci and related numbers *Fibonacci Quarterly* **48** 68-76
[6] Wolfram S 1983 Statistical mechanics of cellular automata *Rev. Mod. Phys.* **55** 601-644
[7] Macfarlane A J 2004 Linear reversible second-order cellular automata and their first-order matrix equivalents *J. Phys. A* **37** 10791-10814
[8] Olver P and Shakiban C 2006 *Applied Linear Algebra* (London: Pearson)
[9] Strang G 1986 *Introduction to Applied Mathematics* (Wellesley, MA: Wellesley Cambridge Press)
[10] Golub G H and van Loan C F 1984 *Matrix Computations* (Baltimore: Johns Hopkins University Press)
[11] Macfarlane A J (in preparation) The *QR* algorithm: explicit expressions for all the elements of all the matrices $\mathrm{A}_r, \mathrm{Q}_r, \mathrm{R}_r$ for all $r \geq 0$

Journal of Physics: Conference Series **254** (2010) 012005 doi:10.1088/1742-6596/254/1/012005

Octonions, E_6, and particle physics

Corinne A Manogue[1] **and Tevian Dray**[2]

[1] Department of Physics, Oregon State University, Corvallis, OR 97331, USA
[2] Department of Mathematics, Oregon State University, Corvallis, OR 97331, USA

E-mail: corinne@physics.oregonstate.edu, tevian@math.oregonstate.edu

Abstract. In 1934, Jordan et al. gave a necessary algebraic condition, the Jordan identity, for a sensible theory of quantum mechanics. All but one of the algebras that satisfy this condition can be described by Hermitian matrices over the complexes or quaternions. The remaining, exceptional Jordan algebra can be described by 3×3 Hermitian matrices over the octonions.

We first review properties of the octonions and the exceptional Jordan algebra, including our previous work on the octonionic Jordan eigenvalue problem. We then examine a particular real, noncompact form of the Lie group E_6, which preserves determinants in the exceptional Jordan algebra.

Finally, we describe a possible symmetry-breaking scenario within E_6: first choose one of the octonionic directions to be special, then choose one of the 2×2 submatrices inside the 3×3 matrices to be special. Making only these two choices, we are able to describe many properties of leptons in a natural way. We further speculate on the ways in which quarks might be similarly encoded.

Received: 4 November 2009

1. Introduction

A personal note from Corinne: During the academic year 1986/87, Tevian and I were living in York, newly married and young postdocs. Tevian was working in the mathematics department there, doing research in general relativity, and I was working in Durham, with David Fairlie, just beginning my research into the octonionic structures reported here. Imagine my pleasure, when I found out that York had its own resident expert on the octonions! I returned to York the following summer, to work with Tony on an attempt to describe the superstring using octonions [1]. I will be forever grateful to him, not only for the generous way in which he shared his vast knowledge and experience in this field, but also for the friendship, respect, collegiality, and mentorship, which he also generously shared.

2. Exceptional Quantum Mechanics

In the Dirac formulation of quantum mechanics, a quantum mechanical state is represented by a *complex* vector v, often written as $|v\rangle$, which is usually normalized such that $v^\dagger v = 1$. In the Jordan formulation [2, 3, 4, 5], the same state is instead represented by the Hermitian matrix $vv^\dagger$, also written as $|v\rangle\langle v|$, which squares to itself and has trace 1. The matrix $vv^\dagger$ is thus the projection operator for the state v, which can also be viewed as a pure state in the density matrix formulation of quantum mechanics. Note that the usual phase freedom in v is no longer present in $vv^\dagger$, which is uniquely determined by the state (and the normalization condition).

A fundamental object in the Dirac formalism is the probability amplitude $v^\dagger w$, or $\langle v|w\rangle$, which is not however measurable; it is the squared norm $|\langle v|w\rangle|^2 = \langle v|w\rangle\langle w|v\rangle$ of the probability amplitude which yields measurable probabilities. One of the basic observations which leads to the Jordan formalism is that these probabilities can be expressed entirely in terms of the Jordan product of projection operators, since

$$\langle v|w\rangle\langle w|v\rangle = (v^\dagger w)(w^\dagger v) \equiv \mathrm{tr}(vv^\dagger \circ ww^\dagger) \tag{1}$$

where $\circ$ denotes the *Jordan product* [2, 3]

$$\mathcal{A} \circ \mathcal{B} = \frac{1}{2}(\mathcal{AB} + \mathcal{BA}) \tag{2}$$

which is commutative but not associative.

Remarkably, the Jordan formulation of quantum mechanics does not require (v and) $\mathcal{A}$ to be complex, but only that the *Jordan identity*

$$(\mathcal{A} \circ \mathcal{B}) \circ \mathcal{A}^2 = \mathcal{A} \circ \left(\mathcal{B} \circ \mathcal{A}^2\right) \tag{3}$$

hold for two Hermitian matrices $\mathcal{A}$ and $\mathcal{B}$. As shown in [3], the Jordan identity (3) is equivalent to power associativity, which ensures that arbitrary powers of Jordan matrices — and hence of quantum mechanical observables — are well-defined.

The Jordan identity (3) is the defining property of a *Jordan algebra* [2], and is clearly satisfied if the operator algebra is associative, which will be the case if the elements of the Hermitian matrices $\mathcal{A}$, $\mathcal{B}$ themselves lie in an associative algebra. Remarkably, one further possibility exists, for which the elements of the Hermitian matrices do *not* lie in an associative algebra. This example is the Albert algebra (also called the exceptional Jordan algebra) $H_3(\mathbb{O})$ of 3×3 octonionic Hermitian matrices [3, 6].[1] In what follows we will restrict our attention to this exceptional case.[2]

3. Quaternions and Octonions

The Hurwitz Theorem states that the real numbers $\mathbb{R}$, complexes $\mathbb{C}$, quaternions $\mathbb{H}$, and octonions $\mathbb{O}$ are the *only* (normed) division algebras (over the real numbers).[3] The quaternions and octonions are extensions of the familiar real and complex numbers. A quaternion is an arbitrary real linear combination of the real identity element 1 and three different square roots of minus one, which are conventionally called $\{i, j, k\}$ and satisfy the multiplication table given in Figure 1. Similarly, the octonions are formed from seven square roots of minus one which we will call $\{i, j, k, k\ell, j\ell, i\ell, \ell\}$, whose multiplication table is summarized in Figure 2. In these multiplication tables, each point corresponds to an imaginary unit. Each line or circle corresponds to a quaternionic triple with the arrow giving the orientation. For example,

$$k\,\ell = k\ell \tag{4}$$
$$\ell\,k\ell = k \tag{5}$$
$$k\ell\,k = \ell \tag{6}$$

and each of these products anticommutes, that is, reversing the order contributes a minus sign.

[1] The 2×2 octonionic Hermitian matrices $H_2(\mathbb{O})$ also form a Jordan algebra, but, even though the octonions are not associative, it is possible to find an associative algebra that leads to the same Jordan algebra [3, 7].

[2] In this case, the equivalence in (1) fails; it is the right-hand side which provides the correct generalization.

[3] A *division algebra* is a *vector space* over a *field* (in this case $\mathbb{R}$) which is also a *ring with identity* under multiplication, and in which $ax = b$ can be uniquely solved for x (unless $a = 0$). A *normed* division algebra satisfies (12) in addition, and is therefore also an *integral domain*, that is, a ring in which $ab = 0$ implies $a = 0$ or $b = 0$.

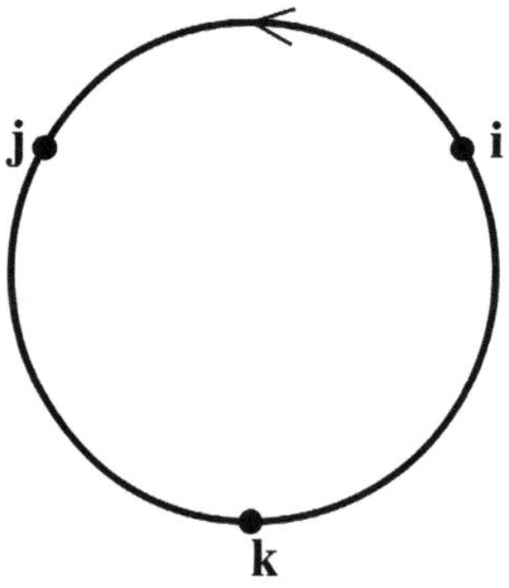

Figure 1. The quaternionic multiplication table.

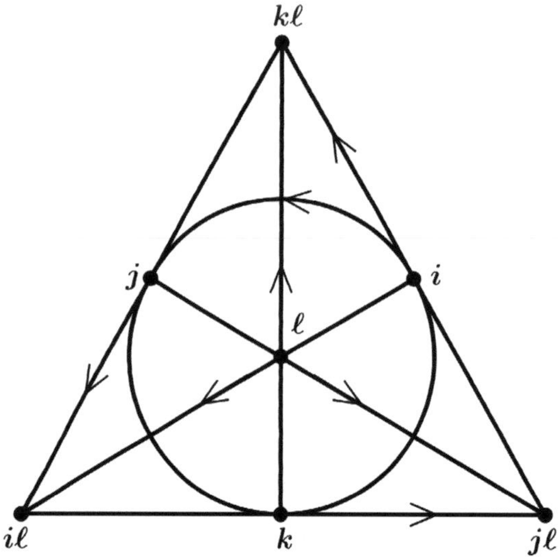

Figure 2. The octonionic multiplication table. Each line segment should be thought of as circle, identical to the quaternionic multiplication table in Figure 1.

We define the *conjugate* $\overline{a}$ of a quaternion or octonion a as the (real) linear map which reverses the sign of each imaginary unit. Thus, if

$$a = a_1 + a_2\, i + a_3\, j + a_4\, k + a_5\, k\ell + a_6\, j\ell + a_7\, i\ell + a_8\, \ell \tag{7}$$

then

$$\overline{a} = a_1 - a_2\, i - a_3\, j - a_4\, k - a_5\, k\ell - a_6\, j\ell - a_7\, i\ell - a_8\, \ell \tag{8}$$

Direct computation shows that

$$\overline{ab} = \overline{b}\,\overline{a} \tag{9}$$

The *norm* $|a|$ of an octonion a is defined by

$$|a|^2 = a\overline{a} = a_1^2 + a_2^2 + a_3^2 + a_4^2 + a_5^2 + a_6^2 + a_7^2 + a_8^2 \tag{10}$$

The only octonion with norm 0 is 0, and every nonzero octonion has a unique inverse, namely

$$a^{-1} = \frac{\overline{a}}{|a|^2} \tag{11}$$

For all the normed division algebras, the norm satisfies the identity

$$|ab| = |a||b| \tag{12}$$

A remarkable property of the octonions is that they are *not* associative! For example, compare

$$(i\,j)(\ell) \;=\; +(k)(\ell) \;=\; +k\ell \tag{13}$$
$$(i)(j\,\ell) \;=\; (i)(j\ell) \;=\; -\,k\ell \tag{14}$$

However, the octonions are *alternative*, that is, products involving no more than 2 independent octonions do associate. The *commutator* of two octonions a, b is given as usual by

$$[a,b] = ab - ba \tag{15}$$

and we define the *associator* of three octonions a, b, c by

$$[a,b,c] = (ab)c - a(bc) \tag{16}$$

which quantifies the lack of associativity. More generally, both the commutator and associator are *antisymmetric*, that is, interchanging any two arguments changes the result by a minus sign; replacing any argument by its conjugate has the same effect, because the real parts don't contribute to the associator.

The units i, j, k, $k\ell$, $j\ell$, $i\ell$, and ℓ are by no means the only square roots of -1. Rather, *any* pure imaginary quaternion or octonion squares to a negative number, so it is only necessary to choose its norm to be 1 in order to get a square root of -1. The imaginary quaternions of norm 1 form a 2-sphere in the 3-dimensional space of imaginary quaternions. The imaginary octonions of norm 1 form a 6-sphere in the 7-dimensional space of imaginary octonions.

Any such unit imaginary quaternion or octonion $\hat{s}$ can be used to construct a complex subalgebra of $\mathbb{H}$ or $\mathbb{O}$, which we will also denote by $\mathbb{C}$, and which takes the form

$$\mathbb{C} = \{a_R + a_s\,\hat{s}\} \tag{17}$$

with $a_R, a_s \in \mathbb{R}$. Regarding $\hat{s}$ as the complex unit, we have the familiar Euler identity

$$e^{\hat{s}\theta} = \cos\theta + \hat{s}\sin\theta \tag{18}$$

so that *any* quaternion or octonion can be written in the form

$$a = re^{\hat{s}\theta} \tag{19}$$

where

$$r = |a| \tag{20}$$

Any *two* unit imaginary octonions $\hat{s}$ and $\hat{t}$ that point in independent directions determine a *quaternionic* subalgebra of $\mathbb{O}$.

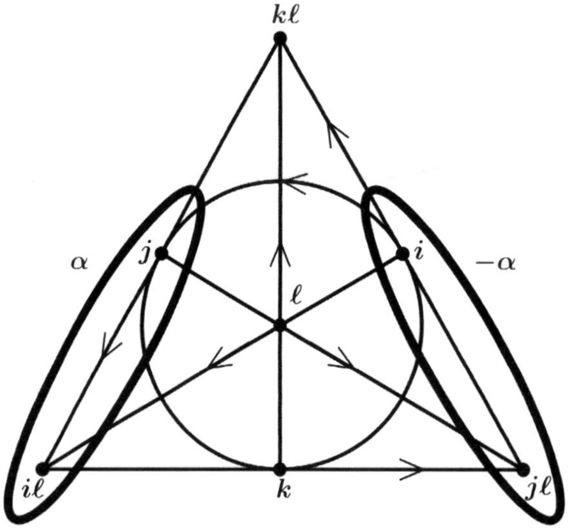

Figure 3. One class of elements of G_2. These transformations are also contained in the preferred $SU(3)$ that fixes ℓ.

4. The Structure of G_2 and $SU(3)$

The freedom to choose an entire 2-sphere or 6-sphere of square roots of minus one within the 3-dimensional space of the pure imaginary quaternions or the 7-dimensional space of pure imaginary octonions leads one to investigate the transformations that preserve the corresponding multiplication table. These transformations form the automorphism group of the corresponding division algebra.

In the case of the quaternions, one can imagine rotating i to any pure imaginary point on the 2-sphere (2 degrees of freedom). Then j can be chosen to be any direction perpendicular to the direction of i, i.e. on the equator of the resulting 2-sphere (1 degree of freedom). The direction of k is determined by the multiplication table. The 3-dimensional automorphism group of the quaternions is therefore seen to be $SO(3)$.

For the octonions, one can again imagine rotating i to any pure imaginary point on the 6-sphere (6 degrees of freedom). Then j must again be perpendicular to i (5 degrees of freedom) and the direction of k is fixed by the multiplication table. But ℓ is now free to be any direction perpendicular to all of the i, j, and k directions (3 degrees of freedom) and the directions of the remaining units are determined by the multiplication table. This 14-dimensional Lie group turns out to be the exceptional group G_2.

Another way of envisioning the transformations in the group G_2 was first shown to us by Sudbery [8]. Consider the octonionic unit $k\ell$ at the top of the multiplication table shown in Figure 3. There are three pairs of octonionic units that form quaternionic subalgebras with $k\ell$, i.e. $\{j, i\ell\}$, $\{j\ell, i\}$, and $\{k, \ell\}$. We call these the pairs that "point to" $k\ell$. If the elements of two of these pairs are rotated into one another oppositely, for instance, if the $\{j, i\ell\}$-plane is rotated by an angle α, and the $\{j\ell, i\}$-plane is rotated by the angle $-\alpha$, then it turns out that the multiplication table is preserved. We have thus constructed a 1-parameter family of automorphisms. There are three ways of pairing up the three pairs of units in this way, but only two are independent. Since there are 7 different units that can be pointed to, the dimension of this group is again 14.

In what follows, we will need not only G_2, but also $SU(3)$, the subgroup of G_2 that fixes one

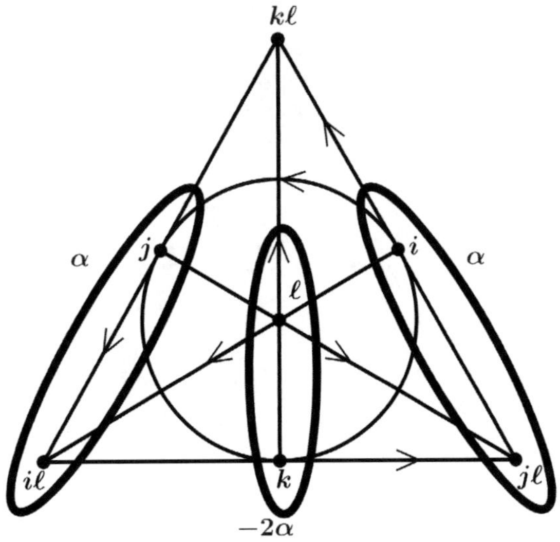

Figure 4. A second class of elements of G_2. These transformations are not contained in the preferred $SU(3)$ that fixes ℓ.

of the octonionic units. Since ℓ is in the middle of our multiplication table, we will, without loss of generality, choose it to be the unit that is fixed. We see that the G_2 transformation in Figure 3 fixes ℓ and is therefore in $SU(3)$, but a G_2 transformation involving either of the other two pairs that point to $k\ell$ will not fix ℓ. To be symmetric, we choose the linear combination of transformations shown in Figure 4 ($\{j, i\ell\}$ and $\{j\ell, i\}$ both rotate by α and $\{k, \ell\}$ rotates by -2α) to be the G_2 transformation that points to $k\ell$ that is *not* in $SU(3)$. If we choose to point in turn to each of the six units that are *not* ℓ, we have six G_2 transformations that are in $SU(3)$ and six that are not. What about the remaining two G_2 transformations? These are transformations that point to ℓ. One such transformation, shown in Figure 5, rotates $\{i\ell, i\}$ by α and $\{j\ell, j\}$ by $-\alpha$. There are three transformations of this type, all of which fix ℓ and are therefore elements of $SU(3)$, but only two are linearly independent. Any two of these transformations complete the 8-dimensional Lie group $SU(3)$.

Yet another way to describe G_2 is in terms of *inner automorphisms*, that is, transformations of the form

$$x \mapsto axa^{-1} \tag{21}$$

Inner automorphisms always preserve an associative multiplication rule, since

$$(axa^{-1})(aya^{-1}) = a(xy)a^{-1} \tag{22}$$

However, this condition is nontrivial over the octonions, since the parentheses cannot be moved. As shown in [9], (22) holds for all $x, y \in \mathbb{O}$ if and only if a is a sixth root of unity. That is, the inner automorphisms of the octonions are given by (21) where

$$a = e^{n\pi\hat{s}/3} \tag{23}$$

where $n \in \mathbb{Z}$ and $\hat{s}$ is any pure imaginary unit octonion. As further discussed in [9], any G_2 transformation can in fact be generated by a finite sequence of nested transformations of the form (21), with a given by (23).

Journal of Physics: Conference Series **254** (2010) 012005 doi:10.1088/1742-6596/254/1/012005

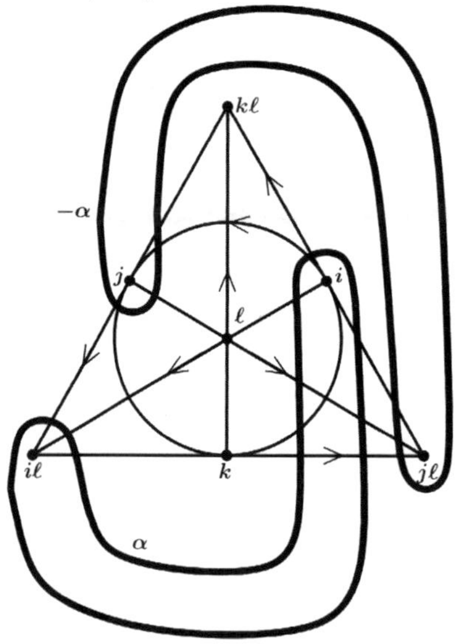

Figure 5. A third class of elements of G_2. These transformations are contained in the preferred $SU(3)$ that fixes ℓ.

5. The Jordan Eigenvalue Problem

In previous work [5], we solved the *Jordan eigenvalue problem*, namely the eigenmatrix problem

$$\mathcal{A} \circ \mathcal{V} = \lambda \mathcal{V} \tag{24}$$

where $\mathcal{A}$ and $\mathcal{V}$ are both 3×3 octonionic Hermitian matrices. Unlike the right eigenvalue problem $\mathcal{A}v = v\lambda$ considered in [10], the Jordan eigenvalue problem (24) admits only real eigenvalues, which do solve the characteristic equation for $\mathcal{A}$, namely [11]

$$-\det(\mathcal{A} - \lambda\mathcal{I}) = \lambda^3 - (\text{tr } \mathcal{A})\,\lambda^2 + \sigma(\mathcal{A})\,\lambda - (\det \mathcal{A})\,\mathcal{I} = 0 \tag{25}$$

where $\mathcal{I}$ denotes the identity matrix, $\sigma(\mathcal{A})$ is defined by

$$\sigma(\mathcal{A}) = \frac{1}{2}\left((\text{tr } \mathcal{A})^2 - \text{tr}(\mathcal{A}^2)\right) = \text{tr}(\mathcal{A} * \mathcal{A}) \tag{26}$$

the operation $*$ denotes the Freudenthal product

$$\mathcal{A} * \mathcal{B} = \mathcal{A} \circ \mathcal{B} - \frac{1}{2}\Big(\mathcal{A}\,\text{tr}(\mathcal{B}) + \mathcal{B}\,\text{tr}(\mathcal{A})\Big) + \frac{1}{2}\Big(\text{tr}(\mathcal{A})\,\text{tr}(\mathcal{B}) - \text{tr}(\mathcal{A} \circ \mathcal{B})\Big)\,\mathcal{I} \tag{27}$$

and the determinant is defined unambiguously by

$$\det(\mathcal{A}) = \frac{1}{3}\,\text{tr}\left((\mathcal{A} * \mathcal{A}) \circ \mathcal{A}\right) \tag{28}$$

The Jordan and Freudenthal products are generalizations of the standard dot and cross products.

Just as in the more familiar complex case, normalized eigenmatrices for each nondegenerate eigenvalue are primitive idempotents, and the degenerate case can be handled using Gram-Schmidt orthogonalization. Furthermore, each primitive idempotent is in fact an element of the *Cayley-Moufang plane* $\mathbb{OP}^2$, which can be characterized as

$$\begin{aligned} \mathbb{OP}^2 &:= \{\mathcal{V} \in H_3(\mathbb{O}) : \mathcal{V} \circ \mathcal{V} = \mathcal{V} \text{ and } \text{tr}\,\mathcal{V} = 1\} \\ &\equiv \{\mathcal{V} \in H_3(\mathbb{O}) : \mathcal{V} * \mathcal{V} = 0 \text{ and } \text{tr}\,\mathcal{V} = 1\} \end{aligned} \tag{29}$$

It is straightforward to show from the first condition in (29) that the components of any element $\mathcal{V} \in \mathbb{OP}^2$ must lie in some quaternionic subalgebra of $\mathbb{O}$, which of course depends on $\mathcal{V}$. Put differently, the associator of the (independent, off-diagonal) components of $\mathcal{V}$, denoted $[\mathcal{V}]$, must vanish. But quaternionic primitive idempotents have (nonunique) "square roots", $\mathcal{V} = \Psi\Psi^\dagger$, so that we can also write

$$\mathbb{OP}^2 = \{\Psi\Psi^\dagger : \Psi \in \mathbb{O}^3, [\Psi] = 0, \Psi^\dagger\Psi = 1\} \tag{30}$$

where $[\Psi]$ denotes the associator of the components of Ψ. We refer to such 3-component octonionic column vectors Ψ as *Cayley spinors.*

Putting this all together, any 3×3 octonionic Hermitian matrix $\mathcal{A}$ can be expressed as the sum of the squares of quaternionic columns, which are orthogonal under the Jordan product, that is

$$\mathcal{A} = \sum_{i=1}^{3} \lambda_i \mathcal{V}_i \tag{31}$$

in terms of primitive idempotents $\mathcal{V}_i = \Psi_i\Psi_i^\dagger \in \mathbb{OP}^2$ and their corresponding eigenvalues λ_i.

6. The Structure of E_6

The automorphism group of the Jordan product (2) (and consequently also of the Freudenthal product (27)) is the exceptional group F_4, and the group which leaves the determinant (28) invariant is a particular real form of the exceptional group E_6. These groups can be interpreted as $F_4 = SU(3,\mathbb{O})$ and $E_6 = SL(3,\mathbb{O})$, as we now show; for further details see [12].

In previous work [9], Manogue and Schray showed how to write the (double cover of the) Lorentz group $SO(9,1)$ as $SL(2,\mathbb{O})$, with the action given by

$$\mathbf{X} \longmapsto \mathbf{M}\mathbf{X}\mathbf{M}^\dagger \tag{32}$$

where $\mathbf{X} \in H_2(\mathbb{O})$, the 2×2 Hermitian matrices with octonionic components. The key to that work was to give an explicit set of basis transformations — the rotations and boosts in coordinate planes — which were *compatible* with the spinor representation in the sense that if $\theta \in \mathbb{O}^2$ transforms like

$$\theta \longmapsto \mathbf{M}\theta \tag{33}$$

then there are no associativity problems in the vector transformation

$$\mathbf{M}(\theta\theta^\dagger)\mathbf{M}^\dagger = (\mathbf{M}\theta)(\mathbf{M}\theta)^\dagger \tag{34}$$

Any such basis transformation $\mathbf{M} \in SL(2,\mathbb{O})$ can be immediately reinterpreted as a 3×3 transformation $\mathcal{M}$ via

$$\mathcal{M} = \begin{pmatrix} \mathbf{M} & 0 \\ 0 & 1 \end{pmatrix} \tag{35}$$

and it is straightforward to verify that any such $\mathcal{M}$ preserves the determinant of $\mathcal{X} \in H_3(\mathbb{O})$ and is therefore an element of E_6.

How many such transformations are there? We first give the basis transformations for the simpler case of $SL(2,\mathbb{C})$, adapted from [9] and rewritten as elements of E_6. When interpreting these transformations, it is helpful to recall that, in this case,

$$\mathbf{X} = \begin{pmatrix} t+z & x-\ell y \\ x+\ell y & t-z \end{pmatrix} \tag{36}$$

We have the three rotations

$$\mathcal{M}_{xy} = \begin{pmatrix} e^{-\ell\theta/2} & 0 & 0 \\ 0 & e^{\ell\theta/2} & 0 \\ 0 & 0 & 1 \end{pmatrix} \tag{37}$$

$$\mathcal{M}_{yz} = \begin{pmatrix} \cos\frac{\theta}{2} & -\ell\sin\frac{\theta}{2} & 0 \\ -\ell\sin\frac{\theta}{2} & \cos\frac{\theta}{2} & 0 \\ 0 & 0 & 1 \end{pmatrix} \qquad \mathcal{M}_{zx} = \begin{pmatrix} \cos\frac{\theta}{2} & -\sin\frac{\theta}{2} & 0 \\ \sin\frac{\theta}{2} & \cos\frac{\theta}{2} & 0 \\ 0 & 0 & 1 \end{pmatrix} \tag{38}$$

and the three boosts

$$\mathcal{M}_{tz} = \begin{pmatrix} e^{\beta/2} & 0 & 0 \\ 0 & e^{-\beta/2} & 0 \\ 0 & 0 & 1 \end{pmatrix} \tag{39}$$

$$\mathcal{M}_{tx} = \begin{pmatrix} \cosh\frac{\beta}{2} & \sinh\frac{\beta}{2} & 0 \\ \sinh\frac{\beta}{2} & \cosh\frac{\beta}{2} & 0 \\ 0 & 0 & 1 \end{pmatrix} \qquad \mathcal{M}_{ty} = \begin{pmatrix} \cosh\frac{\beta}{2} & -\ell\sinh\frac{\beta}{2} & 0 \\ \ell\sinh\frac{\beta}{2} & \cosh\frac{\beta}{2} & 0 \\ 0 & 0 & 1 \end{pmatrix} \tag{40}$$

written in terms of the single imaginary unit ℓ. To generate a 3×3 representation of $SL(2,\mathbb{O})$, start by replacing ℓ in turn by each of the other imaginary units, yielding 18 new transformations, for a total of 24. The remaining 21 transformations in $SL(2,\mathbb{O})$ are precisely the rotations of the imaginary units with each other, which make up an $SO(7)$ (really $Spin(7)$; we are being casual about double covers). As shown in [9], these rotations are obtained by *nesting*, that is by transformations of the form

$$\mathcal{X} \longmapsto \mathcal{M}_2 \left(\mathcal{M}_1 \mathcal{X} \mathcal{M}_1^\dagger\right) \mathcal{M}_2^\dagger \tag{41}$$

where each corresponding $\mathbf{M}$ represents a "flip", that is, a pure imaginary multiple of the $(2\times 2!)$ identity matrix. Thus, a typical $\mathcal{M}$ takes the form

$$\mathcal{M} = \begin{pmatrix} \ell & 0 & 0 \\ 0 & \ell & 0 \\ 0 & 0 & 1 \end{pmatrix} \tag{42}$$

where it is important to note that $\mathcal{M}$ is *not* a multiple of the (3×3) identity matrix.

We are now ready to count the basis transformations of E_6. At first sight, it appears we have three copies of $SL(2,\mathbb{O})$ — simply repeat the embedding (35) with the two other obvious block structures. We call these three copies type I, II, and III. However, this yields $3 \times 45 = 135$ transformations, and, while these transformations do indeed generate all of E_6, it is clear that they can not be a basis, since the dimension of E_6 is only 78.

Let's try again. Each of these three copies of $SL(2,\mathbb{O}) = SO(9,1)$ contains a copy of $SO(8)$. A famous property of $SO(8)$ called *triality* asserts in this context that these three copies of $SO(8)$ in fact consist of the same E_6 transformations (but labeled differently), so we should count these copies only once. But $SO(8)$ has 28 elements, to which we must add 3 copies of the 8 rotations needed to get to $SO(9)$, then 3 copies of the 9 boosts needed to get to $SO(9,1)$, resulting in $28 + 3 \times 8 + 3 \times 9 = 79$ transformations. A bit of thought reveals that the 3 copies of the tz-boost (39) are not independent; removing one of them correctly yields an explicit set of 78 basis transformations for E_6, also justifying the interpretation $E_6 = SL(3,\mathbb{O})$.

It is worth pointing out that, due to triality, only the 14 G_2 transformations need to be written in the nested form (41). Remarkably, the remaining 14 $SO(8)$ transformations can all be expressed using the type I transformation (37) and its types II and III variants. The former are just the usual 7 rotations needed to get from $SO(7)$ to $SO(8)$, but the latter yield an unnested description of the 7 non-G_2 transformations in $SO(7)$, which take the form

$$\mathcal{M}_\ell = \begin{pmatrix} e^{\ell\theta/2} & 0 & 0 \\ 0 & e^{\ell\theta/2} & 0 \\ 0 & 0 & e^{-\ell\theta} \end{pmatrix} \tag{43}$$

Each of these transformations rotates 3 octonionic planes by the same amount, and can therefore be thought of as a "phase" transformation.

What about F_4? Note that we have described $27-1=26$ boosts, and $78-26=52$ rotations. So our E_6 is the real representation with 26 boosts, commonly written as $E_{6(-26)}$, with the number in parentheses denoting the number of boosts minus the number of rotations. It is straightforward to show that F_4 preserves the trace of elements of $H_3(\mathbb{O})$, corresponding to the timelike direction; this is the compact representation of F_4, consisting precisely of the rotation subgroup of this real form of E_6. These considerations justify the interpretation $F_4 = SU(3,\mathbb{O})$.

Returning to the characteristic equation (25), not only does E_6 preserve the determinant, it also preserves the condition $\sigma = 0$. But these two coefficients control the number of nonzero eigenvalues — 3 if $\det \mathcal{A} \neq 0$, 2 if $\det \mathcal{A} = 0 \neq \sigma(\mathcal{A})$, and 1 if $\det \mathcal{A} = 0 = \sigma(\mathcal{A})$. Thus, E_6 preserves the number of nonzero eigenvalues of $\mathcal{A}$, and hence the number of terms in the decomposition (31) with nonzero eigenvalue.

7. Symmetry Breaking and Particle Physics

In order to apply this formalism to elementary particle physics, we break the full symmetry and explore how a Jordan matrix transforms under various subgroups of E_6. One such symmetry breaking occurs when we choose a preferred $SL(2,\mathbb{O})$ subgroup of $SL(3,\mathbb{O})$, as in (35). This leads us to impose a block structure on $H_3(\mathbb{O})$, so that

$$\mathcal{P} = \begin{pmatrix} \mathbf{P} & \psi \\ \psi^\dagger & n \end{pmatrix} \tag{44}$$

where $\mathbf{P} \in H_2(\mathbb{O})$ transforms like a 10-dimensional momentum vector, $\psi \in \mathbb{O}^2$ transforms like a (Majorana-Weyl) spinor, and $n \in \mathbb{R}$ is a scalar. Direct computation shows that

$$\mathcal{P} * \mathcal{P} = \begin{pmatrix} \widetilde{\psi\psi^\dagger} - n\widetilde{\mathbf{P}} & \widetilde{\mathbf{P}}\psi \\ (\widetilde{\mathbf{P}}\psi)^\dagger & \det \mathbf{P} \end{pmatrix} \tag{45}$$

where tilde denotes trace reversal, that is, $\widetilde{\mathbf{P}} = \mathbf{P} - \operatorname{tr}(\mathbf{P})\,\mathbf{I}$. As shown in [14], the massless, momentum-space Dirac equation in 10 dimensions can be written

$$\widetilde{\mathbf{P}}\psi = 0 \tag{46}$$

which implies the nonlinear constraint

$$\det \mathbf{P} = 0 \tag{47}$$

The general solution of (46) and (47) is of the form

$$\psi = \theta\xi \tag{48}$$
$$\mathbf{P} = \theta\theta^\dagger \tag{49}$$

where the components of $\theta \in \mathbb{O}^2$ lie in the complex subalgebra of $\mathbb{O}$ determined by $\mathbf{P}$ and $\xi \in \mathbb{O}$ is arbitrary. Using (45), these equations are seen to be precisely the same as

$$\mathcal{P} * \mathcal{P} = 0 \tag{50}$$

where

$$n = |\xi|^2 \tag{51}$$

Thus, (normalized) solutions of the Dirac equation are precisely elements of $\mathbb{OP}^2$, and therefore the squares of Cayley spinors.

In previous work [14], we discussed solutions of the Dirac equation in the form (46). Remembering that solutions of (50) are quaternionic, and reducing 10 spacetime dimensions to 4 by the simple expedient of choosing a preferred complex subalgebra of $\mathbb{O}$, we used spin eigenstates and particle/antiparticle projection operators as usual to identify particle states within our division algebra formalism. This procedure identified a spin-$\frac{1}{2}$ massive particle with two spin states, namely

$$e_\uparrow = \begin{pmatrix} 1 \\ k \end{pmatrix} \qquad e_\uparrow e_\uparrow^\dagger = \begin{pmatrix} 1 & -k \\ k & 1 \end{pmatrix} \tag{52}$$

$$e_\downarrow = \begin{pmatrix} -k \\ 1 \end{pmatrix} \qquad e_\downarrow e_\downarrow^\dagger = \begin{pmatrix} 1 & -k \\ k & 1 \end{pmatrix} \tag{53}$$

where the direction of the arrow indicates the z-component of the spin, and where the second equality in each case gives the momentum vector. Comparison with (36) shows that the x, y, and z components of the momentum vanish; these states are given at rest. Similarly, there is an analogous antiparticle with two spin states. The procedure also identified a left-handed massless particle, which when moving in the z-direction takes the form

$$\nu_z = \begin{pmatrix} 0 \\ k \end{pmatrix} \qquad \nu_z \nu_z^\dagger = \begin{pmatrix} 0 & 0 \\ 0 & 1 \end{pmatrix} \tag{54}$$

together with a massless particle of the opposite helicity, which when moving in the z-direction takes the form

$$\emptyset_z = \begin{pmatrix} 0 \\ 1 \end{pmatrix} \qquad \emptyset_z \emptyset_z^\dagger = \begin{pmatrix} 0 & 0 \\ 0 & 1 \end{pmatrix} \tag{55}$$

Any other 2-component quaternionic column can be identified as an appropriately rotated and/or boosted superposition of these particles in the usual way. Noting that the first two particles carry an octonionic label (k), it is straightforward to generalize these particles to 3 generations of leptons, labeled by i, j, k; the remaining particle does not have an octonionic label. Thus, an octonionic description of the Dirac equation in 10 dimensions yields a particle spectrum containing precisely 3 generations of leptons, each with a single-helicity, massless neutrino, together with a single "sterile" neutrino of the opposite helicity but with no generation structure.

8. Discussion

We have shown how to break the symmetry group E_6 so that a Lorentzian $3+1$ dimensional momentum space emerges, together with internal symmetries that describe the correct spin/helicity transformations on Cayley spinors to describe leptons. Furthermore, precisely three generations of such leptons exist which respect the octonionic structure of the transformations. Contained naturally within this description of leptons and their symmetries are three massless left-handed neutrinos and a single, sterile, right-handed neutrino.

We have learned several important lessons along the way. First, in spite of the non-commutativity and non-associativity of the octonions, everything that one might want to do can be made to work if one defines everything carefully. Second, when working with the octonions, it is important to make the Lie group structure primary, rather than the Lie algebra structure. Some of the nested group transformations described in (41) cannot be described in terms of the exponentials of any Lie algebra transformations. Finally, in order to keep a complex structure on the Lie algebra from interfering with the octonionic units in the matrices, it is important in the symmetry-breaking process to look at *real* Lie subalgebras rather than complexified ones. We

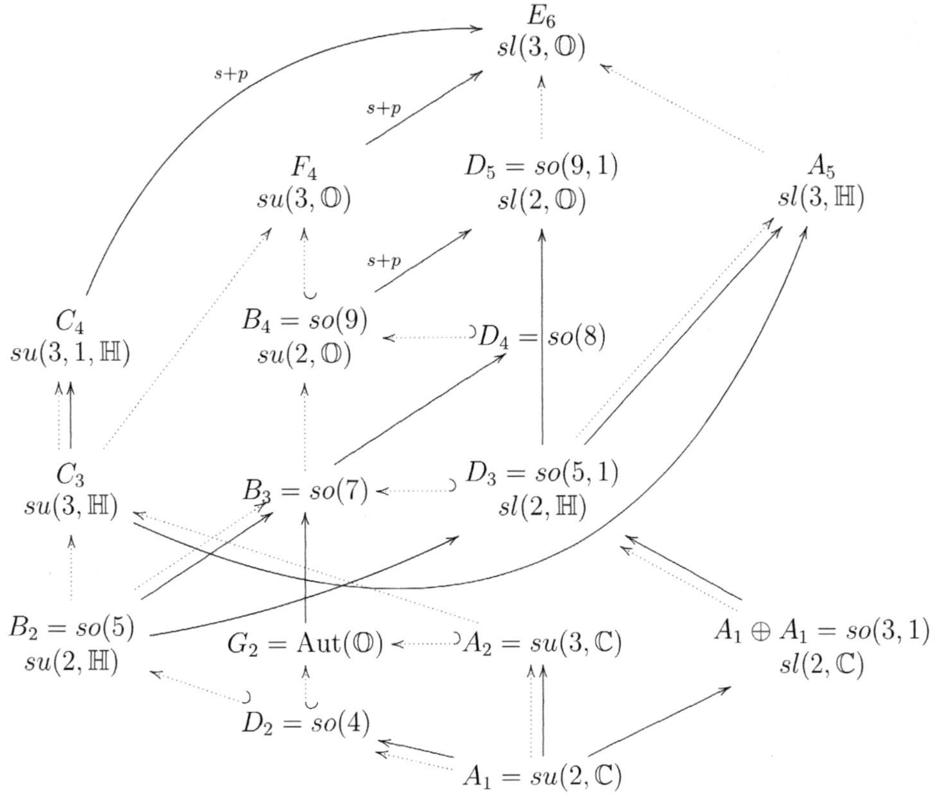

Figure 6. A map of E_6, taken from [13].

call the reader's attention to the recent work of Aaron Wangberg [13], in which the important real forms of the subgroups of E_6 were identified, using a division-algebra perspective. A map of these subgroups, taken from [13], appears in Figure 6.

We conclude with several speculative questions:

(i) In this paper, we have described leptons in terms of "1-squares," i.e. the squares of Cayley spinors. But the most general octonionic hermitian matrix can be a linear combination of up to three such squares. Is it possible to describe the meson and baryon sectors of standard particle physics as "2-squares" and "3-squares," respectively?

(ii) Does the group E_6, broken in the way we have discussed, describe both the standard model interactions and Lorentz transformations in $3+1$-dimensions?

(iii) After the symmetry breaking described in Section 7, some of the Lie group transformations of the form (21), (23) are no longer connected to the identity. Do these discrete transformations correspond to discrete conserved quantities such as charge?

Acknowledgments

We would like to acknowledge the many contributions of colleagues over the years who have contributed to our understanding of this field: David Fairlie, Ed Corrigan, Jörg Schray, Robin Tucker, Jason Janesky, Susumu Okubo, Aaron Wangberg, Jim Wheeler, and, of course, Tony Sudbery. CAM thanks the organizers for support to attend the TonyFest and present this work.

We thank the Department of Physics at Utah State University for support and hospitality during the preparation of this manuscript. This work was supported in part by a grant from the Foundational Questions Institute (FQXi).

References

[1] Sudbery A and Manogue C A 1989 General solutions of the covariant superstring equations of motion *Phys. Rev. D* **40** 4073-4077

[2] Jordan P 1933 Über die Multiplikation quantenmechanischer Größen *Z. Phys.* **80** 285-291

[3] Jordan P, von Neumann J and Wigner E 1934 On an algebraic generalization of the quantum mechanical formalism *Ann. of Math.* **35** 29-64

[4] Gürsey F and Tze C-H 1996 *On the Role of Division, Jordan, and Related Algebras in Particle Physics* (Singapore: World Scientific)

[5] Dray T and Manogue C A 1999 The Exceptional Jordan Eigenvalue Problem *Internat. J. Theoret. Phys.* **38** 2901-2916 (*Preprint* math-ph/9910004)

[6] Albert A A 1934 On a certain algebra of quantum mechanics *Ann. Math.* **35** 65-73

[7] Jacobson N 1968 *Structure and Representations of Jordan Algebras* (*Amer. Math. Soc. Colloq. Publ.* **39**) (Providence: American Mathematical Society)

[8] Sudbery A 1988 private communication

[9] Manogue C A and Schray J 1993 Finite Lorentz transformations, automorphisms, and division algebras *J. Math. Phys.* **34** 3746-3767 (*Preprint* hep-th/9302044)

[10] Dray T and Manogue C A 1998 The octonionic eigenvalue problem *Adv. Appl. Clifford Algebras* **8**, 341-364 (*Preprint* math/9807126)

[11] Freudenthal H 1953 Zur ebenen Oktavengeometrie *Proc. Kon. Ned. Akad. Wet.* **A56**, 195-200

[12] Dray T and Manogue C A (in print) Octonionic Cayley Spinors and E_6 *Comment. Math. Univ. Carolin.* (*Preprint* arXiv:0911.2255)

[13] Wangberg A 2007 *The Structure of* E_6 Ph.D. thesis Oregon State University (*Preprint* arXiv:0711.3447)

[14] Manogue C A and Dray T 2000 Quaternionic spin *Clifford Algebras and their Applications in Mathematical Physics* ed R Abłamowicz and B Fauser (Boston: Birkhäuser) pp 29-46 (*Preprint* hep-th/9910010)

Journal of Physics: Conference Series **254** (2010) 012006 doi:10.1088/1742-6596/254/1/012006

A distributional formulation of electromagnetic fields and sources in the presence of media discontinuities

Robin W Tucker

The Cockcroft Institute and Lancaster University, Lancaster, UK

E-mail: `r.tucker@lancaster.ac.uk`

Abstract.
A self-contained formulation of Maxwell's theory of electromagnetic fields and sources is presented in the language of distributional forms. Properties of a fundamental double-form of bi-degree (p, p) for $p \geq 0$ are reviewed in order to establish a computational framework for analysing equations involving the Hodge-de Rham operator for $p-$forms on space or spacetime and singular sources. A constructive approach to Dirac distributions on (moving) submanifolds embedded in $\mathbf{R}^3$ is developed in terms of (Leray) forms generated by the geometry of the embedding and illustrative examples presented from problems in electrostatics, magnetostatics and radiating point charges. The formulation offers a straightforward analysis of the relativistic jump conditions across static and moving interfaces where certain fields become discontinuous and provides a general methodology for electromagnetic modeling where possibly time dependent sources of certain physical attributes, such as electric charge, electric current and polarization or magnetization, are concentrated on localized regions in space or spacetime.

Received: 13 October 2009

1. Tribute

It is with great pleasure that I dedicate this article to Tony Sudbery. Since our student days we have maintained a most valuable friendship and I am most grateful to Tony for all the things that he has helped me with in Physics and Mathematics. Many years ago I discovered that he was one of the world's experts in the theory of Quaternions over the complex field and this was something I then knew little about. I was using this algebra in generalized theories of gravitation and thanks to Tony's guidance I was able to make sense of this branch of mathematics in the gravitational context. Later he helped me understand Octonions as we worked together on Octonionic Geometry and Simple Supergravity in 11-Dimensions. Today his breadth of expertise roams many new and exciting fields. I count myself fortunate to have such a valuable colleague and friend.

2. Introduction

Mathematical distributions [1] play an all pervading role in the physical sciences. They underpin the theoretical formulation of quantum mechanics and many linear systems in classical field theory and are used to extend the solution space for linear partial differential equations to include non-smooth fields with singularities. They also offer a useful tool in electromagnetic modeling where the sources of certain physical attributes, such as electric charge, electric current, polarization or magnetization, are concentrated on localized regions in space [2], [3], [4], [5], [6], [16], [17] or spacetime. These aspects of distributions often occur together in problems in electromagnetic theory and this article explores this symbiosis in the unifying language of exterior double-forms. Such a formulation has direct relevance to many contemporary design problems in accelerator science where one is confronted with a number of difficult issues in electromagnetic computation. The effective resolution of such problems is often contingent on reliable computationally expensive numerical analysis which can greatly benefit from reliable analytic treatments. The formulation below will be illustrated by a number of applications in applied electro-magnetics.

3. Notation

The formulation exploits the geometric language of exterior differential forms [7] since this is ideally suited to accommodate *local* changes of coordinates that can be used to simplify the description of boundary and initial-value problems and naturally encapsulates intrinsic *global* properties of domains. In the following the distinction between smooth (C^∞) forms on some regular domain and those with possible singularities or discontinuities will be important. Furthermore smooth forms with compact support will play a pivotal role. These will be referred to as *test forms* [8], [9], [10] and distinguished below by a superposed hat. It will also be useful sometimes to notationally distinguish $p-$forms on different manifolds. Thus $\underset{(p)}{\alpha_M}$ will denote a $p-$form [1] on some manifold M while $\underset{(p,q)}{\gamma}(X,Y)$ will denote a double-form of bi-degree (p,q) (see section 4) on the product manifold $X\times Y$. Each $n-$dimensional manifold M will be assumed orientable and endowed with a preferred n-form $\underset{(M)}{\star}1$ induced from a metric tensor g_M. One then has [7] the linear Hodge operator $\underset{(M)}{\star}$ that maps $p-$forms to $(n-p)$-forms on M. If g_M has signature t_g one may write

$$g_M=\sum_{i=1}^{n} e^i_M\otimes e^j_M\,\eta_{ij} \tag{1}$$

where $\eta_{ij}=diag(\pm1,\pm1,\ldots\pm1)$ and

$$\underset{(M)}{\star}1=e^1_M\wedge e^2_M\wedge\ldots\wedge e^n_M \tag{2}$$

with $t_g=det(\eta_{ij})$ and $\{e^i_M\}$ is a set of basis 1−forms in $\Gamma T^\star(M)$.

If α and β are integrable $p-$forms on $\mathcal{U}\subset\mathbf{R}^n$ denote

$$\int_{\mathcal{U}}\alpha\wedge\underset{(\mathbf{R}^n)}{\star}\beta \tag{3}$$

by

$$(\alpha,\beta)_{\mathcal{U}}\equiv(\beta,\alpha)_{\mathcal{U}} \tag{4}$$

Furthermore if one of the forms has compact support on $\mathcal{U}$ (and belongs to the space of Schwartz test functions) an alternative notation is

$$\underset{(p)}{\alpha}^{D}[\underset{(p)}{\hat\phi}]=(\alpha,\hat\phi)_{\mathcal{U}} \tag{5}$$

When $\mathcal{U}=\mathbf{R}^n$ the bracket subscript will be omitted.

[1] i.e. $\alpha_M\in\mathcal{S}\Lambda^p(M)$, the space of sections of the exterior bundle of $p-$forms over M

If $(\alpha, \hat{\phi})$ is well defined (i.e. α is integrable with respect to the bracket $(.,.)$) then one says that α^D is a *regular* Schwartz $p-$form distribution associated with α. Not all distributions are regular and associated with smooth forms.

Suppose Z is a smooth vector field on $\mathbf{R}^n$ then $\widetilde{Z} \equiv g(Z,-)$ denotes a smooth $1-$ form[2]. Using the property

$$\star\star = t_g\,\eta^{n-1} \tag{6}$$

where the involution η on forms satisfies $\eta \underset{(p)}{\alpha} \equiv (-1)^p \underset{(p)}{\alpha}$ then it is straightforward to verify the algebraic identities.

$$\star \underset{(q)}{\alpha}^D[\underset{(n-q)}{\phi}] = t_g\,\eta^{n-1} \underset{(q)}{\alpha}^D[\star^{-1} \underset{(n-q)}{\phi}] \tag{7}$$

$$(\widetilde{Z} \wedge \alpha^D)[\phi] = \alpha^D[i_Z \phi] \tag{8}$$

$$i_Z \alpha^D[\beta] = \alpha^D[\widetilde{Z} \wedge \beta] \tag{9}$$

The operator $\delta \equiv \star^{-1} d \star \eta$ is a formal adjoint of the exterior derivative d with respect to the bracket $(.,.)$ in the sense that

$$d \underset{(p)}{\alpha}[\underset{(p+1)}{\hat{\phi}}] = \alpha[\delta\hat{\phi}] \tag{10}$$

Similarly

$$\delta \underset{(p)}{\alpha}[\underset{(p-1)}{\hat{\psi}}] = \alpha[d\hat{\psi}] \tag{11}$$

These relations are used to define $d\,\alpha^D$ and $\delta\,\alpha^D$. If f is a smooth $0-$form then we denote by $f\,\alpha^D$ the distribution defined by $f\,\alpha^D[\hat{\chi}] = \alpha^D[f\,\hat{\chi}]$.

If one defines the Lie derivative of $\underset{(p)}{\alpha}^D$ with respect to a smooth vector field Z by

$$\mathcal{L}_Z \underset{(p)}{\alpha}^D = (i_Z\, d + d\, i_Z) \underset{(p)}{\alpha}^D$$

then

$$\mathcal{L}_Z \underset{(p)}{\alpha}^D[\underset{(p)}{\hat{\chi}}] = \underset{(p)}{\alpha}^D[\delta\widetilde{Z} \wedge \underset{(p)}{\hat{\chi}} + (\widetilde{Z}^\eta + \widetilde{Z})\,\delta \underset{(p)}{\hat{\chi}} - \nabla_Z \underset{(p)}{\hat{\chi}}] \tag{12}$$

where ∇ denotes covariant differentiation with respect to the Levi-Civita connection for which $\delta \equiv -g^{ab}\, i_{X_a} \nabla_{X_b}$.

In terms of d and δ one defines [3] the Hodge de Rham operator $\Delta = d\,\delta + \delta\, d$. A number of important identities follow by integrating the Leibnitz relation involving smooth forms

$$d(\alpha \wedge \beta) = d\alpha \wedge \beta + (\eta\alpha) \wedge d\beta$$

over regular domains $\mathcal{U}$. These include

$$(d\phi\,, \psi)_{\mathcal{U}} = (\phi\,, \delta\psi)_{\mathcal{U}} + \int_{\partial\mathcal{U}} \phi \wedge \star\psi \tag{13}$$

$$(\delta\phi\,, \psi)_{\mathcal{U}} = (\phi\,, d\psi)_{\mathcal{U}} - \int_{\partial\mathcal{U}} \psi \wedge \star\phi \tag{14}$$

[2] Similarly if χ is a smooth $1-$form on $\mathbf{R}^n$ then $\widetilde{\chi} \equiv g^{-1}(\chi,-)$ denotes a smooth vector field

[3] The traditional Laplacian operator on forms is $-\Delta$

$$(\Delta\phi\,,\psi)_{\mathcal{U}} - (d\phi\,,d\psi)_{\mathcal{U}} - (\delta\phi\,,\delta\psi)_{\mathcal{U}} = \int_{\partial\mathcal{U}} (\delta\phi\,\wedge\star\psi - \psi\,\wedge\star d\phi) \tag{15}$$

$$(\Delta\phi\,,\psi)_{\mathcal{U}} - (\phi\,,\Delta\psi)_{\mathcal{U}} = \int_{\partial\mathcal{U}} (\delta\phi\wedge\star\psi - \delta\psi\wedge\star\phi + \phi\wedge\star d\psi - \psi\wedge\star d\phi) \tag{16}$$

From these elementary identities further useful identities involving singular double-forms can be obtained.

4. Double-Forms on $\mathbf{R}^n$

The notion of a double-form was introduced in order to analyse problems in potential theory in $\mathbf{R}^3$ [11]. Double-forms have since been used by Duff [12, 13] and others as a powerful tool to study exterior equations on more general manifolds. A double-form of bi-degree (p, q) over the product manifold $X \times Y$ may be regarded for each point in X as a p-form valued $q-$form on Y or for each point in Y as a q-form valued $p-$form on X . To illustrate these notions suppose $X = Y = \mathbf{R}^n$. In the natural coordinate system (x^i) for X and (y^i) for Y one has[4]:

$$\underset{(p,q)}{\gamma}(X,Y) = \sum_{i_1<\ldots<i_p} \sum_{j_1<\ldots<j_q} \gamma_{i_1\ldots i_p,j_1\ldots j_p}(X,Y)\,(d\,x^{i_1}\wedge\ldots\wedge d\,x^{i_p})\odot(d\,y^{j_1}\wedge\ldots\wedge d\,y^{j_q}) \tag{17}$$

The *symmetric* product $\odot$ is defined so that:

$$(d\,x^i\odot d\,y^j)\wedge(d\,x^k\odot d\,y^s) = (d\,x^i\odot d\,x^k)\wedge(d\,y^j\odot d\,y^s) \tag{18}$$

and this implies

$$\underset{(p,q)}{\gamma}\wedge\underset{(p',q')}{\gamma'} = (-1)^{pp'+qq'}\underset{(p',q')}{\gamma'}\wedge\underset{(p,q)}{\gamma} \tag{19}$$

for all γ, γ'. The Euclidean metric tensors in natural coordinates and their associated Hodge maps will be designated:

$$g_X = \sum_{i=1}^{n} d\,x^i\otimes d\,x^i \tag{20}$$

$$g_Y = \sum_{i=1}^{n} d\,y^i\otimes d\,y^i \tag{21}$$

and

$$\underset{(X)}{\star}1 = dx^1\wedge dx^2\wedge\ldots\wedge dx^n \tag{22}$$

$$\underset{(Y)}{\star}1 = dy^1\wedge dy^2\wedge\ldots\wedge dy^n \tag{23}$$

The operators $d_X, d_Y, \delta_X, \delta_Y, \underset{(X)}{\star}, \underset{(Y)}{\star}$ then act naturally on forms in the appropriate space.

The *fundamental double-form of bi-degree* (p, p) is defined as

$$\underset{(p,p)}{\gamma}(X,Y) = \sum_{i_1<i_2\ldots<i_p} f(X,Y)(d\,x^{i_1}\wedge\ldots\wedge d\,x^{i_p})_X\odot(d\,y^{i_1}\wedge\ldots\wedge d\,y^{i_p})_Y \tag{24}$$

[4] If $f : \mathbf{R}^n \times \mathbf{R}^n \mapsto \mathbf{R}$ we denote its image by $f(X, Y)$ with a similar notation for double-forms

where

$$f(X,Y) = \begin{cases} \frac{1}{(n-2)V_{n-1}} \frac{1}{|\mathbf{x}-\mathbf{y}|^{n-2}} & n>2 \\ \frac{1}{2\pi} \log \frac{1}{|\mathbf{x}-\mathbf{y}|} & n=2 \end{cases} \tag{25}$$

and [5]

$$V_{n-1} = V_{n-2} \int_0^\pi \sin^{n-2}\theta \, d\theta \tag{26}$$

with $V_1 = 2\pi$, $V_0 = 1$ and $|\mathbf{x}-\mathbf{y}|^2 = \sum_{i=1}^n (x^i - y^i)^2$. This double-form is singular on $X \times Y$ when $\mathbf{x} = \mathbf{y}$ and satisfies

$$\underset{(p,p)}{\gamma}(X,Y) = \underset{(p,p)}{\gamma}(Y,X) \tag{27}$$

$$\Delta_X \underset{(p,p)}{\gamma}(X,Y) = 0 \qquad X \neq Y \tag{28}$$

$$d_X \underset{(p,p)}{\gamma}(X,Y) = \delta_Y \underset{(p+1,p+1)}{\gamma}(X,Y) \tag{29}$$

$$\delta_X \underset{(p+1,p+1)}{\gamma}(X,Y) = d_Y \underset{(p,p)}{\gamma}(X,Y) \tag{30}$$

$$\Delta_X \underset{(p,p)}{\gamma}(X,Y) = \Delta_Y \underset{(p,p)}{\gamma}(X,Y) \tag{31}$$

The singularity at $\mathbf{x} = \mathbf{y}$ in $\underset{(p,p)}{\gamma}(X,Y)$ permeates many of the integrands of integrals that feature below. Such integrals, when they exist, are then defined implicitly as limits. In general integrals $\int_{\mathcal{U}} \alpha$ of n-forms α, over a domain $\mathcal{U} \subset \mathbf{R}^n$, that contain such a singularity are understood as

$$\underset{(\epsilon \to 0)}{Lim} \int_{\mathcal{U}-B_\epsilon} \alpha$$

where B_ϵ is an n-ball of radius ϵ centred on the location of the singularity.

For any smooth $p-$form α on Y the double-form $\underset{(p,p)}{\gamma}(X,Y)$ may be used to define the type-preserving integral operator Γ: $\alpha_X \to (\Gamma\alpha)_Y$ mapping $p-$forms on X to $p-$forms on Y: where

$$(\Gamma\alpha)_Y = \int_X \alpha_X \wedge \underset{(X)}{\star} \underset{(p,p)}{\gamma}(X,Y) = \int_X \underset{(p,p)}{\gamma}(X,Y) \wedge \underset{(X)}{\star} \alpha_X \tag{32}$$

From (13) and (14) it follows that for all test forms $\hat{\phi}_X$, the differential operators d_X and δ_X commute with the integral operator Γ:

$$d_X(\Gamma\hat{\phi}_X) = \Gamma(d_X\hat{\phi}_X) \tag{33}$$

$$\delta_X(\Gamma\hat{\phi}_X) = \Gamma(\delta_X\hat{\phi}_X) \tag{34}$$

If one writes

$$(\Gamma \underset{(p)}{\phi})_Y = \left(\underset{(p,p)}{\gamma}(X,Y), \phi_X \right) \tag{35}$$

[5] V_n is the volume of the unit n-sphere in $\mathbf{R}^n$.

and notes that Γ is a symmetric operator:

$$(\Gamma\alpha,\beta) = (\alpha,\Gamma\beta) = (\Gamma\beta,\alpha)$$

then it follows from Green's identity applied to forms with compact support that:

$$\hat{\psi} = \Gamma\Delta\hat{\psi} \tag{36}$$

Thus

$$(\Gamma\phi,\Delta\hat{\chi}) = (\phi,\Gamma\delta\hat{\chi}) = (\phi,\hat{\chi}) \tag{37}$$

But $(\Gamma\phi,\Delta\hat{\chi}) = (\Delta\Gamma\phi,\hat{\chi})$ hence

$$(\Delta\Gamma\phi,\hat{\chi}) = (\phi,\hat{\chi}) \tag{38}$$

or

$$\Delta\Gamma\phi[\hat{\chi}] = \phi[\hat{\chi}] \tag{39}$$

and Γ is seen to be an inverse of Δ.

For boundary value problems it is useful to introduce singular double-forms on $\mathcal{U}$ that differ from the fundamental double-forms $\underset{(p,p)}{\gamma}(X,Y)$ by regular solutions to Laplace's equation $\Delta \underset{(p)}{\phi} = 0$. Such double-forms inherit the singularity structure of $\underset{(p,p)}{\gamma}(X,Y)$ but can be tailored to satisfy certain conditions on the boundary $\partial\mathcal{U}$.

As an example suppose one requires $p-$form solutions to the generalized Poisson equation

$$\Delta \underset{(p)}{\phi} = \underset{(p)}{\mathcal{S}} \tag{40}$$

that vanish on $\partial\mathcal{U}$. Denote by

$$\mathcal{G}^p_{XY} = \underset{(p,p)}{\gamma}(X,Y) - \widetilde{\underset{(p,p)}{\gamma}(X,Y)} \tag{41}$$

with $\widetilde{\underset{(p,p)}{\gamma}(X,Y)}$ a symmetric *non-singular solution* of $\Delta\, \widetilde{\underset{(p,p)}{\gamma}(X,Y)} = 0$ that coincides with the singular solution $\underset{(p,p)}{\gamma}(X,Y)$ on $\partial\mathcal{U}$. Thus

$$\Delta\mathcal{G}^p_{XY} = 0 \qquad X \neq Y \tag{42}$$

$$\mathcal{G}^p_{XY} = \mathcal{G}^p_{XY} \tag{43}$$

and

$$\mathcal{G}^p_{XY}|_{\partial\mathcal{U}} = 0 \tag{44}$$

Suppose more generally now that $\mathcal{G}^p_{XY}$ is a double-form satisfying chosen conditions on $\partial\mathcal{U}$. For any smooth $p-$form on $\mathbf{R}^n$ define the integral operator $\mathcal{G}$ by

$$(\mathcal{G}\phi)_Y = (\mathcal{G}^p_{XY},\phi_X) \tag{45}$$

If $\mathbf{y} \in \mathcal{U}$ the integral is singular at $\mathbf{x} = \mathbf{y}$ and

$$(\mathcal{G}^p_{XY},\phi_X) = \underset{(\epsilon\to 0)}{Lim}\ (\mathcal{G}^P_{XY},\phi)_{\mathcal{U}-B_\epsilon} \tag{46}$$

One may use (16) together with the limits (derived by explicit computation in spherical polar coordinates with origin at the center of B_ϵ):

$$\underset{(\epsilon\to 0)}{Lim}\int_{S_\epsilon}\left(\delta_X\mathcal{G}^p_{XY}\wedge\underset{(X)}{\star}\phi_X-\phi_X\wedge\underset{(X)}{\star}d_X\mathcal{G}^p_{XY}\right)=\nu\phi_Y$$

$$\underset{(\epsilon\to 0)}{Lim}\int_{S_\epsilon}\left(\delta_X\phi_X\wedge\underset{(X)}{\star}\mathcal{G}^p_{XY}-\mathcal{G}^p_{XY}\wedge\underset{(X)}{\star}d_X\phi_X\right)=0$$

where S_ϵ is the $n-1$ dimensional surface of the ball B_ϵ, to verify

$$\begin{aligned}\nu\phi_Y &= (\mathcal{G}\,\Delta\phi)_Y\\ &-\int_{\partial\mathcal{U}}\left((\delta\phi)_X\wedge\underset{(X)}{\star}\mathcal{G}^p_{XY}-\mathcal{G}^p_{XY}\wedge\underset{(X)}{\star}(d\phi)_X-\delta_X\mathcal{G}^p_{XY}\wedge\underset{(X)}{\star}\phi_X+\phi_X\wedge\underset{(X)}{\star}\mathcal{G}^p_{XY}\right)\end{aligned}\tag{47}$$

In this expression, $\nu=0$ if $\mathbf{y}$ is outside $\mathcal{U}$, $\nu=\frac{1}{2}$ if $\mathbf{y}\in\partial\mathcal{U}$ and $\nu=1$ if $\mathbf{y}\in\mathcal{U}$ If the first term on the right of this equation is written in terms of the source $\underset{(p)}{\mathcal{S}}$ in the equation (40) then (47) offers a representation of the solution of this equation in terms of this source and values of $\delta\phi$, $d\phi$, $\star\phi$ and ϕ on $\partial\mathcal{U}$. It should be stressed however that this representation is not an explicit solution in general since $\delta\phi$, $d\phi$, $\star\phi$ and ϕ cannot in general be assigned values arbitrarily on $\partial\mathcal{U}$. Nevertheless this representation is the cornerstone of many developments in potential theory [2]. The above identities and their subsequent uses are often attributed to Green.

5. Electromagnetic Fields in Spacetime

Maxwell's equations for an electromagnetic field in an arbitrary medium can be written

$$d\,F=0\qquad\text{and}\qquad d\star G=j,\tag{48}$$

where $F\in\mathcal{S}\Lambda^2M$ is the Maxwell 2-form, $G\in\mathcal{S}\Lambda^2M$ is the excitation 2-form and $j\in\mathcal{S}\Lambda^3M$ is the 3-form electric current source[6] on a spacetime M endowed with a metric (signature $(-1,+1,+1,+1)$) and associated Hodge map $\star$. The electric 4-current j describes both (mobile) electric charge and effective (Ohmic) currents in a conducting medium. To close this system, "electromagnetic constitutive relations" relating G and j to F are necessary.

A *unit* future-pointing timelike 4-velocity vector field $U\in\Gamma TM$ may be used to describe an *observer frame* on spacetime and its integral curves model idealized observers. The *electric* field $\mathbf{e}\in\mathcal{S}\Lambda^1M$ and *magnetic induction field* $\mathbf{b}\in\mathcal{S}\Lambda^1M$ associated with F and U are defined by

$$\mathbf{e}=i_UF\qquad\text{and}\qquad c_0\,\mathbf{b}=i_U{\star}F.\tag{49}$$

Thus $i_U\mathbf{e}=0$ and $i_U\mathbf{b}=0$ and since $g(U,U)=-1$

$$F=\mathbf{e}\wedge\widetilde{U}-\star(\,c_0\,\mathbf{b}\wedge\widetilde{U}).\tag{50}$$

Likewise the *displacement* field $\mathbf{d}\in\mathcal{S}\Lambda^1M$ and the *magnetic* field $\mathbf{h}\in\mathcal{S}\Lambda^1M$ associated with G are defined with respect to U by

$$\mathbf{d}=i_UG\,,\qquad\text{and}\qquad \mathbf{h}/\,c_0=i_U\star G\,.\tag{51}$$

[6] All tensors in this article have dimensions constructed from the SI dimensions $[M],[L],[T],[Q]$ where $[Q]$ has the unit of the Coulomb in the MKS system. We adopt $[g]=[L^2],[G]=[j]=[Q],[F]=[Q]/[\epsilon_0\,]$ where the permittivity of free space ϵ_0 has the dimensions $[Q^2\,T^2M^{-1}\,L^{-3}]$ and $c_0=\frac{1}{\sqrt{\epsilon_0\,\mu_0}}$ denotes the speed of light in vacuo. Note that, with $[g]=[L^2]$, for $p-$forms α in n dimensions one has $[\star\alpha]=[\alpha][L^{n-2p}]$

Thus

$$G = \mathbf{d} \wedge \widetilde{U} - \star((\mathbf{h}/c_0) \wedge \widetilde{U}), \tag{52}$$

and $i_U \mathbf{d} = 0$ and $i_U \mathbf{h} = 0$. It follows from these definitions that

$$\star F = c_0 \mathbf{h} \wedge \widetilde{U} + \boldsymbol{E} \tag{53}$$

$$\star G = \frac{\mathbf{h}}{c_0} \wedge \widetilde{U} + \boldsymbol{D} \tag{54}$$

where $\boldsymbol{E} = \star(\mathbf{e} \wedge \widetilde{U})$ and $\boldsymbol{D} = \star(\mathbf{d} \wedge \widetilde{U})$.

6. Time dependent Maxwell Systems in Space

Although the formulation above has a natural setting in a curved spacetime in this article we restrict to a Minkowski spacetime which can be globally foliated by 3-dimensional spacelike hyperplanes. The Minkowski metric induces a metric (and associated Hodge map #) with Euclidean signature on each hyperplane. Furthermore each hyperplane contains events that are deemed simultaneous with respect to a clock attached to any integral curve of a future-pointing unit time-like vector field $U = \frac{1}{c_0}\frac{\partial}{\partial t}$ on spacetime and the Hodge map $\star$ induces a the Hodge map on each hyperplane by the relation

$$\star 1 = \widetilde{U} \wedge \#1$$

The spacetime Maxwell system can now be reduced to a family of exterior systems on $\mathbf{R}^3$. Each member is an exterior system involving forms on $\mathbf{R}^3$ depending parametrically on time t. Let the $g-$orthogonal $3+1$ split of the 4-current 3-form with respect to the foliation be

$$\underset{(3)}{j} = -\underset{(2)}{\mathbf{J}} \wedge dt + \underset{(0)}{\rho}\, \#1, \tag{55}$$

with $i_{\frac{\partial}{\partial t}} \underset{(2)}{\mathbf{J}} = 0$. Then, from (48)

$$d\, j = 0, \tag{56}$$

yields

$$\boldsymbol{d}\, \underset{(2)}{\mathbf{J}} + \underset{(0)}{\dot{\rho}}\, \#1 = 0. \tag{57}$$

Here and below an over-dot denotes (Lie) differentiation with respect to the parameter t and $\dot{\alpha} \equiv \mathcal{L}_{\frac{\partial}{\partial t}} \alpha$. It is convenient to introduce on each hyperplane the (Hodge) dual forms:

$$\underset{(2)}{\mathbf{E}} := \#\underset{(1)}{\mathbf{e}}, \quad \underset{(2)}{\mathbf{D}} := \#\underset{(1)}{\mathbf{d}}, \quad \underset{(2)}{\mathbf{B}} := \#\underset{(1)}{\mathbf{b}}, \quad \underset{(2)}{\mathbf{H}} := \#\underset{(1)}{\mathbf{h}}, \quad \underset{(1)}{\mathbf{j}} := \#\underset{(2)}{\mathbf{J}}$$

so that the $3+1$ split of the spacetime covariant Maxwell equations (48) with respect to $\widetilde{U} = -c_0\, dt$ becomes

$$\boldsymbol{d}\, \underset{(1)}{\mathbf{e}} = -\underset{(2)}{\dot{\mathbf{B}}}, \tag{58}$$

$$\boldsymbol{d}\, \underset{(2)}{\mathbf{B}} = 0, \tag{59}$$

$$\boldsymbol{d}\, \underset{(1)}{\mathbf{h}} = \underset{(2)}{\mathbf{J}} + \underset{(2)}{\dot{\mathbf{D}}}, \tag{60}$$

$$\boldsymbol{d}\underset{(2)}{\mathbf{D}} = \underset{(0)}{\rho}\,\#1. \tag{61}$$

All p-forms ($p \geq 0$) in these equations are independent of dt but have components that may depend parametrically on t.

For a simple isotropic medium at rest in the frame U, $\underset{(1)}{\mathbf{b}} = \mu\,\underset{(1)}{\mathbf{h}}$ and $\underset{(1)}{\mathbf{d}} = \epsilon\,\underset{(1)}{\mathbf{e}}$ where $\epsilon = \epsilon_r\epsilon_0$, $\mu = \mu_r\mu_0$. Thus in terms of $\underset{(1)}{\mathbf{e}}, \underset{(1)}{\mathbf{h}}, \underset{(2)}{\mathbf{E}}, \underset{(2)}{\mathbf{H}}$:

$$\boldsymbol{d}\,\underset{(1)}{\mathbf{e}} = -\mu\underset{(2)}{\dot{\mathbf{H}}}, \tag{62}$$

$$\boldsymbol{d}\,\underset{(2)}{\mathbf{H}} = 0, \tag{63}$$

$$\boldsymbol{d}\,\underset{(1)}{\mathbf{h}} = \epsilon\underset{(2)}{\dot{\mathbf{E}}} + \underset{(2)}{\mathbf{J}}, \tag{64}$$

$$\epsilon\boldsymbol{d}\,\underset{(2)}{\mathbf{E}} = \underset{(0)}{\rho}\,\#1. \tag{65}$$

It is straightforward to show from this exterior system that the fields $\underset{(1)}{\mathbf{e}}$ and $\underset{(1)}{\mathbf{h}}$ satisfy

$$\boldsymbol{\Delta}\,\underset{(1)}{\mathbf{h}} + \epsilon\mu\underset{(1)}{\ddot{\mathbf{h}}} = \underline{\delta}\mathbf{J} + \#\left(\boldsymbol{d}\,\epsilon \wedge \#\epsilon^{-1}\left(\mathbf{J} - \boldsymbol{d}\,\underset{(1)}{\mathbf{h}}\right)\right) \tag{66}$$

$$\boldsymbol{\Delta}\,\underset{(1)}{\mathbf{e}} + \epsilon\mu\underset{(1)}{\ddot{\mathbf{e}}} = -\boldsymbol{d}\left(\frac{\rho}{\epsilon}\right) - \dot{\mathbf{j}} + \#\left(\boldsymbol{d}\,\mu \wedge \#\left(\mu^{-1}\boldsymbol{d}\,\underset{(1)}{\mathbf{e}}\right)\right) \tag{67}$$

Thus for homogeneous media with zero conductivity these equations each reduce to driven wave equations. These equations simplify considerably for non-dispersive simple homogeneous isotropic media where ϵ, μ are constant scalars.

Locally on spacetime one has an equivalence class of 1−forms whose elements A differ by the addition of any exact 1−form. Then since F is closed one may write locally $F = d\,A$. Decomposing A (with respect to a particular $\widetilde{U}$) into 0−form ϕ and 1-form $\mathbf{A}$ potentials yields

$$A = -\phi\,\widetilde{U} - c_0\mathbf{A} \tag{68}$$

with $i_U\mathbf{A} = 0$. With $\widetilde{U} = -c_0\,dt$ and $\star 1 = \widetilde{U} \wedge \#1$ one has in terms of $t-$parameterised forms on $\mathbf{R}^3$

$$\underset{(1)}{\mathbf{e}} = -\boldsymbol{d}\phi - \dot{\mathbf{A}}$$

$$\underset{(1)}{\mathbf{b}} = \#\boldsymbol{d}\,\mathbf{A}$$

The Maxwell system then reduces to

$$\underline{\delta}\,\boldsymbol{d}\phi + \underline{\delta}\dot{\mathbf{A}} = \frac{\rho}{\epsilon} \tag{69}$$

$$\underline{\delta}\,\boldsymbol{d}\,\mathbf{A} + \epsilon\mu\ddot{\mathbf{A}} = \mu\,\mathbf{j} - \epsilon\mu\boldsymbol{d}\,\dot{\phi} \tag{70}$$

In a gauge with

$$\underline{\delta}\mathbf{A} - \epsilon\mu\dot{\phi} = 0$$

the Maxwell system above requires that the potentials must satisfy

$$\boldsymbol{\Delta}\mathbf{A} + \epsilon\mu\,\ddot{\mathbf{A}} = \mu\,\mathbf{j} \tag{71}$$

$$\boldsymbol{\Delta}\phi + \epsilon\mu\,\ddot{\phi} = \frac{\rho}{\epsilon} \tag{72}$$

for sources satisfying (57). Physical solutions correspond to those satisfying physically motivated boundary conditions (in both space and time). These conditions are discussed in later sections. Alternatively in a gauge with

$$\underline{\delta}\mathbf{A} = 0 \tag{73}$$

the Maxwell system becomes

$$\boldsymbol{\Delta}\phi = \frac{\rho}{\epsilon} \tag{74}$$

$$\boldsymbol{\Delta}\mathbf{A} + \epsilon\mu\ddot{\mathbf{A}} = \mu\mathbf{j} - \epsilon\mu\,\boldsymbol{d}\,\dot{\phi} \tag{75}$$

In these equations on $\mathbf{R}^3$ with the Euclidean metric, the Hodge map on all forms satisfies $\#\# = 1$ so $\underline{\delta} = \#\boldsymbol{d}\#\eta$ and $\boldsymbol{\Delta} = \#\boldsymbol{d}\#\eta\boldsymbol{d} + \boldsymbol{d}\,\#\boldsymbol{d}\#\eta$. When attention is devoted to exterior systems on $\mathbf{R}^3$ it is unnecessary to maintain the notational distinction between $\boldsymbol{d}$ and d, $\underline{\delta}$ and δ, $\boldsymbol{\Delta}$ and Δ.

7. Electrostatics in $\mathbf{R}^3$ with Sources in Domains with Boundaries

Integral operators analogous to Γ may be used to solve electrostatic boundary value problems in $\mathbf{R}^3$ involving the Hodge-de Rham operator Δ above and electric fields independent of t. As a basic example consider the Dirichlet problem of finding the electrostatic 0−form potential ϕ on $\mathcal{U} \subset \mathbf{R}^3$ that must satisfy

$$\begin{cases} \Delta\phi = \underset{(0)}{\bar{\rho}} \\ \phi|_{\partial\mathcal{U}} = \bar{\phi} \end{cases} \tag{76}$$

for some data $\bar{\rho} \equiv \frac{\rho}{\epsilon} \in \mathcal{U}, \bar{\phi} \in \partial\mathcal{U}$.

The solution may be represented in terms of the integral operator $\mathcal{G}$ defined by

$$(\mathcal{G}\bar{\rho})_Y = \int_{\mathcal{U}(X)} \bar{\rho}_X \underset{(X)}{\#} \mathcal{G}_{XY} \tag{77}$$

and is given by

$$\phi_Y = (\mathcal{G}\bar{\rho})_Y - \int_{\partial\mathcal{U}} \bar{\phi}_X \underset{(X)}{\#} d_X\mathcal{G}_{XY} \tag{78}$$

where

$$\mathcal{G}_{XY} = \underset{(0,0)}{\gamma}(X,Y) - \underset{(0,0)}{\widetilde{\gamma\,(X,Y)}} \tag{79}$$

with $\underset{(0,0)}{\widetilde{\gamma\,(X,Y)}}$ a symmetric *non-singular solution* of $\Delta \underset{(0,0)}{\widetilde{\gamma\,(X,Y)}} = 0$ that coincides with the singular solution $\underset{(0,0)}{\gamma}(X,Y)$ on $\partial\mathcal{U}$. Thus

$$\Delta\mathcal{G}_{XY} = 0 \qquad X \neq Y \tag{80}$$

Journal of Physics: Conference Series **254** (2010) 012006 doi:10.1088/1742-6596/254/1/012006

$$\mathcal{G}_{XY} = \mathcal{G}_{YX} \tag{81}$$

and

$$\mathcal{G}_{XY}|_{\partial\mathcal{U}} = 0 \tag{82}$$

If the data is smooth it defines *regular* distributions $\bar{\rho}^D, \bar{\phi}^D$ with supports on $\mathcal{U}$ and $\partial\mathcal{U}$ respectively by

$$\bar{\rho}^D[\hat{\chi}] = (\bar{\rho}, \hat{\chi}) \qquad \bar{\phi}^D[\hat{\chi}] = -\int_{\partial\mathcal{U}(X)} \bar{\phi}_X \underset{(X)}{\#} d_X(\mathcal{G}\hat{\chi})_X \tag{83}$$

Now

$$(\mathcal{G}\bar{\rho})[\hat{\chi}] = (\mathcal{G}\bar{\rho}, \hat{\chi}) = (\bar{\rho}, \mathcal{G}\hat{\chi}) = \bar{\rho}[\mathcal{G}\hat{\chi}] \tag{84}$$

Thus for *singular* distributional sources $\bar{\rho}^D$ one then has distributional Dirichlet solutions:

$$\phi^D[\hat{\chi}] = \bar{\rho}^D[\mathcal{G}\hat{\chi}] + \bar{\phi}^D[\hat{\chi}] \tag{85}$$

If the eigenvalues λ_M and Dirichlet eigenfunctions Φ_N of Δ on $\mathcal{U}$ can be found by solving

$$\begin{cases} \Delta\Phi_M = \lambda_M \Phi_M \\ \\ \Phi_M|_{\partial\mathcal{U}} = 0 \end{cases} \tag{86}$$

then a traditional way to satisfy the conditions required of $\mathcal{G}_{XY}$ is to express it as a Fourier expansion in the eigenmodes $\Phi_M(Y)$ for each point of $Y \in \mathcal{U}$.

If the real modes are labeled by a discrete index set M the expansion takes the form

$$\mathcal{G}_{XY} = \sum_M \mathcal{G}_M(X)\Phi_M(Y) \tag{87}$$

If such modes (with support on $\mathcal{U}$) are ortho-normalised so that $(\Phi_M, \Phi_N) = \delta_{MN}$ then

$$\mathcal{G}_N(X) = (\mathcal{G}_{XY}, \Phi_N(Y)) \tag{88}$$

Since the solutions Φ_N are zero on $\partial\mathcal{U}$ it follows that

$$\Phi_N(Y) = \lambda_N \int_{\mathcal{U}} \Phi_N(X) \underset{(X)}{\#} \mathcal{G}_{XY} = \lambda_N(\Phi_N(X), \mathcal{G}_{XY}) = \lambda_N \mathcal{G}_N(Y) \tag{89}$$

or

$$\mathcal{G}_N(X) = \frac{1}{\lambda_N}\Phi_N(X) \qquad \lambda_N \neq 0$$

Thus

$$\mathcal{G}_{XY} = \sum_N \frac{1}{\lambda_N}\Phi_N(X)\,\Phi_N(Y) \tag{90}$$

This series must be regarded as weakly convergent and the summations become integrations when the eigenvalues are continuous.

For example consider $\mathcal{U}$ to be the space between two perfectly conducting parallel plates at $z = 0$ and $z = L$, separated by a distance L in vacuo. Taking cartesian coordinates $\{x, y, z\}$ for

Journal of Physics: Conference Series **254** (2010) 012006 doi:10.1088/1742-6596/254/1/012006

X and $\{x', y', z'\}$ for Y the eigenvalues $N = \{k_x, k_y, n\}$ with $-\infty < k_x < \infty$, $-\infty < k_y < \infty$, $n = 1, 2, \ldots$ and (in complexified form):

$$\Phi_N(x, y, z) = \frac{1}{2\pi}\sqrt{\frac{2}{L}}\exp(ik_x + ik_y)\sin\frac{n\pi z}{L} \tag{91}$$

This gives

$$\mathcal{G}(x, y, z; x', y', z') = \frac{2}{4\pi^2 L}\sum_n^\infty \sin\frac{n\pi z}{L}\sin\frac{n\pi z'}{L}\int_{-\infty}^{\infty}\int_{-\infty}^{\infty}\frac{dk_x\, dk_y\, e^{ik_x(x-x')+ik_y(y-y')}}{k_x^2 + k_y^2 + \frac{n^2\pi^2}{L^2}} \tag{92}$$

If the source $\bar{\rho}$ is not smoothly distributed in $\mathcal{U}$ one needs further technology to evaluate the solution and this will be developed below.

8. Magnetostatics in $\mathbf{R}^3$ with Smooth Sources

Since magnetic charge is absent in Nature a typical problem in magnetostatics is the determination of a static magnetic field from a stationary electric current. The magnetostatic equations are a subset of Maxwell's equations and may be written:

$$\delta \underset{(1)}{\mathbf{h}} = 0 \tag{93}$$

$$d \underset{(1)}{\mathbf{h}} = \underset{(2)}{\mathbf{J}} \tag{94}$$

in terms of forms on $\mathbf{R}^3$. If the current is localized in space (vanishing at ∞) the boundary conditions can be accommodated by using $\underset{(1,1)}{\gamma}(X, Y)$ for the solution. If one sets[7] $\mathbf{h} = \#d\mathbf{A}/\mu_0$ then (93) is immediately satisfied. Writing $\underset{(1)}{\bar{\mathcal{J}}} = -\mu_0\#\mathbf{J}$ one must have from (94)

$$d\#\bar{\mathcal{J}} = 0 \tag{95}$$

and in the gauge with $\delta\mathbf{A} = 0$ equation (94) becomes

$$\Delta\mathbf{A} = \bar{\mathcal{J}} \tag{96}$$

Thus in free space with a smooth source $\bar{\mathcal{J}}$

$$\mathbf{A} = \Gamma\bar{\mathcal{J}} \tag{97}$$

since

$$\Delta\mathbf{A} = \Delta\Gamma\bar{\mathcal{J}} = \bar{\mathcal{J}} \tag{98}$$

i.e.

$$\mathbf{A}_Y = \int_X \underset{(1,1)}{\gamma}(X, Y) \wedge \underset{(X)}{\#\bar{\mathcal{J}}} \tag{99}$$

[7] Since $d\underset{(2)}{\mathbf{B}} = 0$ one has $\underset{(2)}{\mathbf{B}} = d\underset{(1)}{\mathbf{A}}$ in regular domains and in vacuo $\#\mathbf{B} = \mu_0\mathbf{h}$ where μ_0 is the permeability of free space.

Journal of Physics: Conference Series **254** (2010) 012006 doi:10.1088/1742-6596/254/1/012006

with

$$\underset{(1,1)}{\gamma}(X,Y) = \frac{1}{4\pi\,|\mathbf{x}-\mathbf{y}|}\sum_{j=1}^{3} dx^j \odot dx'^j$$

Note if $\bar{\mathcal{J}}$ is bounded

$$\begin{aligned}\delta_Y \mathbf{A}_Y &= \int_X \delta_Y \underset{(1,1)}{\gamma}(X,Y) \wedge \underset{(X)}{\#} \bar{\mathcal{J}}_X = \int_X d_X \underset{(0,0)}{\gamma}(X,Y) \wedge \underset{(X)}{\#} \bar{\mathcal{J}}_X \\ &= -\int_X \underset{(0,0)}{\gamma}(X,Y)\, d_X \underset{(X)}{\#} \bar{\mathcal{J}}_X = 0\end{aligned} \tag{100}$$

Since $\Delta \underset{(1)}{\mathbf{h}} = \delta \mathbf{J}$ one also has directly

$$\underset{(1)}{\mathbf{h}}(Y) = \int_X \underset{(1,1)}{\gamma}(X,Y) \wedge \underset{(X)}{\#} (\delta \mathbf{J})_X. \tag{101}$$

Now for a smooth (regular) source $\bar{\mathcal{J}}$

$$\begin{aligned}\underset{(1)}{\mathbf{A}}[\hat{\psi}] = \Gamma\bar{\mathcal{J}}[\hat{\psi}] &= \int_Y (\Gamma\bar{\mathcal{J}})_Y \wedge \underset{(Y)}{\#} \hat{\psi}_Y \\ &= \int_Y \left(\int_X \underset{(1,1)}{\gamma}(X,Y) \wedge \underset{(X)}{\#} \bar{\mathcal{J}}_X \right) \wedge \underset{(Y)}{\#} \hat{\psi}_Y \\ &= \int_X \int_Y \left(\underset{(1,1)}{\gamma}(X,Y) \wedge \underset{(Y)}{\#} \hat{\psi}_Y \right) \wedge \underset{(X)}{\#} \bar{\mathcal{J}}_X \\ &= \int_X (\Gamma\hat{\psi})_X \wedge \underset{(X)}{\#} \bar{\mathcal{J}}_X \\ &= \int_X \bar{\mathcal{J}}_X \wedge \underset{(X)}{\#} (\Gamma\hat{\psi})_X \\ &\equiv \bar{\mathcal{J}}^D[\Gamma\hat{\psi}]\end{aligned} \tag{102}$$

If $\bar{\mathcal{J}}$ is a smooth 1−form in $\mathbf{R}^3$ the integral $\bar{\mathcal{J}}^D[\Gamma\hat{\psi}]$ furnishes a solution for $\mathbf{A}$. However if the current has support on a curve in $\mathbf{R}^3$ it must be regarded as a *distributional* 1− *form source* not associated with a smooth 1−form. Similarly the charge density in the electrostatic problem may be restricted to a curve or surface in $\mathbf{R}^3$ in which case a distributional source $\bar{\rho}^D$ not associated with a smooth 0−form must be specified.

9. Dirac Distributions on Submanifolds

To define singular distributions with support on (possibly disjoint) submanifolds of a manifold M it is convenient to describe each submanifold parametrically as an embedding in M. Suppose that a distribution has support on a collection of submanifolds $\{S_0^{(r)}\}$, i.e on the chain $\sum_r S_0^{(r)}$. Recall that an $n-k$ dimensional submanifold $S_0^{(r)}$ (possibly with boundary) in an n−dimensional manifold M can also be prescribed in terms of k 0−forms ${f^{(r)}}_1, {f^{(r)}}_2, \ldots, {f^{(r)}}_k$ on M, such that $d{f^{(r)}}_1 \wedge d{f^{(r)}}_2 \ldots \wedge d{f^{(r)}}_k \neq 0$. Such forms generate a local foliation $S^{(r)}$ in the neighborhood of each submanifold $S_0^{(r)}$, where each leaf $S_{\mathbf{c}}^{(r)}$ of the foliation is an $n-k$ dimensional embedding given by ${f^{(r)}}_1 = c_1, {f^{(r)}}_2 = c_2, \ldots, {f^{(r)}}_k = c_k$ for some constants $c_1, c_2, \ldots c_k$ and

$\mathbf{c} = (c_1, c_2, \ldots c_k)$. We may choose the leaf $S_0^{(r)}$ with all these constants zero and denote it $\Sigma^{n-k}_{\mathbf{f}^{(\mathbf{r})}}$ with $\mathbf{f}^{(\mathbf{r})} \equiv (f^{(r)}{}_1, f^{(r)}{}_2, \ldots, f^{(r)}{}_k)$. It is sufficient to establish the distributional framework for a single component of a general chain so henceforth the label (r) will be omitted.

If M is endowed with a metric g (of arbitrary signature and associated Hodge map $\star$), the forms $\{df_j\}$ give rise to a class of forms $\underset{(n-k)}{\Omega_{\mathbf{f}}}$ on M defined with respect to $\star 1$ by

$$\star 1 = df_1 \wedge df_2 \wedge \ldots \wedge df_k \wedge \underset{(n-k)}{\Omega_{\mathbf{f}}} \tag{103}$$

This class restricts to a natural class of measures on $\Sigma^{n-k}_{\mathbf{f}}$. Thus each representative induces the measure

$$\underset{(n-k)}{\omega_{\mathbf{f}}} = \underset{(n-k)}{\Omega_{\mathbf{f}}} |_{\Sigma^{n-k}_{\mathbf{f}}} \tag{104}$$

on the submanifold $\Sigma^{n-k}_{\mathbf{f}}$. Members of the class on M are equivalent if they differ by any combination of the df_j. This *gauge freedom* is of no significance on $\Sigma^{n-k}_{\mathbf{f}}$ since the df_j vanish there under pull-back. Thus it is sufficient to represent the class by an element satisfying

$$i_{\widetilde{df_j}} \underset{(n-k)}{\Omega_{\mathbf{f}}} = 0 \qquad\qquad j = 1 \ldots k. \tag{105}$$

In those situations where the leaves form orthogonal families, i.e. $g^{-1}(df_i, df_j) = 0$ for $i \neq j$ then

$$\underset{(n-k)}{\Omega_{\mathbf{f}}} = \frac{\star(df_1 \wedge df_2 \wedge \ldots \wedge df_k)}{\Pi_{j=1}^k |df_j|^2} \tag{106}$$

where $|df_j|^2 = \widetilde{df_j}(df_j) \neq 0$ and the sign of $|df_j|$ is chosen so that the direction of the vector field $\widetilde{df_j} \equiv g^{-1}(df_j, -)$ is in the direction of increasing values of f_j. Since $df_j \neq 0$ one has on $\Sigma^{n-k}_{\mathbf{f}}$ two fields of "unit normals": $n_j = \pm \frac{\widetilde{df_j}}{|df_j|}$. Then with $n_j = \frac{\widetilde{df_j}}{|df_j|}$

$$\underset{(n-k)}{\Omega_{\mathbf{f}}} = \star\, \Pi_{j=1}^k \left(\frac{n_j}{|df_j|} \wedge \right) \tag{107}$$

For non-orthogonal leaves an $\Omega_{\mathbf{f}}$ can be chosen directly from (103). This condition is equivalent to the equation

$$\pm 1 = \star(df_1 \wedge df_2 \wedge \ldots \wedge df_k \wedge \underset{(n-k)}{\Omega_{\mathbf{f}}}) \tag{108}$$

which yields freedom to choose $\underset{(n-k)}{\omega_{\mathbf{f}}}$ with non-singular coordinate components when restricted to $\Sigma^{n-k}_{\mathbf{f}}$.

The singular Dirac 0−form with support on $\Sigma^{n-k}_{\mathbf{f}}$ is denoted $\underset{(0)}{\boldsymbol{\delta}_{\Sigma^{n-k}_{\mathbf{f}}}}$ and defined by

$$\underset{(0)}{\boldsymbol{\delta}_{\Sigma^{n-k}_{\mathbf{f}}}}[\underset{(0)}{\hat{\phi}}] = \int_{\Sigma^{n-k}_{\mathbf{f}}} \hat{\phi}\, \underset{(n-k)}{\omega_{\mathbf{f}}} \tag{109}$$

Furthermore if Y is a smooth vector field in the vicinity of $\Sigma^{n-k}_{\mathbf{f}}$ one may define the directional derivative $\underset{(0)}{\boldsymbol{\delta}^Y_{\Sigma^{n-k}_{\mathbf{f}}}}$ of $\underset{(0)}{\boldsymbol{\delta}_{\Sigma^{n-k}_{\mathbf{f}}}}$ by:

$$\underset{(0)}{\boldsymbol{\delta}^Y_{\Sigma^{n-k}_{\mathbf{f}}}} \equiv Y\, \underset{(0)}{\boldsymbol{\delta}_{\Sigma^{n-k}_{\mathbf{f}}}} = i_Y\, d\underset{(0)}{\boldsymbol{\delta}_{\Sigma^{n-k}_{\mathbf{f}}}} \tag{110}$$

It follows that

$$\underset{(0)}{\boldsymbol{\delta}^{Y}_{\Sigma^{n-k}_{\mathbf{f}}}}[\underset{(0)}{\hat{\phi}}] = -\int_{\Sigma^{n-k}_{\mathbf{f}}} \underset{(n-k)}{\omega^{Y}_{\mathbf{f}}}(\hat{\phi}) \tag{111}$$

where for orthogonal leaves:

$$\underset{(n-k)}{\omega^{Y}_{\mathbf{f}}}(\hat{\phi}) = \frac{1}{\Pi^{k}_{j=1}|df_j|^2}\, i_{\widetilde{df_k}} \dots i_{\widetilde{df_1}}\, d(\star\hat{\phi}\widetilde{Y})|_{\Sigma^{n-k}_{\mathbf{f}}} \tag{112}$$

Since

$$\delta(\underset{(0)}{f}(\widetilde{Y}\underset{(0)}{\psi})) = f\delta(\widetilde{Y}\psi) - \psi(Yf)$$

$$\begin{aligned}\underset{(n-k)}{\omega^{Y}_{\mathbf{f}}}(\hat{\phi}) &= -(\hat{\phi}\,\delta\widetilde{Y} - Y\hat{\phi})\star\left(\frac{df_1}{|df_1|^2}\wedge\dots\wedge\frac{df_k}{|df_k|^2}\right)\\ &= -(\hat{\phi}\,\delta\widetilde{Y} - Y\hat{\phi})\ \underset{(n-k)}{\Omega_{\mathbf{f}}}\ |_{\Sigma^{n-k}_{\mathbf{f}}}\end{aligned} \tag{113}$$

Additional directional derivatives in the vicinity of $\Sigma^{n-k}_{\mathbf{f}}$ yield the distributions

$$\underset{(0)}{\boldsymbol{\delta}^{Y_1 Y_2 \dots Y_q}_{\Sigma^{n-k}_{\mathbf{f}}}} \equiv Y^1 Y^2 \dots Y^q \underset{(0)}{\boldsymbol{\delta}_{\Sigma^{n-k}_{\mathbf{f}}}}$$

For example

$$\left(Y^1 Y^2 \underset{(0)}{\boldsymbol{\delta}_{\Sigma^{n-k}_{\mathbf{f}}}}\right)[\underset{(0)}{\hat{\phi}}] = \underset{(0)}{\boldsymbol{\delta}_{\Sigma^{n-k}_{\mathbf{f}}}}[\delta(\widetilde{Y_1}\wedge\delta(\hat{\phi}\widetilde{Y_2}))] \tag{114}$$

If one regards any point $p_0 \in M$ as a 0−dimensional space then the original singular Dirac distribution $\underset{(0)}{\delta^{D}_{p_0}}$ with support on p_0 is defined by:

$$\underset{(0)}{\delta^{D}_{p_0}}[\underset{(0)}{\hat{\phi}}] = \hat{\phi}(p_0) \tag{115}$$

It follows that

$$\underset{(0)}{\boldsymbol{\delta}_{\Sigma^{0}_{\mathbf{f}}}} = \underset{(0)}{\omega_{\mathbf{f}}}\,\underset{(0)}{\delta^{D}_{p_0}} \tag{116}$$

The above singular 0−form Dirac distributions can be used to construct singular p−form Dirac distributions with support on submanifolds in various ways. Thus in a spherical polar chart for $\mathbf{R}^3$ a 1−form 0-*layer* Dirac distribution on the unit 2−sphere centred at the origin may be represented as

$$\underset{(0)}{\boldsymbol{\delta}_{\Sigma^{2}_{\mathbf{f}}}}\,(\mu_\theta(\theta,\phi)\,d\theta + \mu_\phi(\theta,\phi)\,d\phi) \tag{117}$$

where μ_θ and μ_ϕ are smooth functions on the sphere. A 1−form 1−*layer* Dirac distribution on the sphere (associated with some vector field Y on $\mathbf{R}^3$) may take the form

$$\underset{(0)}{\boldsymbol{\delta}^{Y}_{\Sigma^{2}_{\mathbf{f}}}}\,(p_\theta(\theta,\phi)\,d\theta + p_\phi(\theta,\phi)\,d\phi) \tag{118}$$

for smooth p_θ and p_ϕ. Similarly a 2−form 0−*layer* Dirac distribution on the sphere takes the form

$$\underset{(0)}{\boldsymbol{\delta}_{\Sigma^2_{\mathbf{f}}}}\, q(\theta,\phi)\, d\theta\wedge d\phi \tag{119}$$

and a 2−form 1−*layer* Dirac distribution takes the form

$$\underset{(0)}{\boldsymbol{\delta}^{Y}_{\Sigma^2_{\mathbf{f}}}} s(\theta,\phi)\, d\theta\wedge d\phi \tag{120}$$

for smooth q and s respectively. More generally in coordinates

$$(\underbrace{\xi^1,\ldots,\xi^k}_{n-p},\underbrace{\sigma_1,\ldots,\sigma_p}_{p})$$

adapted to a p−dimensional submanifold in $\mathbf{R}^n$, one has for $j=0,1,\ldots p$, the j−form, "r-layer" singular Dirac distribution with support on $\Sigma^{n-k}_{\mathbf{f}}$:

$$\underset{(j)}{\mathcal{J}} = \underset{(0)}{\lambda}(\sigma^1\ldots\sigma^p)\,\underset{(0)}{\delta^{Y_1\ldots Y_r}_{\Sigma^{n-k}_{\mathbf{f}}}}\left(\sum_{i_1<i_2<\ldots<i_j} q_{i_1\ldots i_j}(\sigma^1\ldots\sigma^p)\, d\sigma^{i_1}\wedge\sigma^{i_2}\wedge\ldots\wedge d\sigma^{i_j}\right) \tag{121}$$

For a point electrostatic dipole with dipole moment $Z_X = p^i\frac{\partial}{\partial x^i}$ located at $\mathbf{x_0}\in X$ the electrostatic potential distribution ϕ^D_Y is given by the singular distributional source ρ^D_Y where

$$\phi^D_Y[\hat{\chi}] = \rho^D_Y[\hat{\chi}] = \left(\mathcal{L}_{Z_X}\underset{(0,0)}{\gamma}(X,Y)\right)\delta^D_{\mathbf{x_0}}[\hat{\chi}] = \mathbf{p}\cdot\nabla\left(\frac{1}{|\mathbf{y}-\mathbf{x_0}|}\right)\hat{\chi}(\mathbf{x_0}) \tag{122}$$

10. Line and Surface Currents in $\mathbf{R}^3$

The source of a smooth magnetostatic field is the smooth current 2-form J in $\mathbf{R}^3$ and the total electric current passing through a surface $S\subset\mathbf{R}^3$ is $I[S]=\int_S J$, measured in amps in MKS units. In some circumstances this current density may be concentrated in the vicinity of 1-dimensional submanifolds $\Sigma_1\subset\mathbf{R}^3$ or 2-dimensional submanifolds $\Sigma_2\subset\mathbf{R}^3$. Electric current filaments on material curves in space (wires) and current sheets in material surfaces in space are idealizations of such localized sources and may be modeled by singular distribution-valued 1-forms with support on curves and surfaces respectively. In these situations new physical current densities are introduced so that the total current in a segment of wire or a region of a surface is finite and measurable. These singular distributional sources can be mathematically modeled in terms of the distributions $\delta_{\Sigma^r_{\mathbf{f}}}$ for $r=1,2$ and a vector field W in $\mathbf{R}^3$ with support on $\Sigma^r_{\mathbf{f}}$.

For a wire described by the curve with image $\Sigma^1_{\mathbf{f}}$ introduce the distributional 1-form

$$\underset{(1)}{\mathcal{I}^D_{\Sigma^1_{\mathbf{f}}}} = I_0\,\underset{(1)}{\widetilde{W}}\,\underset{(0)}{\delta_{\Sigma^1_{\mathbf{f}}}} \tag{123}$$

where I_0 is a smooth function on the wire. Hence

$$\mathcal{I}^D_{\Sigma^1_{\mathbf{f}}}[\underset{(1)}{\hat{\psi}}] = \int_{\mathbf{R}^3}\underset{(1)}{\hat{\psi}}\wedge J$$

and with $\mathcal{J} = \# J$:

$$\begin{aligned}\mathcal{I}^D_{\Sigma^1_{\mathbf{f}}}[\hat{\psi}] &= \delta_{\Sigma^1_{\mathbf{f}}}[I_0\, i_W \hat{\psi}] = \int_{\Sigma^1_{\mathbf{f}}} I_0\,(i_W \hat{\psi})\, \underset{(1)}{\omega_{\mathbf{f}}} = \int_{\mathbf{R}^3} \hat{\psi} \wedge J = \int_{\mathbf{R}^3} \hat{\psi} \wedge \# \mathcal{J} = \mathcal{J}^D[\hat{\psi}] \\ &= \int_{\Sigma^1_{\mathbf{f}}} \hat{\psi}\, I_0\,(i_W \underset{(1)}{\omega_{\mathbf{f}}}) \equiv \int_{\Sigma^1_{\mathbf{f}}} \hat{\psi}\, \underset{(0)}{\breve{I}}\end{aligned} \tag{124}$$

since $\hat{\psi} \wedge \underset{(1)}{\omega_{\mathbf{f}}}$ is zero on $\Sigma^1_{\mathbf{f}}$. It does not make sense to ascribe a physical dimension to $\mathcal{I}^D_{\Sigma^1_{\mathbf{f}}}$ since it is a functional not a value. However if $\{\mathcal{J}^D[\hat{\psi}]\}$ denotes the physical dimension of $\mathcal{J}^D[\hat{\psi}]$ one has in MKS units

$$\{\mathcal{J}^D[-]\} = \{J\}\{-\} = \text{amp}\,\{-\}$$

and $\{\mathcal{J}\} = \{\# J\} = \text{amp m}^{-1}$ with $\{I_0\}\{\widetilde{W}\}\{\underset{(1)}{\omega_{\mathbf{f}}}\} = \{J\} = $ amp. The (smooth) 0-form density

$$\underset{(0)}{\breve{I}}\,(s) \equiv I_0(i_W \underset{(1)}{\omega_{\mathbf{f}}}) \tag{125}$$

belonging to $\Gamma\Lambda^0\Sigma^1_{\mathbf{f}}$ is the total current (amps) flowing along the wire at any point $s \in \Sigma^1_{\mathbf{f}}$

In a similar way, for the 1-form distribution on $\mathbf{R}^3$:

$$\underset{(1)}{\mathcal{K}^D_{\Sigma^2_{\mathbf{f}}}} = \kappa_0\, \widetilde{W}\, \underset{(0)}{\delta_{\Sigma^2_{\mathbf{f}}}} \tag{126}$$

one has

$$\begin{aligned}\mathcal{K}^D_{\Sigma^2_{\mathbf{f}}}[\hat{\psi}] &= \int_{\mathbf{R}^3} \hat{\psi} \wedge J = \underset{(0)}{\delta_{\Sigma^2_{\mathbf{f}}}}[\kappa_0 i_W \hat{\psi}] \\ &= \int_{\Sigma^2_{\mathbf{f}}} \kappa_0 (i_W \hat{\psi}) \underset{(2)}{\omega_{\mathbf{f}}} = \int_{\Sigma^2_{\mathbf{f}}} \hat{\psi} \wedge i_W(\kappa_0 \underset{(2)}{\omega_{\mathbf{f}}}) = \int_{\Sigma^2_{\mathbf{f}}} \hat{\psi} \wedge \underset{(1)}{\breve{\kappa}}\end{aligned} \tag{127}$$

since $\kappa_0(i_W \hat{\psi})\, \underset{(2)}{\omega_{\mathbf{f}}} = \kappa_0 \hat{\psi} \wedge (i_W \underset{(2)}{\omega_{\mathbf{f}}})$ on $\Sigma^2_{\mathbf{f}}$. Here

$$\underset{(1)}{\breve{\kappa}} = \kappa_0\, i_W (\underset{(2)}{\omega_{\mathbf{f}}}) \in \Gamma\Lambda^1\Sigma^2_{\mathbf{f}}$$

is a smooth 1-form (measured in amps) on the surface $\Sigma^2_{\mathbf{f}}$. If $C_1 \in \mathbf{R}^3$ is any space-curve lying on this surface then $\int_{C_1} \underset{(1)}{\breve{\kappa}}$ is the total current in amps crossing this curve in the direction W, on $\Sigma^2_{\mathbf{f}}$. This may be compared with the definition above of the total current (in amps), $\int_S J$, crossing the surface $S \in \mathbf{R}^3$ in the direction $\widetilde{\# J}$. If $C_1^\star \frac{\partial}{\partial s}$ is a *unit* tangent vector [8] to C_1 then $i_{\frac{\partial}{\partial s}} \underset{(1)}{\breve{\kappa}}$ is the *surface current density* in amp m^{-1}.

It is worth noting that $I_0 \in \Gamma\Lambda^0\mathbf{R}^3$, $\kappa_0 \in \Gamma\Lambda^0\mathbf{R}^3$, $W \in \Gamma T\mathbf{R}^3$ and $\Sigma^1_{\mathbf{f}}$, $\Sigma^2_{\mathbf{f}}$ are the primary objects used to define the distributions $\mathcal{I}^D_{\Sigma^1_{\mathbf{f}}}$ and $\mathcal{K}^D_{\Sigma^2_{\mathbf{f}}}$ in $\mathbf{R}^3$ in terms of which are defined the smooth 0-forms $\breve{I} \in \Gamma\Lambda^0\Sigma^1_{\mathbf{f}}$ and 1-forms $\breve{\kappa} \in \Gamma\Lambda^1\Sigma^2_{\mathbf{f}}$.

[8] i.e. in terms of the Euclidean metric tensor g in $\mathbf{R}^3$, $g(C_1^\star \frac{\partial}{\partial s}, C_1^\star \frac{\partial}{\partial s}) = 1$

11. Magnetostatic Fields from a Singular Distributional Stationary Current Source

Suppose a uniform electric current confined to a wire in $\mathbf{R}^3$ is modeled in terms of a 1−form 0−*layer* Dirac distribution $\delta_{\Sigma^1_{\mathbf{f}}}$ with support on the wire. A distributional source $\bar{\mathcal{J}}^D$ will generate a distributional 1−form $\mathbf{A}^D$. Thus define $\mathbf{A}^D$ by

$$\mathbf{A}^D[\hat{\psi}] = \bar{\mathcal{J}}^D[\Gamma\hat{\psi}] \tag{128}$$

Suppose $\bar{\mathcal{J}}^D$ has a support on the locus described by $\mathbf{f} = (f_1, f_2)$ with $f_1 = r - a$, and $f_2 = z$ in cylindrical coordinates (r, ϕ, z). This will be used to model a uniform stationary current source in a circular loop of radius a if

$$\bar{\mathcal{J}}^D = \lambda \underset{(0)}{\boldsymbol{\delta}_{\Sigma^1_{\mathbf{f}}}}\, d\phi \tag{129}$$

with some constant λ. Thus

$$\begin{aligned}\bar{\mathcal{J}}^D[\Gamma\hat{\psi}] &= \lambda \underset{(0)}{\boldsymbol{\delta}_{\Sigma^1_{\mathbf{f}}}}[i_{\widetilde{d\phi}}(\Gamma\hat{\psi})] \\ &= \lambda \int_{\Sigma^1_{\mathbf{f}}} \left(i_{\widetilde{d\phi}}(\Gamma\hat{\psi})_Y\right) \underset{(1)}{\omega^Y_{\mathbf{f}}}\end{aligned} \tag{130}$$

Since

$$(\Gamma\hat{\psi})_Y = \int_X \underset{(1,1)}{\gamma}(X, Y) \wedge \# \underset{(X)}{\hat{\psi}_X} \tag{131}$$

and

$$i_{\widetilde{d\phi(Y)}}(\Gamma\hat{\psi})_Y = \int_X \left(i_{\widetilde{d\phi(Y)}} \underset{(1,1)}{\gamma}(X, Y)\right) \wedge \# \underset{(X)}{\hat{\psi}_X} \tag{132}$$

then

$$\mathbf{A}^D[\hat{\psi}] = \int_X \underset{(1)}{\mathcal{A}_X} \wedge \# \underset{(X)}{\hat{\psi}_X} \tag{133}$$

where

$$\mathcal{A}_X = \lambda \int_{\Sigma^1_{\mathbf{f}}(Y)} \left(i_{\widetilde{d\phi(Y)}} \underset{(1,1)}{\gamma}(X, Y)\right) \underset{(1)}{\omega^Y_{\mathbf{f}}} \tag{134}$$

In cylindrical polar coordinates, $\mathbf{x} = (r\cos\phi, r\sin\phi, z)$, $\mathbf{y} = (r'\cos\phi', r'\sin\phi', z')$

$$|\mathbf{x} - \mathbf{y}|^2 \equiv R^2(r, \phi, z; r', \phi', z') = r^2 + r'^2 + z^2 + z'^2 - 2zz' - 2rr'\cos(\phi - \phi')$$

and the Euclidean metric tensor is

$$g = dr \otimes dr + r^2\, d\phi \otimes d\phi + dz \otimes dz \tag{135}$$

with

$$g^{-1} = \frac{\partial}{\partial r} \otimes \frac{\partial}{\partial r} + r^{-2} \frac{\partial}{\partial \phi} \otimes \frac{\partial}{\partial \phi} + \frac{\partial}{\partial z} \otimes \frac{\partial}{\partial z} \tag{136}$$

Hence $\widetilde{d\phi'} = \frac{1}{r'^2}\frac{\partial}{\partial \phi'}$. With $dx' = dr'\cos\phi' - r'\sin\phi'\, d\phi'$, $dy' = dr'\sin\phi' + r'\cos\phi'\, d\phi'$, $i_{\frac{\partial}{\partial\phi'}}\, dx' = -r'\sin\phi'$, $i_{\frac{\partial}{\partial\phi'}}\, dy' = r'\cos\phi'$, $i_{\frac{\partial}{\partial\phi'}}\, dz' = 0$

$$i_{\widetilde{d\phi(Y)}} \underset{(1,1)}{\gamma}(X, Y)|_{r'=a, z'=0} = \frac{1}{aR}\{dr\,\sin(\phi - \phi') + r\, d\phi\,\cos(\phi - \phi')\} \tag{137}$$

Further with

$$\underset{(Y)}{\#1} = r'\,d\,r' \wedge d\,z \wedge d\,\phi' = d\,f_1 \wedge d\,f_2 \wedge \underset{(1)}{\omega_{\mathbf{f}}^Y}$$

$$\underset{(1)}{\omega_{\mathbf{f}}^Y} = a\,d\,\phi'$$

and one has [9]

$$\mathcal{A}_X(r,\phi,z) = r\,d\,\phi \frac{1}{4\pi}\int_0^{2\pi} \frac{\lambda\cos(\phi-\phi')}{(r^2+a^2+z^2-2ar\cos(\phi-\phi'))^{1/2}}\,d\,\phi' \tag{138}$$

or with $\Psi = \phi' - \phi$ and noting that the integrand is an even periodic function of Ψ:

$$\mathcal{A}_X(r,\phi,z) = -r\,d\,\phi \int_0^{2\pi} \frac{\lambda\cos\Psi\,d\,\Psi}{(r^2+a^2+z^2-2ar\cos\Psi)^{1/2}} \tag{139}$$

with magnitude independent of ϕ. For a constant current $\breve{I}$ in the circular loop the constant $\lambda = 4\pi\,\mu_0\,\breve{I}a$. In MKS units the physical unit for $\underset{(2)}{J}$ and hence $\underset{(1)}{\mathbf{h}}$ is the ampere and that for $\bar{\mathcal{J}}$ is μ_0 ampere/m. The methodology here illustrated for a circular planar coil is directly applicable to any open or closed current carrying conductor of arbitrary shape in space and possibly composed of piecewise smooth connected segments.

To verify that when $\mathbf{A}^D[\hat{\psi}] = \int_X \mathcal{A} \wedge \#\hat{\psi}$ the distribution $\mathbf{A}^D$ lies in the gauge satisfying $\delta\mathbf{A}^D = 0$ one must compute

$$\begin{aligned}
\delta\mathbf{A}^D[\underset{(0)}{\hat{\xi}}\,] &= \mathbf{A}^D[d\underset{(0)}{\hat{\xi}}\,] \\
&= \int_X \mathcal{A} \wedge \underset{(X)}{\#}\, d\underset{(0)}{\hat{\xi}} = \int_X d\underset{(0)}{\hat{\xi}} \wedge \underset{(X)}{\#}\mathcal{A} \\
&= -\int_X \underset{(0)}{\hat{\xi}}\, d\underset{(X)}{\#}\mathcal{A} = -\int_X \underset{(0)}{\hat{\xi}}\, \underset{(X)}{\#}\delta\mathcal{A} \\
&= -\lambda \int_X \underset{(0)}{\hat{\xi}}\,\underset{(X)}{\#} \int_{\Sigma^1_{\mathbf{f}}(Y)} i_{\widetilde{d\phi(Y)}}\,\delta_X \underset{(1,1)}{\gamma}(X,Y)\,\underset{(1)}{\omega_{\mathbf{f}}^Y}
\end{aligned} \tag{140}$$

But

$$\delta_X \underset{(1,1)}{\gamma}(X,Y) = d_Y \underset{(0,0)}{\gamma}(X,Y)$$

$$i_{\widetilde{d\phi(Y)}}\, d_Y \underset{(0,0)}{\gamma}(X,Y) = \frac{\partial}{\partial\phi'}\underset{(0,0)}{\gamma}(X,Y)$$

so

$$\delta\mathbf{A}^D[\underset{(0)}{\hat{\xi}}\,] = 0$$

since $\underset{(1)}{\omega_{\mathbf{f}}^Y} = a\,d\,\phi'$ and $\Sigma^1_{\mathbf{f}}(Y)$ is a closed curve. This is physically equivalent to the statement that the current in the coil is “conserved”.

[9] The term in the integrand proportional to $\sin(\phi - \phi')$ integrates to zero.

12. Magnetostatic Fields from a Steady Helical Line Current

The derivation above of the potential for the magnetic field due to a circular line current was somewhat labored. However no further steps are required to find the potentials due to more complex geometries once one parametrises the source in terms of the geometry of its support in space.

Suppose a helical wire with pitch $p > 0$ and radius a is the space curve $(r = a, \phi = \frac{\sqrt{1-p^2}}{a}\sigma, z = p\,\sigma)$ in cylindrical polar coordinates. The parameter σ is the arc-length parameter so $0 < \sigma < L$ for a helix of length L. It is convenient to introduce $P > 0$ with $a^2\,P^2 + p^2 = 1$. Then the unit tangent to the helix is the direction $W = p\frac{\partial}{\partial z} + a\,P\frac{\partial}{\partial \phi}$. If one chooses to describe the helix in terms of the foliating set $\mathbf{f} = \{f_1 = r - a, f_2 = z - \frac{p}{P}\,\phi\}$ this yields a Leray form $\Omega_{\mathbf{f}} = -r\,d\,\phi$. Then, in this case, the source $\mathcal{J}^D = \lambda\,\widetilde{W}\,\boldsymbol{\delta}_{\underset{(0)}{\Sigma^1_{\mathbf{f}}}}$ with support on the helical wire yields the magnetostatic potential 1−form

$$\mathcal{A}(r,\phi,z) = \mathcal{A}_r(r,\phi,z)\,d\,r + \mathcal{A}_\phi(r,\phi,z)\,r\,d\,\phi + \mathcal{A}_z(r,\phi,z)\,d\,z \tag{141}$$

where

$$\mathcal{A}_r = \frac{\lambda}{4\pi}\int_0^L \frac{d\,\sigma}{R}\,\sin\left(P\sigma - \phi\right) \tag{142}$$

$$\mathcal{A}_\phi = -\frac{\lambda\,a^2\,P^2}{4\pi}\int_0^L \frac{d\,\sigma}{R}\,\cos\left(P\sigma - \phi\right) \tag{143}$$

$$\mathcal{A}_z = -\frac{\lambda\,a\,p\,P}{4\pi}\int_0^L \frac{d\,\sigma}{R} \tag{144}$$

with

$$R^2 = r^2 + a^2 + z^2 + \sigma^2\,p^2 - 2\,z\,p\,\sigma - 2a\,r\,\cos(P\sigma - \phi) \tag{145}$$

For a constant current $\breve{I}_0$ in the helix the constant $\lambda = 4\pi\,\mu_0\breve{I}_0 a$. The potential $\mathcal{A}$ yields a global description of the field in terms of elliptic integrals that depend on the geometry of the helix specified by the parameters L, a and p or P. Such parameters offer natural scales that are useful for defining dimensionless variables that in turn can be used to generate multipole or asymptotic expansions of the above integrals. The practical generation of high intensity uniform magnetic fields by carefully designing coils with complex helical windings is of paramount importance in the construction of undulators and free electron lasers.

13. Time Dependent Electromagnetic fields and Smooth Sources

When the smooth electromagnetic sources ρ and $\mathbf{J}$ depend on t they may generate time dependent electromagnetic fields that must satisfy (62), (63), (64), (65). These equations offer a well-posed initial-boundary value problem that is traditionally approached by finding time-dependent potentials ϕ and $\mathbf{A}$ that satisfy the source-driven wave-equations (71), (72) (subject to initial and boundary conditions) in the Lorentz gauge. When the sources are distributional one seeks distributional potentials subject to similar conditions. The distributional formulation given above for real static field configurations generalizes without difficulty to the time dependent situation. The essential modification is to define functionals on the space of t dependent complex test forms on $\mathbf{R^3}$ and exploit the Fourier transform [14] of such forms. Thus if $\hat{\psi}$ is a t-dependent test p-form on $X = \mathbf{R^3}$ its Fourier transform is the ω-dependent test p-form $\mathcal{F}\hat{\psi}$ on $X = \mathbf{R^3}$ defined by

$$(\mathcal{F}\hat{\psi})(X,\omega) = \frac{1}{\sqrt{2\pi}}\int_R \hat{\psi}(X,t)\,\exp(i\omega t)\,d\,t \tag{146}$$

with inverse

$$\hat{\psi}(X,t) = \frac{1}{\sqrt{2\pi}} \int_R (\mathcal{F}\hat{\psi})(X,\omega) \exp(-i\omega t)\, d\omega \tag{147}$$

For fields in the vacuum the fundamental t-dependent (p,p) type double-form is the real part of the Fourier transform of [10]

$$\underset{(p,p)}{K_\omega}(X,Y) = \exp\left(-i\frac{\omega}{c_0}|\mathbf{x}-\mathbf{y}|\right) \underset{(p,p)}{\gamma}(X,Y). \tag{148}$$

It is useful, then, to define the singular complex-linear integral operator $\mathbf{\Gamma}_\omega$ on all $p-$forms ψ by:

$$(\mathbf{\Gamma}_\omega \psi)(Y,\omega) = \int_X \underset{(p,p)}{K_\omega}(X,Y) \wedge \underset{(X)}{\#} \psi(X,\omega) \tag{149}$$

Suppose that $\bar{\mathcal{J}}$ is a smooth $t-$dependent $p-$form source (regular at spatial infinity) that enters into the equation

$$\Delta_X C(X,t) + \frac{1}{{c_0}^2}\ddot{C}(X,t) = \bar{\mathcal{J}}(X,t) \tag{150}$$

for the time-dependent $p-$form C on $X = \mathbf{R^3}$. Then

$$(\mathcal{F}C)(X,\omega) = (\Delta_X - \frac{\omega^2}{{c_0}^2})^{-1} (\mathcal{F}\bar{\mathcal{J}})(X,\omega) \tag{151}$$

Since the singularity structure of $\underset{(p,p)}{K_\omega}(X,Y)$ is the same as that for $\underset{(p,p)}{\gamma}(X,Y)$ it is straightforward to show that a particular solution, regular at spatial infinity, is given by

$$(\mathcal{F}C)(X,\omega) = \int_Y \underset{(p,p)}{K_\omega}(X,Y) \wedge \underset{(Y)}{\#} (\mathcal{F}\bar{\mathcal{J}})(Y,\omega) \tag{152}$$

For each ω the function $\mathcal{F}C$ may be associated with a functional by applying the operation $\int_X \underset{(X)}{\#}\hat{\alpha}(X,\omega)$ to define:

$$(\mathcal{F}C)[\hat{\alpha}](\omega) = \int_X (\mathcal{F}C)(X,\omega) \wedge \underset{(X)}{\#}\hat{\alpha}(X,\omega) \tag{153}$$

But from (152) this may be expressed

$$(\mathcal{F}C)[\hat{\alpha}](\omega) = \int_Y (\mathcal{F}\bar{\mathcal{J}})(Y,\omega) \wedge \underset{(Y)}{\#} (\mathbf{\Gamma}_\omega\hat{\alpha})(Y,\omega) \equiv (\mathcal{F}\bar{\mathcal{J}})^D[\mathbf{\Gamma}_\omega\hat{\alpha}](\omega) \tag{154}$$

Thus from (148) and Fourier inversion

$$C(X,t) = \int_Y \underset{(p,p)}{\gamma}(X,Y) \wedge \underset{(Y)}{\#} \bar{\mathcal{J}}\left(Y, t - \frac{|\mathbf{x}-\mathbf{y}|}{c}\right) \tag{155}$$

and

$$C[\hat{\alpha}](t) = \frac{1}{\sqrt{2\pi}} \int_R dt' \int_Y \bar{\mathcal{J}}(Y,t') \wedge \underset{(Y)}{\#} \int_X \underset{(p,p)}{\gamma}(X,Y) \wedge \underset{(X)}{\#} (\mathcal{F}\hat{\alpha})\left(X, t - t' - \frac{|\mathbf{x}-\mathbf{y}|}{c}\right) \tag{156}$$

[10] Note that the phases in (147) and (148) are chosen so that for $0 < r < \infty$, $t > 0$ $\exp(\pm i\frac{\omega}{c_0}r).\exp(\mp i\omega t) = \exp(\pm i\frac{\omega}{c_0}(r - c_0 t))$ describes a radially outgoing wave.

or

$$C[\hat{\alpha}](t) = \frac{1}{\sqrt{2\pi}} \int_R dt' \int_Y \bar{\mathcal{J}}\left(Y, t - t' - \frac{|\mathbf{x}-\mathbf{y}|}{c}\right) \wedge \underset{(Y)}{\#} \int_X \underset{(p,p)}{\gamma}(X,Y) \wedge \underset{(X)}{\#} (\mathcal{F}\hat{\alpha})\left(X, t'\right) \quad (157)$$

Motivated by (155) the distribution C^D associated with the smooth $p-$form $C(X,t)$ is defined by

$$C^D[\hat{\beta}] = \int_R dt \int_X C(X,t) \wedge \underset{(X)}{\#} \hat{\beta}(X,t) \quad (158)$$

Hence

$$C^D[\hat{\beta}] = \int_R dt \int_X \int_Y \underset{(p,p)}{\gamma}(X,Y) \wedge \underset{(Y)}{\#} \bar{\mathcal{J}}\left(Y, t - \frac{|\mathbf{x}-\mathbf{y}|}{c}\right) \wedge \underset{(X)}{\#} \hat{\beta}(X,y) \quad (159)$$

or with $t \mapsto t' - t - \frac{|\mathbf{x}-\mathbf{y}|}{c}$ and interchange of X and Y

$$C^D[\hat{\beta}] = \int_R dt' \int_X (\mathbf{\Gamma}_{\omega R}\hat{\beta})(X,t') \wedge \underset{(X)}{\#} \bar{\mathcal{J}}(X,t') \quad (160)$$

where

$$(\mathbf{\Gamma}_R\hat{\beta})(X,t') \equiv \int_Y \underset{(p,p)}{\gamma}(X,Y) \wedge \underset{(Y)}{\#} \hat{\beta}\left(Y, t' + \frac{|\mathbf{x}-\mathbf{y}|}{c}\right)$$

Thus finally

$$C^D[\hat{\beta}] = \int_R dt' \bar{\mathcal{J}}(X,t') \wedge \underset{(X)}{\#} (\mathbf{\Gamma}_R\hat{\beta})(X,t') \equiv \bar{\mathcal{J}}^D[\mathbf{\Gamma}_R\hat{\beta}] \quad (161)$$

where $\bar{\mathcal{J}}^D$ is the distribution associated with the *smooth* time-dependent $p-$form $\bar{\mathcal{J}}$. If one now contemplates a distributional source $\bar{\mathcal{J}}^D$ that is not associated with a smooth time-dependent $p-$form on $\mathbf{R}^3$ then (161) offers a particular distributional solution C^D to (150) based on a retarded extension of the fundamental solution $\underset{(p,p)}{\gamma}(X,Y)$.

The above solution is immediately applicable to the problem of finding the retarded electromagnetic potentials $\mathbf{A}$ and ϕ in free space, where ${c_0}^2\,\epsilon\mu = 1$, for *any* smooth time dependent source. Thus from section (6) one sees that with a smooth source $\bar{\mathcal{J}} = -\frac{\varrho}{\epsilon}$ in equation (150) the smooth $0-$form solution ϕ associated with C^D describes a free-space scalar potential solution to (72) and that with a smooth source $\bar{\mathcal{J}} = \mu\#\mathbf{J}$ in equation (150) the smooth $1-$form solution $\mathbf{A}$ associated with C^D describes a free-space vector potential solution to (71).

In the absence of losses due to conduction the *convective current* spacetime $3-$form is given as $j = \rho_0 \star \widetilde{V}$ where the unit future-pointing time-like vector field V convects *proper-charge density* ρ_0. In a local spacetime co-basis $\{c_0\, dt, e^1, e^2, e^3\}$ adapted to U the source velocity $\widetilde{V} = \gamma\left(-c_0\, dt + \frac{\widetilde{\mathbf{v}}}{c}\right)$ where $\gamma^{-2} = 1 - \frac{|\mathbf{v}|^2}{c_0^2}$, $|\mathbf{v}|^2 = g(\mathbf{v},\mathbf{v})$ and $\widetilde{\mathbf{v}} = \sum_{k=1}^3 v_k\, e^k$ is the instantaneous time-dependent (Newtonian) $3-$velocity $1-$form in $\mathbf{R}^3$. Since

$$j = -\mathbf{J} \wedge dt + \rho\#1 = \rho_0 \star \widetilde{V} \quad (162)$$

one finds immediately

$$\#\mathbf{J} = -\rho\widetilde{\mathbf{v}}$$

The analysis above also offers distributional solutions to (71) and (72) in terms of scalar and $1-$form distributional sources. The electromagnetic field associated with an arbitrarily moving charge point source is space can now be modeled in terms of a moving Dirac distribution in space.

14. Time Dependent Electromagnetic fields and Distributional Sources: Moving Dirac Distributions in $\mathbf{R}^3$

The above section formulates a distributional solution to (150) associated with a $t-$ dependent p-form distributional source $\bar{\mathcal{J}}^D$ on $\mathbf{R}^3$. $0-$layer singular distributions associated with electromagnetic sources constrained to moving curves or surfaces in space can be constructed in terms of a family of distributions $\underset{(0)}{\delta_{\Sigma_{\mathbf{f}}^{3-k}\times R}}$ defined for $k = 1, 2$ by

$$\underset{(0)}{\delta_{\Sigma_{\mathbf{f}}^{3-k}\times R}}[\underset{(0)}{\hat{\beta}}] = \int_R \underset{(0)}{\boldsymbol{\delta}_{\Sigma_{\mathbf{f}}^{3-k}}}[\underset{(0)}{\hat{\beta}}]\, dt \tag{163}$$

Then

$$\widetilde{W}\, \underset{(0)}{\delta_{\Sigma_{\mathbf{f}}^{3-k}\times R}}[\underset{(1)}{\hat{\beta}}] = \int_R \widetilde{W}(t)\, \underset{(0)}{\boldsymbol{\delta}_{\Sigma_{\mathbf{f}}^{3-k}}}[\underset{(1)}{\hat{\beta}}]\, dt \tag{164}$$

for some $t-$dependent $1-$form $\widetilde{W}$ in $\mathbf{R}^3$ with support on $\Sigma_{\mathbf{f}}^{3-k}$.

For the case $k = 3$ the support of $\underset{(0)}{\boldsymbol{\delta}_{\Sigma_{\mathbf{f}}^{3-k}}}$ is a moving point $p_0(t)$ in $X = \mathbf{R}^3$. Then the worldline history of the moving point is encoded into the support of a distribution $\underset{(0)}{\delta_{p_0\times R}}$ where

$$\underset{(0)}{\delta_{p_0\times R}}[\hat{\beta}] = \int_R \hat{\beta}(p_0(t), t)\, dt \tag{165}$$

If the history of $p_0 \in X$ is parameterized by t it is the spacetime curve

$$t' \mapsto (\mathbf{x}, t) = (\mathbf{x}_0(t'), t') \tag{166}$$

Then

$$\underset{(0)}{\delta_{p_0\times R}}[\underset{(0)}{\hat{\beta}}] = \int_R \underset{(0)}{\hat{\beta}}\left(\mathbf{x}_0(t'), t'\right)\, dt' \tag{167}$$

and

$$\widetilde{W}\, \underset{(0)}{\delta_{p_0\times R}}[\underset{(1)}{\hat{\beta}}] = \int_R \underset{(0)}{(i_{W(t')}\hat{\beta})}\left(\mathbf{x}_0(t'), t'\right)\, dt' \tag{168}$$

One may now easily calculate

$$\begin{aligned}\underset{(0)}{\delta_{p_0\times R}}[\mathbf{\Gamma}_R\hat{\beta}] &= \int_R (\mathbf{\Gamma}_R\hat{\beta})(\mathbf{x}_0(t'), t')\, dt' \\ &= \int_R dt' \int_Y \underset{(0)}{\gamma}(\mathbf{x}_0(t'), Y)\ \underset{(Y)}{\#}\hat{\beta}\left(Y, t + \frac{|\mathbf{x}_0(t') - \mathbf{y}|}{c_0}\right)\end{aligned} \tag{169}$$

Changing variable $t' \mapsto t = t' + \frac{|\mathbf{x}_0(t')-\mathbf{y}|}{c_0}$ at fixed Y with Jacobian $\mathcal{Q}$ defined by

$$dt = \mathcal{Q}^{-1}(Y, t')\, dt'$$

yields

$$\underset{(0)}{\delta_{p_0\times R}}[\mathbf{\Gamma}_R\hat{\beta}] = \int_R dt \int_Y \left(\mathcal{Q}(Y, t')\, \underset{(0)}{\gamma}(\mathbf{x}_0(t'), Y)\right)\ \underset{(Y)}{\#}\hat{\beta}(Y, t) \tag{170}$$

where $t' \equiv \hat{t}'(Y,t)$ solves the equation $t' = t - \frac{|\mathbf{x}_0(t')-\mathbf{y}|}{c_0}$. Thus

$$\mathcal{Z}(Y,t) = \mathcal{Q}(Y,t') \underset{(0)}{\gamma}(\mathbf{x}_0(t'), Y) \tag{171}$$

is the 0−form associated with the distributional source $\underset{(0)}{\delta_{p_0 \times R}}[\mathbf{\Gamma}_R\hat{\beta}]$ at the field point Y at the instant t. It is straightforward to calculate the Jacobian in terms of $\mathbf{x}_0(t)$ and its derivative $\dot{\mathbf{x}}_0(t)$:

$$\mathcal{Q}^{-1}(Y,t) = \left(1 + \frac{1}{c_0}\frac{(\mathbf{x}_0(t') - \mathbf{y})\cdot(\dot{\mathbf{x}}_0(t'))}{|\mathbf{x}_0(t') - \mathbf{y}|}\right) \tag{172}$$

The instantaneous Newtonian 3−velocity of the point support is $\mathbf{v}(t) \equiv \dot{\mathbf{x}}_0(t)$ and if one introduces the Euclidean unit vector

$$\mathbf{n}(Y,t) \equiv \frac{\mathbf{y} - \mathbf{x}_0(t')}{|\mathbf{y} - \mathbf{x}_0(t')|}$$

connecting the field point $\mathbf{y}$ at time t to the source point $\mathbf{x}_0(t')$at the earlier time $t' = \hat{t}'(Y,t)$, then the Jacobian inverse takes the form

$$\mathcal{Q}^{-1}(Y,t) = \left(1 - g\left(\frac{\mathbf{v}(t')}{c_0}, \mathbf{n}(Y,t)\right)\right) \tag{173}$$

where $t' \equiv \hat{t}'(Y,t)$ solves the equation $t' = t - \frac{|\mathbf{x}_0(t')-\mathbf{y}|}{c_0}$ and g denotes the Euclidean metric tensor on $\mathbf{R}^3$

By contrast to the smooth electromagnetic sources $\bar{\mathcal{J}}$ discussed in section (13) one models a moving point source (with electric charge q and 3−velocity $\mathbf{v}(t)$) for the scalar potential ϕ^D by

$$\underset{(0)}{\bar{\mathcal{J}}^D} = -\frac{q}{\epsilon_0}\underset{(0)}{\delta_{p_0 \times R}} \tag{174}$$

and for the potential 1−form $\mathbf{A}^D$ by

$$\underset{(1)}{\bar{\mathcal{J}}^D} = -q\mu_0\,\widetilde{\mathbf{v}}(t)\,\underset{(0)}{\delta_{p_0 \times R}} \tag{175}$$

It then follows immediately from (161) that the distribution ϕ^D may be associated with

$$\underset{(0)}{\mathcal{P}}(Y,t) = -\frac{q}{\epsilon_0}\,\mathcal{Q}(Y,t)\underset{(0,0)}{\gamma}(\mathbf{x}(t'), Y) \tag{176}$$

while the distribution $\mathbf{A}^D$ may be associated with

$$\underset{(1)}{\mathcal{A}}(Y,t) = -q\,\mu_0\,\mathcal{Q}(Y,t)\,i_{\mathbf{v}(t')}\underset{(1,1)}{\gamma}(\mathbf{x}(t'), Y) \tag{177}$$

where $t' = \hat{t}'(Y,t)$ as above. These are the classic Lienard-Weichert potentials for a moving point charge [2]. It is of interest to verify from $\#\mathbf{J}^D = -\widetilde{\mathbf{v}}\rho^D$ that the distributions $\mathbf{A}^D$ and ϕ^D satisfy the gauge condition $\delta\,\mathbf{A}^D + \epsilon_0\mu_0\dot{\phi}^D = 0$. It may also be noted that by using the appropriate fundamental double-forms in place of $\underset{(0,0)}{\gamma}$ and $\underset{(1,1)}{\gamma}$ these potentials maintain their structure for solutions in $\mathcal{D} \subset Y$ that satisfy appropriate boundary conditions on ∂D.

15. Distributional Analysis of Fixed Junction Conditions in $\mathbf{R}^3$

The formulation in this section begins by assuming that all electromagnetic sources are contained within a static region $\mathcal{U} \subset \mathbf{R}^3$. This region is partitioned into compact static regions some of which are empty, $\mathcal{U} = \bigcup_I U_I$ where each U_I is the image of some 3− chain. Some regions will serve to support distinct physical media characterized by electromagnetic constitutive properties such as permittivity, permeability and conductivity. Others may contain the vacuum or smooth volume distributions of electric charge or current. Some components of some electromagnetic fields in the vicinity of adjacent domains containing such sources will not be continuous. All regions may also include singular sources distributed on points, curves or surfaces, where electromagnetic fields may be singular. In this manner one can apply Stoke's theorem to regular forms on regular domains and account for discontinuities and singularities in terms of different types of singular distributions. Let $\alpha(U)$ denote a smooth $p-$form with bounded components on some regular domain $U \subset \mathbf{R}^3$ in the partition of $\mathcal{U}$. Denote typical test $p-$forms with arbitrary support on $\mathbf{R}^3$ by $\hat{\Lambda}$ and those with support containing $\mathcal{W} \subset \mathbf{R}^3$ by $\hat{\Lambda}(\mathcal{W})$. It will be assumed that all $p-$chains $[0,1]^p \mapsto \mathbf{R}^3$ inherit a fixed orientation from the orienting 3-form $\#1$ on $\mathbf{R}^3$. Then $\int_\Sigma \alpha(U)$ denotes the integral of the pull-back of the smooth $p-$form $\alpha(U)$ to the domain of the *oriented* $p-$chain Σ. Following convention it is convenient not to distinguish notationally between the domain and image of any $p-$chain. When U' and U'' lie on opposite sides of a common interface between regions and the integral $\int_\Sigma (\alpha(U') - \alpha(U'')) \neq 0$ where $\Sigma \subset \partial U'$, and $\Sigma \subset \partial U''$, the interface is termed a non-zero *discontinuity $p-$chain* for α.

Singular distributions of electric charge and point electric multipoles are taken to reside on a finite set of fixed isolated points P_I (0-chains), fixed (possibly intersecting curves) (1−chains) C_I and fixed(possibly intersecting) surfaces (2−chains) S_I.

Now for any smooth $p-$form $\alpha(U_I)$ on any domain $U_I \subset \mathcal{U}$ and test $(p+1)-$form $\hat{\Lambda}$ on $\mathbf{R}^3$ one has from (13) the identity:

$$(\alpha(U_I), \delta\hat{\Lambda})_{U_I} = (d\alpha(U_I), \hat{\Lambda})_{U_I} - \int_{\partial U_I} \alpha(U_I) \wedge \#\hat{\Lambda} \tag{178}$$

In general ∂U_I is a sum of disjoint 2−chains that may include parts of $\partial\mathcal{U}$.

With these preliminaries one may analyze the content of the distributional Maxwell equation

$$d\,D^D = \rho^D \tag{179}$$

where the total distributional Maxwell displacement 2−form D^D on $\mathbf{R}^3$ is defined for all test 2−forms $\hat{\beta}$ by

$$D^D[\hat{\beta}] = \sum_{U_I} (D(U_I), \hat{\beta})_{U_I} + \sum_{P_I} \mathcal{D}^D_{P_I}[\hat{\beta}] + \sum_{C_I} \mathcal{D}^D_{C_I}[\hat{\beta}] + \sum_{S_I} \mathcal{D}^D_{S_I}[\hat{\beta}] \tag{180}$$

and the distributional electric charge density 3−form ρ^D on $\mathbf{R}^3$ is defined for all test 3−forms $\hat{\gamma}$ by

$$\begin{aligned} \rho^D[\hat{\gamma}] = {} & \sum_{U_I} (\rho(U_I), \hat{\gamma})_{U_I} + \sum_q \sum_{P_I} Q_{P_I} \underset{(0)}{\boldsymbol{\delta}^{Y_1 Y_2 \ldots Y_q}_{\Sigma^0_{\mathbf{f}}}}[\#\hat{\gamma}] \\ & + \sum_q \sum_{C_I} \underset{(1)}{\lambda^q_{C_I}} \wedge d\check{f}_{I_1} \wedge d\check{f}_{I_2} \underset{(0)}{\boldsymbol{\delta}^{Y_1 Y_2 \ldots Y_q}_{\Sigma^1_{\mathbf{f}}}}[\hat{\gamma}] \\ & + \sum_q \sum_{S_I} \underset{(2)}{\sigma^q_{S_I}} \wedge df_{I_1} \underset{(0)}{\boldsymbol{\delta}^{Y_1 Y_2 \ldots Y_q}_{\Sigma^2_{\mathbf{f}}}}[\hat{\gamma}] \end{aligned} \tag{181}$$

The first term on the right hand side above represents a sum over all regular volume distributions of charge density[11], the second term represents a sum over all point multipoles in $\mathcal{U}$ (including

[11] $\rho(U_I)$ may be zero on some U_I.

point charges), the third a sum over all line-distributions of charge on curves C_I (given by the loci $\check{f}_{I_1} = 0, \check{f}_{I_2} = 0$) in $\mathcal{U}$ and the last term a sum over all surface distributions of charge on surfaces S_I (given by the locus $f_I = 0$) in $\mathcal{U}$. Each Q_{P_I} is a constant describing the electric charge of point mono-poles or the moments of point multi-poles, $\lambda^q_{C_I}$ is a 1−form on C_I containing a line-charge density and each $\sigma^q_{S_I}$ is a 2−form on S_I containing a surface charge density. The distributional equation implies that

$$d\,D^D[\hat{\gamma}] = \rho^D[\hat{\gamma}] \tag{182}$$

for *all* test forms $\hat{\gamma}$ on $\mathbf{R}^3$. By restricting the support of $\hat{\gamma}$ the quantities $D, \mathcal{D}^D_{P_I}, \mathcal{D}^D_{C_I}, \mathcal{D}^D_{S_I}$ can be related to sources with support U_I, P_I, C_I, S_I respectively. Now

$$d\,D^D[\hat{\gamma}] = D^D[\delta\hat{\gamma}] = \sum_{U_I} (D(U_I), \delta\hat{\gamma})_{U_I} + \sum_{P_I} \mathcal{D}^D_{P_I}[\delta\hat{\gamma}] + \sum_{C_I} \mathcal{D}^D_{C_I}[\delta\hat{\gamma}] + \sum_{S_I} \mathcal{D}^D_{S_I}[\delta\hat{\gamma}]$$

and from (178) one has

$$\begin{aligned}\sum_{U_I} (D(U_I), \delta\hat{\gamma})_{U_I} &= \sum_{U_I} (d\,D(U_I), \hat{\gamma})_{U_I} - \sum_{U_I} \int_{\partial U_I} D(U') \wedge \#\hat{\gamma} \\ &= \sum_{U_I} (d\,D(U_I), \hat{\gamma})_{U_I} - \sum_{U'} \int_{\mathcal{U}} D(U') \wedge \#\hat{\gamma} \\ &\quad + \sum_j \int_{I_j(U''', U'')} (D(U''') - D(U'')) \wedge \#\hat{\gamma}\end{aligned}$$

where $\{U'\}$ is the set of regular domains that contain all or part of $\mathcal{U}$ in $\partial U'$, $\{U''''\}$ is the set of regular domains that contain all or part of $\mathcal{U}$ in $\partial U''''$ and $\{U''', U''\}$ is the set of regular domains that have common interfaces $I_j(U''', U'')$. If one chooses the support of $\hat{\gamma}$ to lie wholly or partly in a subset of U_I that excludes all singular sources one collects the set of regular differential equations

$$d\,D(U_I) = \rho(U_I) \tag{183}$$

on each such subset U_I. If one chooses the support of $\hat{\gamma}$ to contain the interface $I_j(U''', U'')$ which is part of the support $f = 0$ of a surface charge with density σ_j (and no other singular source) one has:

$$\begin{aligned}\int_{I_j(U''', U'')} (D(U''') - D(U'')) \wedge \#\hat{\gamma} &= \left(\underset{(0)}{\boldsymbol{\delta}_{\Sigma^2_{\mathbf{f}}}}\, \sigma_j \wedge df \right) [\hat{\gamma}] \\ &= \underset{(0)}{\boldsymbol{\delta}_{\Sigma^2_{\mathbf{f}}}} [i_{\widetilde{\sigma_j}}\, i_{\widetilde{df}}\, \hat{\gamma}] = \int_{I_j(U''', U'')} \omega_{\mathbf{f}}\, \hat{\Phi}\end{aligned}$$

where $\hat{\Phi} = i_{\widetilde{\sigma_j}}\, i_{\widetilde{df}}\, \hat{\gamma}$. Now $\omega_{\mathbf{f}} = \#\,df/|df|^2$ and $\hat{\Phi}\,\#\,df = i_{\widetilde{df}}\, i_{\widetilde{\sigma_j}}\hat{\gamma}\,\#\,df = df \wedge i_{\widetilde{\sigma_j}}\hat{\gamma}$. But

$$i_{\widetilde{df}} \left(i_{\widetilde{\sigma_j}}\hat{\gamma} \wedge \#\,df \right) = \left(i_{\widetilde{df}} i_{\widetilde{\sigma_j}} \right) \#\,df = |df|^2\,\#\, i_{\widetilde{\sigma_j}}\hat{\gamma} = -df \wedge \left(i_{\widetilde{df}} \#\, i_{\widetilde{\sigma_j}}\hat{\gamma} \right)$$

The last term pulls back to zero on $f = 0$ so:

$$\left(i_{\widetilde{df}} i_{\widetilde{\sigma_j}}\hat{\gamma} \right) \frac{\#\,df}{|df|^2} = \#\, i_{\widetilde{\sigma_j}}\hat{\gamma} = \sigma_j \wedge \hat{\gamma} \quad mod \quad df$$

yielding

$$\int_{I_j(U''',U'')} (D(U''') - D(U'') - \sigma_j) \wedge \#\hat{\gamma} = 0$$

and one infers the pull-back relations:

$$\left(I_j(U''',U'')\right)^* \left(D(U''') - D(U'') - \sigma_j(I_j(U''',U''))\right) = 0 \tag{184}$$

on $I_j(U''',U'')$. Note the correlation between the pull-back discontinuity $(I_j(U''',U''))^* (D(U''') - D(U''))$ and the chosen orientation of the chain $I_j(U''',U'')$. If the interface is part of the surface $f = 0$ in $\mathbf{R}^3$ this pull-back relation can also be expressed in terms of $d\,f$ and forms injected into $\mathbf{R}^3$:

$$([D] \wedge d\,f)|_{I_j} = (\sigma_j \wedge d\,f)|_{I_j} \tag{185}$$

where I_j is the interface discontinuity relative to D, $[D] = D(U''') - D(U'')$ for f increasing from U''' to U'' and σ_j is the surface charge 2-form, now injected into $\mathbf{R}^3$.

Similarly if B^D is the distributional Maxwell magnetic induction 2−form on $\mathbf{R}^3$ defined for all test 2−forms $\hat{\beta}$ on $\mathbf{R}^3$ by

$$B^D[\hat{\beta}] = \sum_{U_I} (B(U_I), \hat{\beta})_{U_I} + \sum_{P_I} \mathcal{B}^D_{P_I}[\hat{\beta}] + \sum_{C_I} \mathcal{B}^D_{C_I}[\hat{\beta}] + \sum_{S_I} \mathcal{B}^D_{S_I}[\hat{\beta}] \tag{186}$$

the distributional Maxwell equation (in the absence of magnetic charge)

$$d\,B^D = 0 \tag{187}$$

implies

$$d\,B(U_I) = 0 \tag{188}$$

on each subset of $\mathcal{U}$ free of singularities and

$$\left(I_j(U''',U'')\right)^* \left(B(U''') - B(U'')\right) = 0 \tag{189}$$

or

$$([B] \wedge d\,f)|_{I_j} = 0 \tag{190}$$

where I_j is the interface discontinuity relative to B and $[B] = B(U''') - B(U'')$ for f increasing from U''' to U''.

The total distributional Maxwell magnetic field 1−form h^D is defined for all test 1−forms $\hat{\alpha}$ on $\mathbf{R}^3$ by

$$h^D[\hat{\alpha}] = \sum_{U_I} (h(U_I), \hat{\alpha})_{U_I} + \sum_{C_I} \mathcal{H}^D_{C_I}[\hat{\alpha}] + \sum_{S_I} \mathcal{H}^D_{S_I}[\hat{\alpha}] \tag{191}$$

where each $h(U_I)$ is a smooth 1−form on U_I. The distributional Maxwell equation

$$d\,h^D = J^D + \dot{D}^D \tag{192}$$

involves the time derivative of the total Maxwell displacement distribution, $\dot{D}^D$, defined by $\dot{D}^D = \mathcal{L}_{\partial_t} D^D$ and, for all test 2−forms $\hat{\beta}$, the distributional electric current 2−form is taken to be:

$$J^D[\hat{\beta}] = \sum_{U_I} (J(U_I), \hat{\beta})_{U_I} + \sum_{C_I} \underset{(1)}{\check{\lambda}_{C_I}} \, \underset{(0)}{\boldsymbol{\delta}_{\Sigma^1_{\mathbf{f}}}}[\#\hat{\beta}] + \sum_{S_I} \underset{(1)}{\kappa_{S_I}} \, \underset{(0)}{\boldsymbol{\delta}_{\Sigma^2_{\mathbf{f}}}}[\#\hat{\beta}] \tag{193}$$

The distributions $\mathcal{H}^D_{C_I}, \mathcal{H}^D_{S_I}$ are associated with current line-sources with density $\check{\lambda}_{C_I}$ and current surface sources with density κ_{S_I}. Each $\check{\lambda}_{C_I}$ is a 1−form on C_I and each κ_{S_I} is a 1-form on S_I. It follows from

$$d\,h^D[\hat{\beta}] = J^D[\hat{\beta}] + \dot{D}[\hat{\beta}]$$

that

$$d\,h(U_I) = J(U_I) + \dot{D}(U_I) \tag{194}$$

on each subset of U_I free of singular sources and

$$\left(I_j(U''', U'')\right)^* \left(h(U''') - h(U'') - \kappa_j(I_j(U''', U''))\right) = 0 \tag{195}$$

on each $I_j(U''', U'')$ containing a surface current 1− form κ_j. By injecting the surface current 1−form κ_j into $\mathbf{R}^3$ one may write the above junction condition as

$$([h] \wedge d\,f)|_{I_j} = (\kappa_j \wedge d\,f)|_{I_j} \tag{196}$$

where I_j is the interface discontinuity relative to h and $[h] = h(U''') - h(U'')$ for f increasing from U''' to U''.

Finally if e^D denotes the total distributional Maxwell electric field 1−form defined for all test 1−forms $\hat{\alpha}$ on $\mathbf{R}^3$ by

$$e^D[\hat{\alpha}] = \sum_{U_I} (e(U_I), \hat{\alpha})_{U_I} + \sum_{P_I} \mathcal{E}^D_{P_I}[\hat{\alpha}] + \sum_{C_I} \mathcal{E}^D_{C_I}[\hat{\alpha}] + \sum_{S_I} \mathcal{E}^D_{S_I}[\hat{\alpha}] \tag{197}$$

where each $e(U_I)$ is a smooth 1−form on U_I then the distributional Maxwell equation

$$d\,e^D = -\dot{B}^D$$

where $\dot{B}^D = \mathcal{L}_{\partial_t} B^D$ yields

$$d\,e(U_I) = -\dot{B}(U_I) \tag{198}$$

on each U_I free of singular sources (where $\dot{B} = \mathcal{L}_{\partial_t} B$) and

$$\left(I_j(U''', U'')\right)^* \left(e(U''') - e(U'')\right) = 0 \tag{199}$$

on each $I_j(U''', U'')$. One may write this junction condition as

$$([e] \wedge d\,f)|_{I_j} = 0 \tag{200}$$

where I_j is the interface discontinuity relative to e and $[e] = e(U''') - e(U'')$ for f increasing from U''' to U''.

In all these boundary conditions, if U_I denotes an idealized perfect conductor then $e(U_I), b(U_I), D(U_I), H(U_I)$ are all zero. However if U_I denotes an uncharged idealized perfect insulator then $\sigma_0 = 0$ on its interface with another uncharged idealized perfect insulator. If Σ is a charged conducting surface its distribution of surface charge cannot be arbitrarily prescribed unlike the distribution of volume charge or surface charge that can exist in a non-conducting medium.

Journal of Physics: Conference Series **254** (2010) 012006 doi:10.1088/1742-6596/254/1/012006

16. Distributional Analysis of Moving Junction Conditions

In the previous section the distributional Maxwell system has been written as an exterior system on $\mathbf{R}^3$ in terms of differential forms dependent on time regarded as an evolution parameter. The partition of $\mathcal{U}$ was fixed relative to this time parameter. Although this system is fully compatible with a relativistic description of electromagnetism it was not expressed in a manifestly Lorentz covariant form. For many relativistic problems and situations where dynamic interface phenomena are relevant it is more natural to reformulate the distributional descriptions in terms of the spacetime structures introduced in section 4. In this section an outline of this approach will be given. The partitioning of spacetime by a collection of immersions into spacetime is best approached in the language of algebraic topology which is not attempted here and attention is restricted to a simple partition of spacetime by the history of a single compact region of space with source on its spatial boundary.

Let Φ be a leaf in a local foliation defining a hypersurface in flat spacetime. Assume further that $d\Phi \neq 0$ and $g(d\Phi, d\Phi) \equiv d\Phi \cdot d\Phi) > 0$. The unit time-like vector field U on spactime defines a (local) frame associated with a laboratory observer while a unit time-like vector field W on spacetime will be associated below with the observed motion of an interface specified by Φ. At each event p on the hypersurface one also has a unit space-like normal $N = \frac{\widetilde{d\Phi}}{\sqrt{d\Phi \cdot d\Phi}}$. In the following it proves convenient to make notationally explicit the dependence of various fields on U. Hence, any superscripted bold face form $\boldsymbol{\xi}^U$ on spacetime is orthogonal to U in the sense $i_U \boldsymbol{\xi}^U = 0$.

In this context the distributional Maxwell system in spacetime can be written in a self-evident notation as

$$d\,F^D = 0 \tag{201}$$

$$d \star G^D = j^D \tag{202}$$

By arguments in complete analogy to those in the previous section, for a single hypersurface each equation yields an equation of the type

$$\left(\underset{(2)}{\Lambda} \wedge d\,\Phi\right)|_p = \left(\underset{(2)}{\mathcal{K}} \wedge d\,\Phi\right)|_p \tag{203}$$

for some smooth 2−forms Λ and $\mathcal{K}$ in the neighborhood of the hypersurface and p on the hypersurface. Let the orthogonal decomposition of these forms with respect to U be:

$$\Lambda = \underset{(1)}{\mathbf{l}^U} \wedge \tilde{U} + \underset{(2)}{\mathbf{S}^U}$$

$$\mathcal{K} = \underset{(1)}{\boldsymbol{\kappa}^U} \wedge \tilde{U} + \underset{(2)}{\boldsymbol{\sigma}^U}$$

$$d\,\Phi = \boldsymbol{\alpha}^U\,\tilde{U} + \boldsymbol{\beta}^U$$

Thus in the neighborhood of the hypersurface one has

$$\boldsymbol{\alpha}^U\,\mathbf{S}^U - \mathbf{l}^U \wedge \boldsymbol{\beta}^U = \boldsymbol{\alpha}^U\,\boldsymbol{\sigma}^U - \boldsymbol{\kappa}^U \wedge \boldsymbol{\beta}^U$$

$$\mathbf{S}^U \wedge \boldsymbol{\beta}^U = \boldsymbol{\sigma}^U \wedge \boldsymbol{\beta}^U$$

One now relates $\boldsymbol{\alpha}^U$ and $\boldsymbol{\beta}^U$ to a particular space-like unit vector N^U and the instantaneous normal component v_{N^U} of the 3−velocity of an interface with respect to N^U in the frame U. The space-like direction $d\Phi$ induces a time-like vector field W on the hypersurface $d\Phi = 0$ satisfying

$i_W\, d\,\Phi = 0$. The vector field W is unique up to a scaling so with $d\,\Phi \neq 0$ one normalises W so that $W \cdot W = -1$ and writes

$$W = \gamma\left(U - \frac{\mathbf{v}^U}{c_0}\right)$$

where $\gamma^2 = \frac{1}{1-\frac{\mathbf{v}^U\cdot\mathbf{v}^U}{c_0^2}}$. The Newtonian 3-velocity field $\mathbf{v}^U$ of W relative to the frame U is given in terms of U and W by:

$$\frac{\mathbf{v}^U}{c_0} = -\frac{(1+\tilde{U}\wedge i_U)W}{W\cdot U}$$

Let $\Sigma_3^{d\,\Phi}$ denote the hypersurface given by $\Phi = 0$ and Σ_3^U a local hypersurface defined by U. In a local chart with $U = \partial_t/c_0$, Σ_3^U is the locus $t = t_0$ for some positive constant t_0. Within $\Sigma_3^{d\,\Phi}$ one may identify a codimension 2 submanifold Σ_2 given by $\Phi = 0,\ t - t_0 = 0$. The instantaneous space-like unit normal to Σ_2 in $\Sigma_3^{d\,\Phi}\bigcap\Sigma_3^U$ is denoted N^U. In terms of Φ and U one has

$$\widetilde{N}^U = \frac{\left(1+\tilde{U}\wedge i_U\right)d\,\Phi}{\sqrt{d\Phi\cdot d\Phi}}$$

and then

$$\begin{aligned}\boldsymbol{\alpha}^U &= \frac{v_{N^U}\sqrt{d\Phi\cdot d\Phi}}{c_0}\\ \boldsymbol{\beta}^U &= \widetilde{N}^U\sqrt{d\Phi\cdot d\Phi}\end{aligned}$$

where $v_{N^U} = \widetilde{\mathbf{v}^U}\cdot\widetilde{N}^U$ and W is the solution of $W\cdot d\,\Phi = 0,\ W\cdot W = -1$. Thus the generic interface conditions become in terms of normal interface properties in the frame U

$$\left(\frac{v_{N^U}}{c_0}\mathbf{S}^U - \mathbf{l}^U\wedge\widetilde{N}^U\right)|_p = \left(\frac{v_{N^U}}{c_0}\boldsymbol{\sigma}^U - \boldsymbol{\kappa}^U\wedge\widetilde{N}^U\right)|_p \tag{204}$$

$$(\mathbf{S}^U\wedge\widetilde{N}^U)|_p = (\boldsymbol{\sigma}^U\wedge\widetilde{N}^U)|_p \tag{205}$$

for any event $p\in\Sigma_3^{d\,\Phi}\bigcap\Sigma_3^U$.

If $\mathbf{S}^U$ is written in terms of the 1-form $\mathbf{s}^U$ using the induced Hodge map $\#$ on Σ_3^U defined by $\star 1 = \tilde{U}\wedge\#1$:

$$\mathbf{S}^U = \#\mathbf{s}^U$$

then with the orthogonal split relative to $\widetilde{N}^U$

$$\mathbf{s}^U = \mathbf{s}_N^U\widetilde{N}^U + \mathbf{s}_{/\!/}^U$$

where $i_N\mathbf{s}_{/\!/}^U = 0$ and $\mathbf{s}_N^U = i_N\mathbf{s}^U$ one has

$$\mathbf{S}^U\wedge\widetilde{N}^U = \mathbf{s}_N^U\#1$$

i.e.

$$\mathbf{S}^U\wedge\widetilde{N}^U = -(i_{N^U}\#\mathbf{S}^U)i_U\star 1$$

The distributional equation $d\,F^D = 0$ yields the generic condition (203) with $\mathcal{K} = 0$ and $\Lambda|_p = [F]$ where $[F]$ denotes the discontinuity of F across the hypersurface $\Sigma_3^{d\,\Phi}$. This condition can now be decomposed into conditions on the instantaneous discontinuities $[\mathbf{e}^U]$ and $[\mathbf{b}^U]$ at some event $p\in\Sigma_3^{d\,\Phi}\bigcap\Sigma_3^U$ by writing $\Lambda|_p = [F] = [\mathbf{e}^U]\wedge\tilde{U} - c_0[\mathbf{B}^U]$. Then

$\mathbf{l}^U = [\mathbf{e}^U]$, $\mathbf{S}^U = -c_0[\mathbf{B}^U]$ and $\mathbf{B}^U = \#\mathbf{b}^U$. The condition $\mathbf{S}^U \wedge \widetilde{N}^U = 0$ implies $[\mathbf{B}^U] \wedge \widetilde{N}^U = 0$, i.e $[\mathbf{b}^U_N]\,\#1 = 0$ or

$$[\mathbf{b}^U_N] = 0 \tag{206}$$

Also $\frac{v_{NU}}{c_0}\mathbf{S}^U - \mathbf{l}^U \wedge \widetilde{N}^U = 0$ implies $\frac{v_{NU}}{c_0}[-c_0\mathbf{B}^U] = [\mathbf{e}^U] \wedge \widetilde{N}^U = 0$, i.e. $v_{NU}\,[\mathbf{B}^U] + [\mathbf{e}^U] \wedge \widetilde{N}^U = 0$ or [12] (on applying #):

$$v_{NU}\,[\mathbf{b}^U] + \#([e^u] \wedge \widetilde{N}^U) = 0 \tag{207}$$

In a similar way the distributional equation $d \star G^D = j^D$ yields the generic condition (203) with $\mathcal{K} \neq 0$ and $\Lambda|_p = [\star G]$ where $[\star G]$ denotes the discontinuity of $\star G$ across the hypersurface $\Sigma_3^{d\Phi}$. This condition can now be decomposed into conditions on the instantaneous discontinuities $[\mathbf{d}^U]$ and $[\mathbf{h}^U]$ at some event $p \in \Sigma_3^{d\Phi} \bigcap \Sigma_3^U$ by writing $\Lambda|_p = [\star G] = [\mathbf{h}^U/c] \wedge \tilde{U} + [\mathbf{D}^U]$. In this case $\mathbf{l}^U = [\mathbf{h}^U/c_0]$ and $\mathbf{S}^U = [\mathbf{D}^U]$ and the orthogonal decomposition of $\mathcal{K}$ is written

$$\mathcal{K} = \boldsymbol{\kappa}^U \wedge \tilde{U} + \boldsymbol{\sigma}^U$$

Then

$$\mathbf{S}^U \wedge \widetilde{N}^U = \boldsymbol{\sigma}^U \wedge \widetilde{N}^U = (\sigma_0^U\,\#1) \wedge \widetilde{N}^U = \sigma_0^U\,\#1$$

in terms of the surface charge density 0−form σ_0^U on the interface. Hence $[D] \wedge \widetilde{N}^U = \sigma_0^U\,\#1$ or with $\mathbf{d}^U = \mathbf{d}^U_N \widetilde{N}^U + \mathbf{d}^U_{/\!/}$:

$$[\mathbf{d}^U_N] = \sigma_0 \tag{208}$$

Finally $\frac{v_{NU}}{c_0}\mathbf{S}^U - \mathbf{l}^U \wedge \widetilde{N}^U = \frac{v_{NU}}{c_0}[\mathbf{D}^U] - \frac{[\mathbf{h}^U]}{c_0} \wedge \widetilde{N}^U$ and with $\boldsymbol{\kappa}^U = \kappa_0 \widetilde{N}^U + \boldsymbol{\kappa}^U_{/\!/}$ one has the moving interface condition[13]

$$v_{NU}\,[\mathbf{d}^U] - \#([\mathbf{h}^U \wedge \widetilde{N}^U]) = v_{NU}\,\sigma_0^U \widetilde{N}^U - c_0\,\#\,(\boldsymbol{\kappa}^U_{/\!/} \wedge \widetilde{N}^U) \tag{209}$$

The component of this condition in the direction $\widetilde{N}^U$ reproduces the condition (208) above on the normal component of $[\mathbf{d}^U]$. Thus for an arbitrarily moving interface relative to an arbitrary observer field U the local interface conditions on the fields $\mathbf{e}^U, \mathbf{b}^U, \mathbf{d}^U, \mathbf{h}^U$ depend on the local normal velocity component of the interface, the direction of the interface normal and the values of the surface charge and current at that instant.

17. Conclusions

Using properties of the fundamental double-form of bi-degree (p, p) associated with the Hodge-de Rham operator Δ on differential forms, a distributional framework for analyzing the Maxwell system with singular sources has been developed. A set of $r-$layer Dirac singular distributions with supports on (moving) embeddings in $\mathbf{R}^3$ has been constructed and finds application in electromagnetic source modeling. The framework has been illustrated by explicitly calculating the fields associated with particular singular sources including those due to an arbitrarily moving point charge in free space. With the aid of $r-$layer Dirac singular distributions many fields with more complex distributional sources can be readily reduced to quadratures once one parametrises such sources in terms of the geometry of their support in space. Further applications can be found in [16]. The local jump conditions on Maxwell fields across arbitrarily moving interfaces have been deduced directly from the covariant Maxwell equations for media on spacetime

[12] In Gibb's 3−vector notation: $v_{NU}\,[\mathbf{b}^\mathbf{U}] = (\mathbf{N}^U \times [\mathbf{e}^\mathbf{U}])$
[13] In Gibb's 3−vector notation: $\mathbf{N}^U \cdot [\mathbf{d}^U] = \sigma_0^U$ and $v_{NU}\,[\mathbf{d}^U] + (\mathbf{N}^U \times [\mathbf{h}^U]) = v_{NU}\,\sigma_0^U \mathbf{N}^U + c_0(\mathbf{N}^U \times \boldsymbol{\kappa}^U)$

Acknowledgments
The author is grateful to S Goto and colleagues at the Cockcroft Institute for valuable discussions and to the EPSRC for a Springboard Fellowship and financial support for this research which is part of the Alpha-X collaboration.

References
[1] Schwartz L 1965 *Méthodes mathématiques pour les sciences physiques* (Paris: Hermann et Cie)
[2] Jackson J D ^{3}1999 *Classical Electrodynamics* (New York: Wiley)
[3] Van Bladel J 1991 *Singular Electromagnetic Fields and Sources* (IEEE Press)
[4] Skinner R and Weil J A 1989 An Introduction to Generalised functions and their Applications to static electromagnetic point dipoles, including hyperfine interactions *Am. J. Phys.* **57** 777
[5] Glasser M L 1970 The Potential of a Point Charge between Capacitor Plates *Am. J. Phys.* **36** 415
[6] Labinac V, Erceg N and Koinik-Karuza D 2006 Magnetic Field of a Cylindrical Coil *Am. J. Phys.* **74** 621
[7] Benn I M and Tucker R W 1987 *An introduction to Spinors and Geometry with Applications in Physics* (IOP Publishing Ltd)
[8] Gel'fand I M and Shilov G E 1966 *Generalised Functions* Vol. 1 (New York, Academic Press)
[9] Shilov G E 1968 *Generalised Functions and Partial Differential Equations* (New York: Gordon and Breach)
[10] Barros-Neto J 1973 *An Introduction to the Theory of Distributions* (New York: Dekker)
[11] Nickerson H K, Spencer D C and Steenrod NE 1959 *Advanced Calculus*)Princeton: Van Nostrand)
[12] Duff G E D 1952 Differential Forms on Manifolds with Boundary *Annals of Math.* **56** 115
[13] Duff G E D 1953 Boundary Value Problems associated with the Tensor Laplacian *Canadian J. Math.* **5** 196
[14] Taylor M E 1999 *Partial Differential Equations I* (Amsterdam: Springer)
[15] Cooper R K 1969 On the Equivalence of Certain Green's Function Expressions *Am. J. Phys.* **37** 1032
[16] Tucker R W 2009 Differential form valued forms and distributional electromagnetic sources *J. Math. Phys.* **50** 033506
[17] Kaiser G 2009 Huygens' principle in classical electrodynamics: a distributional approach *Preprint* arXiv:0906.4167v1

From SICs and MUBs to Eddington

Ingemar Bengtsson
Fysikum, Stockholms Universitet, S-106 91 Stockholm, Sweden
E-mail: ingemar@physto.se

Abstract. This is a survey of some very old knowledge about Mutually Unbiased Bases (MUB) and Symmetric Informationally Complete POVMs (SIC). In prime dimensions the former are closely tied to an elliptic normal curve symmetric under the Heisenberg group, while the latter are believed to be orbits under the Heisenberg group in all dimensions. In dimensions 3 and 4 the SICs are understandable in terms of elliptic curves, but a general statement escapes us. The geometry of the SICs in 3 and 4 dimensions is discussed in some detail.

Received: 28 October 2009

1. Introduction

Two problems that have attracted much attention in the quantum information community are the MUB and SIC problems for Hilbert spaces of finite dimension N. In the MUB problem [1, 2] one looks for $N+1$ orthonormal bases that are mutually unbiased, in the sense that

$$|\langle e_m|f_n\rangle|^2 = \frac{1}{N}\ , \qquad 0 \leq m, n \leq N-1\ , \tag{1}$$

whenever the vector $|e_m\rangle$ belongs to one basis and the vector $|f_n\rangle$ to another. In the SIC problem [3, 4] one looks for a symmetric and informationally complete POVM, which translates to the problem of finding N^2 unit vectors $|\psi_i\rangle$ such that

$$|\langle \psi_i|\psi_j\rangle|^2 = \frac{1}{N+1}\ , \qquad 0 \leq i, j \leq N^2-1\ , \tag{2}$$

whenever $i \neq j$. These problems are hard. For the MUB problem an elegant solution exists whenever N is a power of a prime [5]. For the SIC problem quite ad hoc looking analytic solutions are known for eighteen different dimensions; these are described (and in some cases derived) by Scott and Grassl, who also give full references to the earlier literature [6]. The belief in the community is that a complete set of $N+1$ MUB does not exist for general N, while the SICs do.

Since the problems are so easy to state, it is not surprising that they have been posed independently in many different branches of science. One purpose of this article is to describe what nineteenth century geometers had to say about them. A story told by Eddington [7] is relevant here:

Some years ago I worked out the structure of this group of operators in connection with Dirac's theory of the electron. I afterwards learned that a great deal of what I had written was to be found in a treatise on

Kummer's quartic surface. There happens to be a model of Kummer's quartic surface in my lecture-room, at which I had sometimes glanced with curiosity, wondering what it was all about. The last thing that entered my head was that I had written (somewhat belatedly) a paper on its structure. Perhaps the author of the treatise would have been equally surprised to learn that he was dealing with the behaviour of an electron.

We will see what Eddington saw as we proceed. Meanwhile, let us observe that when N is a prime the MUB are the eigenbases of the $N+1$ cyclic subgroups of the Heisenberg group, while there is a conjecture (enjoying very considerable numerical support [6]) that the SICs can always be chosen to be special orbits of this group. When N is a power of a prime the solution of the MUB problem shifts a little, since the MUBs now consist of eigenvectors of the cyclic subgroups of the Heisenberg group defined over a finite field rather than over the ring of integers modulo N. Concerning SICs that are orbits under the Heisenberg group there is a link to the MUB problem: If the dimension N is a prime the SIC Bloch vectors, when projected onto any one of the MUB eigenvalue simplices, have the same length for all the $N+1$ MUB [8, 9].

In mathematics elliptic curves provide the natural home for the Heisenberg group, so it seems natural to investigate if elliptic curves can be used to illuminate the MUB and SIC problems. In dimensions 3 [10] and 4 they certainly can, as we will see, but in higher dimensions I am not so sure. There will be some comments and formulas that I could not find in the books and papers I studied, but keeping Eddington's example in mind I do not claim originality for them.

2. Two pieces of background information

We had better define the Heisenberg group properly. A defining non-unitary representation is given by the upper triangular matrices

$$g(\gamma,\alpha,\beta) = \begin{pmatrix} 1 & \alpha & \gamma \\ 0 & 1 & \beta \\ 0 & 0 & 1 \end{pmatrix} . \tag{3}$$

Here the matrix elements belong to some ring. In the original Weyl-Heisenberg group [11] they are real numbers, but here we are more interested in the case that they belong to the ring of integers modulo N. We denote the resulting group by $H(N)$. It is generated by two elements X and Z obeying

$$ZX = qXZ\ , \qquad X^N = Z^N = \mathbf{1}\ , \qquad q = e^{\frac{2\pi i}{N}}\ . \tag{4}$$

For $N = 2$ we can use the Pauli matrices to set $X = \sigma_X$, $Z = \sigma_Z$, which makes it possible to remember the notation. We will consider the group projectively, so for our purposes it can often be regarded as a group of order N^2.

Because q is a primitive Nth root of unity the unitary representation in which Z is diagonal is unique up to permutations [11]. It is known as the clock and shift representation. If the components of any vector are denoted x_a the action is given by

$$\begin{array}{ll} X: & x_0 \to x_{N-1} \to x_{N-2} \to \ldots \to x_1 \to x_0 \\ Z: & x_a \to q^a x_a \end{array} \ , \qquad 0 \leq a \leq N-1\ . \tag{5}$$

The unitary automorphism group of the Heisenberg group plays prominent roles in quantum information theory [12, 13], and is often called the Clifford group. In the older literature the Heisenberg group is sometimes called the Clifford collineation group, and the Clifford group is called the Clifford transform group [14]. Although we will discuss it in detail for the case $N = 3$

later on, we will mostly be concerned with automorphisms of order 2. In the clock and shift representation such an automorphism acts according to

$$A: \quad x_a \leftrightarrow x_{-a} \; . \; . \tag{6}$$

Adding this generator leads us to consider an extended group which is twice as large as $H(N)$. In quantum information language the involution A is generated by one of Wootters' phase point operators [2]. Finally there is the curious conjecture [3] that the SIC vectors are always left invariant by a unitary automorphism of the Heisenberg group having order 3. No one knows why this should be so, but it does appear to be true [15, 6], and in four dimensions we will see exactly how it happens.

What is special about the case when N is prime is that $H(N)$ then admits $N+1$ cyclic subgroups of order N, forming a flower with $N+1$ petals with only the unit element in common. Correspondingly there are $N+1$ eigenbases, and they necessarily form a complete set of MUB [16]. In prime power dimensions $N = p^k$ the known complete set of MUB is the set of eigenbases of the cyclic subgroups of a Heisenberg group defined over a Galois field. The only case we will discuss is when $N = 4$, for which the Galois Heisenberg group is the tensor product $H(2) \otimes H(2)$.

Another piece of background information is that SICs and MUBs look natural in Bloch space, which is the N^2-1 dimensional space of Hermitean operators of trace 1, considered as a vector space with the trace inner product and with the maximally mixed state at the origin. Density matrices form a convex body in Bloch space. A SIC is simply a regular simplex in Bloch space, inscribed into this convex body. But it is not easy to rotate the simplex while keeping the body of density matrices fixed, because the symmetry group of this body is only $SU(N)/Z_N$, a rather small subgroup of $SO(N^2-1)$ as soon as $N > 2$. This is why the SIC problem is hard. An orthonormal basis is a regular simplex with only N corners, spanning some $(N-1)$-plane through the origin in Bloch space. Two bases are mutually unbiased if the corresponding $(N-1)$-planes are totally orthogonal, from which it immediately follows that no more than $N+1$ MUB can exist.

Any pure state corresponds to a Bloch vector of a definite length. Given a complete set of MUB we can project this vector onto the $N+1$ different $(N-1)$-planes defined by the MUB. Should it happen that these projected vectors all have the same length the vector is as it were unbiased with respect to the MUB, and is then—for some reason—called a Minimum Uncertainty State [17, 18]. The condition on a state vector to be unbiased in this sense is easily worked out using the Euclidean metric on Bloch space in conjunction with Pythagoras' theorem. Choose any one of the MUB as the computational basis, and express the Hilbert space components of a unit vector with respect to that basis as

$$x_a = \sqrt{p_a}e^{i\mu_a} \; , \qquad \sum_{n=0}^{N-1} p_a = 1 \; . \tag{7}$$

If the corresponding Bloch vector projected onto the $(N-1)$-plane spanned by the computational basis has the length appropriate to a Minimum Uncertainty State it must be true that

$$\sum_{a=0}^{N-1} p_a^2 = \frac{2}{N+1} \; . \tag{8}$$

This is simple enough, but there is the complication that this has to be done for all the $N+1$ MUB, which will give an additional set of N constraints on the phases ensuring that the vector has the appropriate length when projected to the other MUB planes. We spare the reader from the details, but we repeat that all Heisenberg covariant SIC vectors are Minimum Uncertainty States whenever N is a prime. Examining the proof of this interesting statement shows that something similar is true also when no complete set of MUB is available: In any eigenbasis of a cyclic subgroup of $H(N)$ of order N eq. (8) will hold for any vector belonging to a Heisenberg covariant SIC [8, 9]. This is true regardless of how many bases of this kind there are.

3. The syzygetic Hesse pencil

We now descend to the complex projective plane, and begin by introducing the language used by nineteenth century geometers. Points are represented by ket or column vectors in $\mathbf{C}^3$, or more precisely by one-dimensional subspaces, while lines are represented by two-dimensional subspaces. Using the scalar product in Hilbert space we can equally well represent the lines by bra or row vectors orthogonal to the subspaces they represent, so that the relation

$$\langle Y|X\rangle = 0 \tag{9}$$

means that the point X lies on the line Y. The two-dimensional subspace representing the line consists of all vectors whose scalar product with the bra vector $\langle Y|$ vanishes. Since there is a one-to-one correspondence $|X\rangle \leftrightarrow \langle X|$ between bras and kets there is also a one-to-one correspondence between points and lines. Clearly eq. (9) implies that

$$\langle X|Y\rangle = 0 \ , \tag{10}$$

which says that the point Y lies on the line X. This is known as the duality between points and lines in the projective plane.

We will study complex plane curves defined by homogeneous polynomials in three variables. Linear polynomials define two-dimensional subspaces, that is to say two real-dimensional subsets of the complex plane, and by the above they define projective lines. Intrinsically they are spheres, namely Bloch spheres, because $\mathbf{CP}^1 = \mathbf{S}^2$. Quadratic polynomials or quadrics define conic sections, and over the complex numbers the intrinsic geometry of a conic section is again that of a sphere. The set of spin coherent states is an example [19]. To the next order in complication we choose a cubic polynomial. We require the curve to transform into itself under the Heisenberg group in the clock and shift representation (5). Up to an irrelevant overall constant the most general solution for the cubic is then

$$P = x_0^2 + x_1^2 + x_2^2 + tx_0x_1x_2 \ . \tag{11}$$

Here t is a complex number parametrising what is known as the syzygetic Hesse pencil of cubics. Intrinsically each cubic is a torus rather than a sphere. We observe that the polynomial is automatically invariant under the additional involution A given above in (6).

Hesse [20], and before him Plücker [21], studied this family of curves in detail. Their first object was to determine the inflection points. They are given by those points on the curve for which the determinant of its matrix of second derivatives—its Hessian—vanishes. In the present case this is a cubic polynomial as well; in fact

$$H = \det \partial_i\partial_j P = (6^3 + 2t^3)xyz - 6t^2(x^3 + y^3 + z^3) \ . \tag{12}$$

This is again a member of the Hesse pencil of cubics. In astronomy a "syzygy" occurs when three planets lie on a line, so we can begin to appreciate why the pencil is called "syzygetic". The inflection points are given by $P = H = 0$. By Bézout's theorem two cubics in the complex projective plane intersect in nine points, hence there are nine inflection points. They coincide for all cubics in the pencil, and are given by

$$\begin{bmatrix} 0 & 0 & 0 & -1 & -q & -q^2 & 1 & 1 & 1 \\ 1 & 1 & 1 & 0 & 0 & 0 & -1 & -q & -q^2 \\ -1 & -q & -q^2 & 1 & 1 & 1 & 0 & 0 & 0 \end{bmatrix} \ . \tag{13}$$

This is recognisable as a set of nine SIC vectors covariant under the Heisenberg group [3, 4]. We can normalise our vectors if we want to, but in the spirit of projective geometry we choose not to.

There are four singular members of the Hesse pencil, defined by values of the parameter t such that there are non-zero solutions to $P = P_{,x} = P_{,y} = P_{,z} = 0$. These values are

$$t = \infty \quad \text{and} \quad t^3 = -3^3 \; . \tag{14}$$

If $t = \infty$ the polynomial reduces to $xyz = 0$. In this case the singular cubic consists of three projective lines that make up a triangle. The remaining three singular cases will give rise to three other triangles. Therefore the syzygetic pencil singles out 4 special triangles in the projective plane, given by their 12 vertices

$$\Delta^{(0)} = \begin{bmatrix} 1 & 0 & 0 \\ 0 & 1 & 0 \\ 0 & 0 & 1 \end{bmatrix} , \qquad \Delta^{(1)} = \begin{bmatrix} 1 & q^2 & q^2 \\ q^2 & 1 & q^2 \\ q^2 & q^2 & 1 \end{bmatrix} ,$$
$$\Delta^{(2)} = \begin{bmatrix} 1 & q & q \\ q & 1 & q \\ q & q & 1 \end{bmatrix} , \qquad \Delta^{(\infty)} = \begin{bmatrix} 1 & 1 & 1 \\ 1 & q & q^2 \\ 1 & q^2 & q \end{bmatrix} , \tag{15}$$

where $q = e^{2\pi i/3}$. The columns, labelled consecutively by $0, 1, 2$, can indeed be regarded as 12 points or by duality as 12 lines. The four triangles are referred to as the inflection triangles.

What gives the triangles their name is the remarkable fact that the nine inflection points lie by threes on their twelve edges. Hesse calls this a "*schönen Lehrsatz*", and attributes it to Plücker [21]. It is not hard to verify. After a small calculation one finds that the orthogonalities between the columns in the four triangles and the vectors representing the inflection points are as follows:

	$\Delta_0^{(0)}$	$\Delta_1^{(0)}$	$\Delta_2^{(0)}$	$\Delta_0^{(1)}$	$\Delta_1^{(1)}$	$\Delta_2^{(1)}$	$\Delta_0^{(2)}$	$\Delta_1^{(2)}$	$\Delta_2^{(2)}$	$\Delta_0^{(\infty)}$	$\Delta_1^{(\infty)}$	$\Delta_2^{(\infty)}$
X_0	•			•			•			•		
X_1	•					•		•			•	
X_2	•				•				•			•
Y_0		•			•			•		•		
Y_1		•		•					•		•	
Y_2		•				•	•					•
Z_0			•			•			•	•		
Z_1			•		•		•				•	
Z_2			•	•				•				•

Thus we have

$$\langle \Delta_0^{(0)} | X_0 \rangle = \langle \Delta_0^{(0)} | X_1 \rangle = \langle \Delta_0^{(0)} | X_2 \rangle = 0 \tag{16}$$

and so on. Recalling the interpretation of the vanishing scalar products we see by inspection of the table that Hesse's beautiful theorem is true.

We have verified that there exists a configuration of 9 points and 12 lines such that each point belongs to four lines, and each line goes through three points. This is denoted $(9_4, 12_3)$, and is known as the Hesse configuration. Using the duality between points and lines we have also proved the existence of the configuration $(12_3, 9_4)$. From an abstract point of view such a configuration is a combinatorial object known as a finite affine plane [22]. In the language of quantum information theory the inflection triangles form a complete set of four MUB, while the inflection points form a SIC.

We can now expand on our discussion of group theory in section 2. First, every plane cubic can be regarded as a commutative group in a natural way. This is not surprising, given that the curve is intrinsically a torus—that is a group manifold. The idea relies on Bézout's theorem, which this time assures us that any line intersects the cubic in three points—two of which coincide if the line is a tangent, and all of which coincide if the line is a line of inflection. An arbitrary point on the cubic is taken to be the identity element, and denoted O. To add two arbitrary points A and B on the cubic, draw the line between them and locate its third intersection point P with the cubic. Then draw the line between O and P and again locate the third intersection point C. By definition then $A + B = C$. All the group axioms are obeyed, although it is non-trivial to prove associativity.

Now choose the origin to sit at one of the inflection points. With Hesse's construction in hand one sees that the nine inflection points form a group of order nine, which is precisely the projective Heisenberg group. This is also the torsion group of the curve, meaning that it contains all group elements of finite order. Because they are group elements of order 3 the inflection points are also called 3-torsion points.

Next we ask for the group of transformations transforming the cubics of the Hesse pencil among themselves. Recall that the parameter t in the Hesse cubic (11) can serve as a complex coordinate on a sphere. The four singular members of the pencil defines a tetrahedron on that sphere. Transformations within the pencil act as Möbius transformations on the complex number t. Moreover they must permute the singular members of the Hesse pencil among themselves. This means that they form a well known subgroup of $SO(3)$, namely the symmetry group A_4 of the regular tetrahedron. It enjoys the isomorphism

$$A_4 \sim PSL(2, \mathbf{F}_3) \ , \tag{17}$$

where $\mathbf{F}_3$ is the field of integers modulo 3. The group $SL(2, \mathbf{F}_3)$ consists of unimodular two by two matrices with integer entries taken modulo three; here only its projective part enters because the subgroup generated by the matrix $-\mathbf{1}$ gives rise to the involution A and does not act on t, although it does permute the inflection points among themselves. The full symmetry group of the pencil is a semi-direct product of the Heisenberg group and $SL(2, \mathbf{F}_3)$. This is the affine group on a finite affine plane. It is known as the Hessian group [23], or as the Clifford group.

There are many accounts of this material in the literature, from geometric [24], undergraduate [25], and modern [26] points of view. It forms a recurrent theme in Klein's history of nineteenth century mathematics [27]. The fact that the inflection points form a SIC was first noted by Lane Hughston [10].

4. The elliptic normal curve in prime dimensions

Felix Klein and the people around him put considerable effort into the description of elliptic curves embedded into projective spaces of dimension higher than 2. They proceeded by means of explicit parametrisations of the curve using Weierstrass' σ-function [28, 29]. As far as we are concerned now, we only need to know that the symmetries they built into their curves is again the Heisenberg group supplemented with the involution $x_a \leftrightarrow x_{-a}$ coming from the Clifford group. An analysis of this group of symmetries leads directly to "*une configuration très-remarquable*" originally discovered by Segre [30]. We will present it using some notational improvements that were invented later [31, 9].

Since $N = 2n - 1$ is odd, the integer n serves as the multiplicative inverse of 2 among the integers modulo N. It is then convenient to write the Heisenberg group elements as

$$D(i,j) = q^{nij}X^iZ^j \quad \Rightarrow \quad D(i,j)D(k,l) = q^{n(jk-il)}D(i+k,j+l) = q^{jk-il}D(k,l)D(i,j) \ . \tag{18}$$

Let us also introduce explicit matrix representations of the group generators:

$$D(i,j) = q^{nij+bj}\delta_{a,b+i} \ , \qquad A = \delta_{a+b,0} \ . \tag{19}$$

Note that the spectrum of the involution A consists of n eigenvalues 1 and $n-1$ eigenvalues -1. Hence A splits the vector space into the direct sum

$$\mathcal{H}_N = \mathcal{H}_n^{(+)} \oplus \mathcal{H}_{n-1}^{(-)} \ . \tag{20}$$

It is these subspaces that we should watch. In fact there are altogether N^2 subspaces of dimension n singled out in this way, because there are N^2 involutions

$$A_{ij} = D(i,j)AD(i,j)^\dagger \ . \tag{21}$$

The eigenvectors of the various cyclic subgroups can be collected into the $N+1$ MUB

$$\triangle_{am}^{(k)} = \begin{cases} \delta_{am} & , \quad k = 0 \\ \frac{1}{\sqrt{N}}q^{\frac{(a-m)^2}{2k}} & , \quad 1 \leq k \leq N-1 \\ \frac{1}{\sqrt{N}}q^{am} & , \quad k = \infty \end{cases} \ . \tag{22}$$

Here k labels the basis, m the vectors, and a their components. For $N=3$ this coincides with form (15) given earlier. Note that $N-1$ MUB have been written as circulant matrices, which is a convenient thing to do.

The key observation is that the zeroth columns in the MUB all obey—we suppress the index labelling components—

$$A\triangle_0^{(k)} = \triangle_0^{(k)} \ . \tag{23}$$

Hence this set of $N+1$ vectors belongs to the n-dimensional subspace $\mathcal{H}_n^{(+)}$ defined by the involution A. We can go on to show that each of the n-dimensional eigenspaces defined by the N^2 involutions A_{ij} contain $N+1$ MUB vectors. Conversely, each MUB vector belongs to N subspaces. We have found the Segre configuration

$$\Big(N(N+1)_N, N^2_{N+1}\Big) \tag{24}$$

containing N^2+N points and N^2 $(n-1)$-planes in projective $(N-1)$-space, always assuming that N is an odd prime.

The intersection properties of the Segre configuration are remarkable. Two n-spaces in $2n-1$ dimensions typically intersect in a single ray. With a total of N^2 such subspaces to play with we expect many vectors to arise in this way. But let ψ be such a vector. A minor calculation shows that

$$\psi = A_{ij}A_{kl}\psi = D(2i-2k, 2j-2k)A_{kl}^2\psi = D(2i-2k, 2j-2k)\psi \ . \tag{25}$$

Thus ψ must be an eigenvector of some element in the Heisenberg group, and hence the intersection of any two n-spaces is always one of the $N(N+1)$ eigenvectors in the configuration. In the other direction things are a little more complicated. Two vectors belonging to the same basis are never members of same eigenspace, while two vectors of two different MUB belong to a unique common eigenspace. Using projective duality we obtain the dual configuration

$$\left(N^2_{N+1}, N(N+1)_N\right) \tag{26}$$

consisting of N^2 $(n-1)$-spaces and N^2+N hyperplanes. The intersection properties are precisely those of a finite affine plane [22].

These are the facts that so delighted Segre. A hundred years later they delighted Wootters [2]—although he phrased the discussion directly in terms of the phase point operators A_{ij} rather than in terms of their eigenspaces. A systematic study of prime power dimensions in Segre's spirit appears not to have been made, although there are some results for $N = 9$ [14].

But where is the SIC? It is hard to tell. When the dimension $N = 2n - 1 = 3$ we observe that $n - 1 = 1$, so the dual Segre configuration involves N^2 vectors, and these are precisely the SIC vectors (13). When $N \geq 5$ the Segre configuration contains not even a candidate set of N^2 vectors. But at least, as a byproduct of the construction, we find a set of $2n$ equiangular vectors in any n dimensional Hilbert space such that $2n - 1$ is an odd prime. Explicitly they are

$$\begin{bmatrix} \sqrt{2n-1} & 1 & 1 & \dots & 1 \\ 0 & \sqrt{2} & \sqrt{2}q^{1\cdot 1^2} & \dots & \sqrt{2}q^{(2n-2)\cdot 1^2} \\ 0 & \sqrt{2} & \sqrt{2}q^{1\cdot 2^2} & \dots & \sqrt{2}q^{(2n-2)\cdot 2^2} \\ \vdots & \vdots & \vdots & & \vdots \\ 0 & \sqrt{2} & \sqrt{2}q^{1\cdot(n-1)^2} & \dots & \sqrt{2}q^{(2n-2)(n-1)^2} \end{bmatrix} . \tag{27}$$

Such sets are of some interest in connection with pure state quantum tomography [32].

The elliptic curve itself has not been much in evidence in this section. It is still there in the background though, and in any dimension it contains N^2 distinguished N-torsion points. A study of the explicit expression for the Heisenberg covariant elliptic curve shows that each of its torsion points belong to one of the N^2 eigenspaces $\mathcal{H}^{(-)}_{n-1}$ [29], and with the single exception of the $N = 3$ example (13) the known SICs never sit in such a subspace, so the torsion points are not SICs. This is discouraging, but we will find some consolation when we proceed to examine the $N = 4$ case.

5. The SIC in 4 dimensions

In an $N = 4$ dimensional Hilbert space there is a parting of the ways, in the sense that the MUB and the SIC are defined using two different versions of the Heisenberg group. The elliptic curve stays with $H(4)$. Using an argument concerning line bundles and employing ingredients such as the Riemann-Roch theorem, it can be shown that an elliptic normal curve in projective 3-space (not confined to any projective plane) is the non-singular intersection of two quadratic polynomials. If we insist that it is transformed into itself by the Heisenberg group in its clock and shift representation (5), it follows [29] that these quadratic polynomials are

$$Q_0 = x_0^2 + x_2^2 + 2ax_1x_3 , \qquad Q_1 = x_1^2 + x_3^2 + 2ax_0x_2 . \tag{28}$$

The extra symmetry under the involution A, defined in (6), again appears automatically. We can diagonalise these quadratic forms by means of a unitary transformation of our Hilbert space. In the new coordinates we have

$$Q_0 = z_0^2 + iz_1^2 + a(iz_2^2 + z_3^2) , \qquad Q_1 = iz_2^2 - z_3^2 + a(z_0^2 - iz_1^2) . \tag{29}$$

Note that $Q_0 = Q_1 = 0$ implies

$$z_0^4 + z_1^4 + z_2^4 + z_3^4 = 0 \ . \tag{30}$$

Hence the elliptic curve lies on a quartic surface.

The new basis that we have introduced has a natural interpretation in terms of the involution A. First of all, by acting on A with the Heisenberg group as in eq. (21) we obtain only four involutions altogether, rather than N^2 as in the odd prime case. Their spectra are $(1, 1, 1, -1)$, and in the new basis they are all represented by diagonal matrices. Hence each basis vector is inverted by one involution, and left invariant by the others. In projective 3-space they correspond to four reference points, and one can show that the 16 tangents of the 16 torsion points on the curve divide into 4 sets of 4 each coming together at one of the 4 reference points [29]. Each such set is an orbit under the subgroup of elements of order 2.

In our preferred basis the generators of the Heisenberg group appear in the form

$$Z = e^{\frac{i\pi}{4}} \begin{pmatrix} 0 & 1 & 0 & 0 \\ -i & 0 & 0 & 0 \\ 0 & 0 & 0 & -i \\ 0 & 0 & -1 & 0 \end{pmatrix} , \qquad X = e^{\frac{i\pi}{4}} \begin{pmatrix} 0 & 0 & 1 & 0 \\ 0 & 0 & 0 & 1 \\ -i & 0 & 0 & 0 \\ 0 & i & 0 & 0 \end{pmatrix} . \tag{31}$$

Finding a set of 16 SIC-vectors covariant under the Heisenberg group is now a matter of simple guesswork. One answer, ignoring overall phases and normalisation, is

$$\begin{bmatrix} x & x & x & x & i & i & -i & -i & i & i & -i & -i & i & i & -i & -i \\ 1 & 1 & -1 & -1 & x & x & x & x & i & -i & i & -i & 1 & -1 & 1 & -1 \\ 1 & -1 & 1 & -1 & 1 & -1 & 1 & -1 & x & x & x & x & -i & i & i & -i \\ 1 & -1 & -1 & 1 & -i & i & i & -i & -1 & 1 & 1 & -1 & x & x & x & x \end{bmatrix} , \tag{32}$$

where

$$x = \sqrt{2 + \sqrt{5}} \ . \tag{33}$$

All scalar products have the same modulus because

$$(x^2 - 1)^2 = |x + 1 + i(x - 1)|^2 \ . \tag{34}$$

Thanks to our change of basis, this is significantly more memorable than the standard solutions [3, 4] (and it was in fact arrived at, without considering the Heisenberg group at all, by Belovs [33]). The whole set is organised into 4 groups, where each group sits at a standard distance from the 4 basis vectors that are naturally singled out by the elliptic curve. The normalised vectors obey eq. (8) for a Minimum Uncertainty State, even though our basis is unusual.

The otherwise mysterious invariance of the SIC vectors under some element of the Clifford group of order 3 is now easy to see. We focus on the group of vectors

$$\begin{bmatrix} x & x & x & x \\ 1 & 1 & -1 & -1 \\ 1 & -1 & 1 & -1 \\ 1 & -1 & -1 & 1 \end{bmatrix} . \tag{35}$$

They form an orbit under the subgroup of elements of order 2. When we project them to the subspace orthogonal to the first basis vector we have 4 equiangular vectors in a 3 dimensional subspace. Each projected vector will be invariant under a rotation of order 3 belonging to the symmetry group of this tetrahedron. An example leaving the first vector invariant is

$$R = \begin{pmatrix} 1 & 0 & 0 & 0 \\ 0 & 0 & 0 & 1 \\ 0 & 1 & 0 & 0 \\ 0 & 0 & 1 & 0 \end{pmatrix} . \tag{36}$$

It is straightforward to check that the rotation R belongs to the Clifford group, and is indeed identical to one of "Zauner's unitaries" [3].

Each of the four involutions A admit a "square root" belonging to the Clifford group, such as

$$F = \begin{pmatrix} 1 & 0 & 0 & 0 \\ 0 & 0 & 1 & 0 \\ 0 & 1 & 0 & 0 \\ 0 & 0 & 0 & i \end{pmatrix} \quad \Rightarrow \quad F^2 = A = \begin{pmatrix} 1 & 0 & 0 & 0 \\ 0 & 1 & 0 & 0 \\ 0 & 0 & 1 & 0 \\ 0 & 0 & 0 & -1 \end{pmatrix} . \tag{37}$$

Acting with these unitaries on the SIC (32) will give a set of altogether 16 different SICs, collectively forming an orbit under the Clifford group [3, 15].

Note that the 16 SIC points in projective space do not actually sit on the elliptic curve. In this sense the step from $N = 3$ to $N = 4$ is non-trivial. In an arbitrary even dimension $N = 2n$ the involution A, see (6), has a spectrum consisting of $n+1$ eigenvalues 1 and $n-1$ eigenvalues -1. When $N = 4$ this singles out a unique ray but in higher dimensions this is not so, so generalising to arbitrary even dimension will not be easy.

6. Minimum Uncertainty States in four dimensions

Eddington and his surface have not yet appeared. The group on whose twofold cover his Fundamental Theory hinged was not the Heisenberg group over the ring of integers modulo 4, but a different Heisenberg group of the form $H(2) \otimes H(2)$ [7]. This group can be represented by real matrices, and is in fact the group which gives rise to the complete set of MUB in 4 dimensions. What can we do with it?

There does not exist a SIC which is covariant under Eddington's group. In fact the group $H(2)^{\otimes k}$ admits such an orbit only if $k = 1$ or $k = 3$ [34]. As a substitute we can look for an orbit of 16 Minimum Uncertainty States with respect to the maximal set of MUB. Such an orbit does exist, and is given by the 16 vectors

$$\begin{bmatrix} x & x & x & x & \alpha & \alpha & -\alpha & -\alpha & \alpha & \alpha & -\alpha & -\alpha & \alpha & \alpha & -\alpha & -\alpha \\ \alpha & \alpha & -\alpha & -\alpha & x & x & x & x & \alpha & -\alpha & \alpha & -\alpha & \alpha & -\alpha & \alpha & -\alpha \\ \alpha & -\alpha & \alpha & -\alpha & \alpha & -\alpha & \alpha & -\alpha & x & x & x & x & \alpha & -\alpha & -\alpha & \alpha \\ \alpha & -\alpha & -\alpha & \alpha & \alpha & -\alpha & -\alpha & \alpha & \alpha & -\alpha & -\alpha & \alpha & x & x & x & x \end{bmatrix} , \tag{38}$$

where

$$x = \sqrt{2+\sqrt{5}} , \qquad \alpha = e^{ia} , \qquad \cos a = \frac{\sqrt{5}-1}{2\sqrt{2+\sqrt{5}}} . \tag{39}$$

I omit the lengthy proof that these 16 vectors really are Minimum Uncertainty States [35]. Although this is not a SIC, in a way it comes close to being one. Like the SICs (13) and (32), it can be arrived at using the following procedure: Introduce a vector $(x, e^{i\mu_1}, \ldots, e^{i\mu_{N-1}})^{\mathrm{T}}$, and adjust the value of x so that the normalised vector solves eq. (8) for a Minimum Uncertainty State. Next introduce a complex Hadamard matrix, that is to say a unitary matrix all of whose matrix elements have the same modulus. Such matrices exist in any dimension, although their classification problem is unsolved if the dimension exceeds 5 [36]. By multiplying with an overall factor $\sqrt{N}$, and then multiplying the columns with phase factors, we can ensure that all matrix elements in the first row equal 1. Replace these elements with x. Next multiply the rows with phase factors until one of the columns equals the vector we introduced. The result is a set of N vectors with all mutual scalar products taking the value that characterises a SIC. Next permute the entries of the original vector cyclically, and afterwards try to adjust the phases μ_a so that the resulting N vectors are again equiangular with the mutual scalar products characterising a SIC. Extending the new vectors using an Hadamard matrix in the same way as before then gives N equiangular vectors each of which belongs to a separate group of N equiangular vectors. Before we can say that we have constructed a SIC we must check that all scalar products between pairs of vectors not belonging to the same group take the SIC values. The vectors (38) fail to form a SIC only because the last step fails.

Finally we come back to Eddington's lecture room. In the treatise that he read [37] it is explained that an orbit of $H(2) \otimes H(2)$ gives a realisation of the Kummer configuration 16_6, consisting of 16 points and 16 planes in projective 3-space, such that each point belongs to 6 planes and each plane contains 6 points. The above set of Minimum Uncertainty States realises this configuration. As an example, the 6 vectors

$$\begin{bmatrix} -\alpha & -\alpha & -\alpha & -\alpha & -\alpha & -\alpha \\ x & x & \alpha & -\alpha & \alpha & -\alpha \\ \alpha & -\alpha & x & x & -\alpha & \alpha \\ -\alpha & \alpha & -\alpha & \alpha & x & x \end{bmatrix} \tag{40}$$

are orthogonal to the row vector

$$\begin{pmatrix} x & \alpha & \alpha & \alpha \end{pmatrix} , \tag{41}$$

or in other words the corresponding 6 points belong to the corresponding plane. This is a purely group theoretical property and does not require the vectors to be Minimum Uncertainty States. Still, Eddington's story suggests that our 16 special vectors may have some use, somewhere.

Acknowledgments

I thank Subhash Chaturvedi for telling me about the Segre configuration, at a point in time when neither of us knew about Segre. Both of us give our best wishes to Tony!

References

[1] Ivanović I D 1981 Geometrical description of quantal state determination *J. Phys. A* **14** 3241
[2] Wootters W K 1987 A Wigner-function formulation of finite-state quantum mechanics *Ann. Phys.* **176** 1
[3] Zauner G 1999 *Quantendesigns. Grundzüge einer nichtkommutativen Designtheorie* PhD thesis (Universität Wien)
[4] Renes J M, Blume-Kohout R, Scott A J and Caves C M 2004 Symmetric informationally complete quantum measurements *J. Math. Phys.* **45** 2171
[5] Wootters W K and Fields B D 1989 Optimal state-determination by mutually unbiased measurements *Ann. Phys.* **191** 363
[6] Scott A J and Grassl M 2009 SIC-POVMs: A new computer study *Preprint* arXiv:0910.5784
[7] Eddington A S 1935 *New Pathways in Science* (Cambridge: Cambridge University Press)
[8] Khatirinejad M 2008 On Weyl-Heisenberg orbits of equiangular lines *J. Algebr. Comb.* **28** 333
[9] Appleby D M, Dang H B and Fuchs C A 2007 Physical significance of Symmetric Informationally-Complete sets of quantum states *Preprint* arXiv:0707.2071
[10] Hughston L P 2007 talk at the Perimeter Institute
[11] Weyl H 1932 *Theory of Groups and Quantum Mechanics* (New York: Dutton)
[12] Fivel D I 1995 Remarkable phase oscillations appearing in the lattice dynamics of Einstein-Podolsky-Rosen states *Phys. Rev. Lett.* **74** 835

[13] Gottesman D 1998 A theory of fault-tolerant quantum computation *Phys. Rev. A* **57** 127

[14] Horadam A F 1957 A locus in [8] invariant under a group of order 51840×81 *Quart. J. Math. Oxford* **8** 241

[15] Appleby D M 2005 SIC-POVMs and the extended Clifford group *J. Math. Phys.* **46** 052107

[16] Bandyopadhayay S, Boykin P O, Roychowdhury V and Vatan F 2002 A new proof for the existence of a mutually unbiased bases *Algorithmica* **34** 512

[17] Wootters W K and Sussman D M 2007 Discrete phase space and minimum-uncertainty states *Preprint* arXiv:0704.1277

[18] Appleby D M 2009 SIC-POVMs and MUBs: Geometrical relationships in prime dimensions in *Proceedings of the Växjö Conference on Foundations of Probability and Physics - 5* ed L. Accardi et al (New York: AIP Conf. Proc. 1101)

[19] Brody D C and Hughston L P 2001 Geometric quantum mechanics *J. Geom. Phys.* **38** 19

[20] Hesse O 1844 Über die Wendepuncte der Curven dritter Ordnung *J. Reine Angew. Math.* **28** 97

[21] Plücker J 1835 *System der analytischen Geometrie, auf neue Betrachtungsweisen gegründet* (Berlin: Duncker und Humblot)

[22] Dolgachev I 2004 Abstract configurations in algebraic geometry, in *Proc. of the Fano Conference* (Univ. Torino)

[23] Jordan C 1877 Mémoire sur les équations différentielles linéaires á intégrale algébraique *J. Reine Angew. Math.* **84** 89

[24] Grove C C 1906 *The syzygetic pencil of cubics with a new geometrical development of its Hesse Group,* G_{216} PhD thesis (Johns Hopkins University, Baltimore, Md.)

[25] Gibson C G 1998 *Elementary Geometry of Algebraic Curves* (Cambridge: Cambridge University Press)

[26] Artebani M and Dolgachev I (in press) The Hesse pencil of plane cubic curves *L'Enseignement Mathématique*

[27] Klein F 1926 *Vorlesungen über die Entwicklung der Mathematik im 19. Jahrhundert* (Berlin: Springer)

[28] Bianchi L 1880 Ueber die Normalformen dritter und fünfter Stufe des elliptischen Integrals erster Gattung *Math. Ann.* **17** 234

[29] Hulek K 1986 Projective geometry of elliptic curves *Asterisque* **137** 1

[30] Segre C 1886 Remarques sur les transformations uniformes des courbes elliptiques en elles-mèmes *Math. Ann.* **27** 296

[31] Gross D 2006 Hudson's theorem for finite dimensional quantum systems *J. Math. Phys.* **47** 122107

[32] Flammia S T, Silberfarb A, and Caves C M 2005 Minimal informationally complete measurements for pure states *Found. Phys.* **35** 1985

[33] Belovs A 2008 *Welch Bounds and Quantum State Tomography* Master's Thesis (Univ. of Waterloo)

[34] Godsil C and Roy A 2009 Equiangular lines, mutually unbiased bases, and spin models *Euro. J. Combin.* **30** 246

[35] Bengtsson I and Ericsson Å 2009 (unpublished)

[36] Tadej W and Życzkowski K 2006 A concise guide to complex Hadamard matrices *Open Syst. Inf. Dyn.* **13** 133

[37] Hudson R W H T 1905 *Kummer's Quartic Surface* (Cambridge: Cambridge University Press)

Mutually unbiased bases and semi-definite programming

Stephen Brierley[1] **and Stefan Weigert**[2]

[1] QuIC, Ecole Polytechnique, Universit Libre de Bruxelles, CP 165, 1050 Brussels, Belgium
[2] Department of Mathematics, University of York, York YO10 5DD, UK

E-mail: steve.brierley@ulb.ac.be, stefan.weigert@york.ac.uk

Abstract. A complex Hilbert space of dimension six supports at least *three* but not more than *seven* mutually unbiased bases. Two computer-aided *analytical* methods to tighten these bounds are reviewed, based on a discretization of parameter space and on Gröbner bases. A third algorithmic approach is presented: the non-existence of more than three mutually unbiased bases in composite dimensions can be decided by a *global* optimization method known as semidefinite programming. The method is used to confirm that the *spectral matrix* cannot be part of a complete set of seven mutually unbiased bases in dimension six.

Received: 1 June 2010

This paper is dedicated to Tony Sudbery.

1. Introduction

Two orthonormal bases $\mathcal{B}^0 = \{|\psi_j^0\rangle, j = 1\dots d\}$ and $\mathcal{B}^1 = \{|\psi_j^1\rangle, j = 1\dots d\}$ of $\mathbb{C}^d, d \in \mathbb{N}$, are mutually unbiased (MU) if the modulus of the inner product of any vectors not in the same basis is constant,

$$|\langle\psi_i^0|\psi_j^1\rangle| = \frac{1}{\sqrt{d}}, \quad \text{for all} \quad i, j = 1\dots d. \tag{1}$$

In this paper, we will examine three computer-aided analytical methods which could be used to prove the conjecture that no more than three MU bases exist in the space $\mathbb{C}^6$ [1]. This open problem concerning the geometry of the state space of a low-dimensional quantum system has attracted considerable interest in recent years; see [2] for a thorough review of the properties and uses of MU bases in complex Hilbert spaces of dimension d.

First, we review and illustrate an approach which is based on *discretising the parameter space* when searching for vectors that constitute MU bases. This simplification results in only a *finite*—but possibly large—number of states which must be checked. The results obtained will be exact once rigorous error bounds have been established [3]. Second, we show that the constraints defining an MU constellation can be expressed in terms of coupled polynomial equations. Thus, constructing a *Gröbner basis* may show that they have no solution. Finally, we propose a new approach by writing the problem in a form suitable to use *semidefinite programming.* This allows one to algorithmically identify the (non-) existence of solutions by means of global optimization techniques.

We will illustrate each of the proposed methods by reproducing known results for MU bases in low dimensions [4]. For the sake of simplicity, each of the algorithms will be used to prove the non-existence of four MU bases in dimension two. All three techniques have yet to be applied successfully to the problem in dimension six: the *computational* resources required to implement them were beyond those available.

Throughout this paper, we will use the concept of *MU constellations* [5]: the MU constellation $\{d-1,\lambda,\mu,\nu\}_d$ with $0 \leq \lambda,\mu,\nu \leq d-1$, for example, consists of four sets of orthonormal vectors containing $d-1,\lambda,\mu$, and ν vectors, respectively, with vectors taken from different sets having the required overlap $1/\sqrt{d}$ (cf. Eqs. (1)). It is sufficient to list at most $(d-1)$ vectors in each set since a suitable d^{th} vector can be constructed from them.

The non-existence of any MU constellation smaller than $\{5,5,5,5\}_6 \equiv \{5^4\}_6$ proves the non-existence of four (or more) MU bases in dimension six. The smallest number of vectors sufficient to parameterize a MU constellation $\{d-1,\lambda,\mu,\nu\}_d$ is given by $s = \lambda+\mu+\nu-1$ (the underlying equivalence relations between MU constellations can be found in Appendix A of [5]). The number s effectively determines the number of variables and equations needed to define a MU constellation and, *a fortiori*, the computational resources required to run any algorithm solving the problem. For example, the MU constellation $\{5,3,3,3\}_6 \equiv \{5,3^3\}_6$ contains $s=8$ free vectors depending on 40 real parameters, whereas the full set of seven MU bases in dimension six involves $s=29$ vectors depending on 145 real parameters. In order to reduce the problem as much as possible, it seems imperative to work with smallest possible constellations such as $\{5,3^3\}_6$ or $\{5^2,4,1\}_6$ which have *not* been found by the extensive numerical searches presented in [5].

2. Discretizing the parameter space

One of the few rigorous results about MU bases in $\mathbb{C}^6$ is due to Jaming et al. [3]. They have shown that a particular class of matrices, the so-called *Fourier family* [6], cannot be a member of a quadruple of MU bases. The Fourier family consists of (6×6) matrices $F(x_1,x_2)$ depending on two continuous parameters x_1 and x_2; for $x_1=x_2=0$, the general expression reduces to the standard Fourier matrix $F_{jk}=\omega^{jk}/\sqrt{6}, j,k=0\ldots 5$, where $\omega=\exp(2\pi i/6)$. We now describe the idea of this approach which, as outlined in [7], could lead to a no-go theorem in composite dimensions such as $d=6$.

Let us assume that there exists a MU constellation C of the form $\{d-1,\lambda,\mu,\nu\}_d$; we will attempt to find a contradiction. The constellation C can be parameterised by p_d phases $\alpha \equiv (\alpha_1,\ldots,\alpha_{p_d})^T$, according to Eq. (8) of [5]. The first step is to *approximate* this MU constellation by another set of vectors with components being restricted to N^{th} roots of unity only. Such an approximation is easily achieved by dividing the interval $[0,2\pi)$ into N non-overlapping intervals $I_j=[(2j-1)\pi/N,(2j+1)\pi/N) \pmod{2\pi}$, with $j=1,\ldots,N$. If the phase α_k lies in the interval I_j, we approximate it by the mid-point of I_j, $\alpha_k \to \widetilde{\alpha}_{k,j} \equiv 2\pi j_k/N$. The mapping $\alpha \to \widetilde{\alpha}$ thus sends the vectors of a given MU constellation C to a different collection of vectors (denoted by $\widetilde{C}$) which do not necessarily satisfy the MU conditions (1). Effectively, the p_d-dimensional continuous parameter space α has been replaced by a grid consisting of a *finite* number of N^{p_d} states. The accuracy of the approximation of the MU constellation C by $\widetilde{C}$ is determined by the value of N, and additional flexibility results from partitioning the range of each variable α_k individually.

The key step forward made by Jaming et al. [3] has been to establish rigorous bounds on the errors of the scalar products introduced by the discretization. These bounds allow one to conclude that no MU constellation exists if none can be found for a sufficiently accurate discrete approximation. Thus, the search for a contradiction has effectively been reduced to checking a *finite* set of constellations which can be done exhaustively, in principle. If N is small, the search is easy to perform; however, the error bounds are not tight for small values of N. Finer partitions are necessary which, in turn, come at a computational cost since the number of grid points is proportional to N^{p_d}. A hierarchical refinement of the procedure may be used to reduce the amount of computational resources [7].

2.1. Dimension two

It is not difficult to illustrate this algorithm by searching for the MU constellation $\{1^4\}_2$, i.e. a set of four MU bases in the space $\mathbb{C}^2$—which is known *not* to exist. If the constellation $\{1^4\}_2$ existed, the four vectors defining it could be written in the form

$$\begin{pmatrix} 1 \\ 0 \end{pmatrix}, \frac{1}{\sqrt{2}}\begin{pmatrix} 1 \\ 1 \end{pmatrix}, \frac{1}{\sqrt{2}}\begin{pmatrix} 1 \\ e^{i\alpha} \end{pmatrix}, \frac{1}{\sqrt{2}}\begin{pmatrix} 1 \\ e^{i\beta} \end{pmatrix}, \tag{2}$$

depending on $p_2 = 2$ real parameters $\alpha(\equiv \alpha_1)$ and $\beta(\equiv \alpha_2)$. The conditions resulting from Eqs. (1) lead to three constraints,

$$\left|\cos\frac{\alpha}{2}\right| = \left|\cos\frac{\beta}{2}\right| = \left|\cos\frac{\alpha-\beta}{2}\right| = \frac{1}{\sqrt{2}} \tag{3}$$

and it is obvious that they have no solution.

To see the discretization procedure in action, we now show the inconsistency of Eqs. (3) by checking only a finite number of cases. Assume that there are values of α and β which give rise to a solution of Eqs. (3), and write them as $\alpha = \alpha_j + \Delta\alpha_j$ and $\beta = \beta_{j'} + \Delta\beta_{j'}$, where α_j is the Nth root of unity closest to α and differing by $\Delta\alpha_j$ from it, etc. The maximum distance between the actual value of α and an Nth root is π/N. Using Taylor's theorem for a differentiable function $f(x)$,

$$f(x+h) = f(x) + hf'(x+\theta(x)h), \quad \text{for some } \theta(x) \in (0,1), \tag{4}$$

the condition $|\cos(\alpha/2)| = 1/\sqrt{2}$ implies that

$$\frac{1}{\sqrt{2}} - \frac{\pi}{2N} \leq \left|\cos\frac{\alpha_j}{2}\right| \leq \frac{1}{\sqrt{2}} + \frac{\pi}{2N}, \tag{5}$$

where the triangle inequality

$$|a_1| - |a_2| \leq |a_1 + a_2| \leq |a_1| + |a_2|, \tag{6}$$

and the relations $|\Delta\alpha_j| \leq \pi/N$ have been used. The other two constraints lead to inequalities which are obtained by substituting either $\beta_{j'}$ or $\alpha_j - \beta_{j'}$ for α_j in (5). Thus, one needs to check N^2 inequalities corresponding to the indices $j, j' = 1 \ldots N$. For increasing values of N, the numbers $\alpha_j, \beta_{j'}$, *and* $\alpha_j - \beta_{j'}$ must approach odd multiples of $\pi/2$ which is impossible. Thus, for a sufficiently large *finite* value of N, the inequalities have no solutions in terms of Nth roots of unity and in some intervals centered around them—thus covering the *entire* parameter space.

2.2. Dimension six

Jaming et al. have used this method to exclude the entire Fourier family, $F(x_1, x_2)$, from a complete set of MU bases [3]. This was achieved by discretising the fundamental region of the Fourier family using $N = 180$ and two other complete bases using $N' = 19$. The restriction to consider sets of the form $\{I, F(x_1, x_2), \mathcal{B}^2, \mathcal{B}^3\}$ with two orthonormal bases $\mathcal{B}^2$ and $\mathcal{B}^3$, reduces the computational complexity considerably but it is not dictated by an inherent limitation of the method. Interestingly, the non-existence of a finite projective plane was shown by an exhaustive search [8]. Since the existence of a complete set of MU bases shares some properties with the existence of finite projective planes [9], discretization appears a promising avenue.

3. Using Gröbner bases

The second method we review relies on a technique to algorithmically search for solutions of coupled polynomial equations. The approach is based on the construction of a *Gröbner basis* [10] for these equations, a tool developed in commutative algebra. It has been applied successfully in [11] to show that no two MU bases of a particular form can be supplemented by a third basis plus a single vector which is MU to those of the three bases. Further rigorous results excluding many candidate bases from a complete set of MU bases have been reported in [12]. We first formulate the approach for arbitrary dimensions d, then work through an explicit example in the space $\mathbb{C}^2$, and finally discuss the prospects of applying this approach in dimension six.

Any MU constellation of type $\{d-1,\lambda,\mu,\nu\}_d$ will correspond to a point on a $s(d-1)$-dimensional hypertorus, with $s=\lambda+\mu+\nu-1$ [5]. In a parameterization analogous to the one given in (2), only phase factors of modulus one each depending on a single real parameter will occur. Upon expressing each phase factor as $e^{i\alpha_j}=x_j+iy_j$, supplemented by the conditions $x_j^2+y_j^2=1$, we are led to a parameterization which requires $2s(d-1)$ real variables $\mathbf{x}\equiv(x_1,x_2,\ldots,y_{2s(d-1)})\in\mathbb{R}^{2s(d-1)}$.

The conditions for vectors, orthonormal in sets with $d-1,\lambda,\mu$ and ν elements, to be MU now turn into N multivariate polynomial equations,

$$p_j(\mathbf{x})=0\,,\qquad j=1\ldots N\,, \tag{7}$$

with

$$N=\frac{1}{2}(s+1)(s-1)+\frac{1}{2}(\lambda^2+\mu^2+\nu^2)+s(d-1)\,. \tag{8}$$

Any solution of the Eqs. (7) gives rise to a MU constellation of type $\{d-1,\lambda,\mu,\nu\}_d$. Geometrically speaking, we wish to describe the variety

$$V\equiv\{\mathbf{x}\in\mathbb{R}^{2s(d-1)}:p_1(\mathbf{x})=0,\ldots,p_N(\mathbf{x})=0\}\,, \tag{9}$$

which is a subset of the space $\mathbb{R}^{2s(d-1)}$. No MU constellation of the form $\{d-1,\lambda,\mu,\nu\}_d$ exists if and only if the associated variety is empty, $V=\emptyset$. The set V has an algebraical description in terms of an ideal $I=\langle p_1,\ldots,p_N\rangle$ which is generated by the polynomials $p_1(\mathbf{x})$ to $p_N(\mathbf{x})$. This ideal I consists of all linear combinations a of the polynomials $p_j(\mathbf{x})$ with coefficients r_j polynomial in the variables $\mathbf{x}$,

$$a=\sum_{j=1}^{S}r_j(\mathbf{x})p_j(\mathbf{x})\,. \tag{10}$$

It is important to note that the ideal corresponds to the variety over the algebraic closure of the coefficient field, here the *complex* numbers.

Having re-cast the problem of identifying constellations of MU vectors in terms of ideals, it becomes possible to apply methods from commutative algebraic geometry. The equations (7) have no (real or complex-valued) solutions if the polynomial 1 is contained in the ideal $\langle p_1,\ldots,p_N\rangle$. The construction of a Gröbner basis G of the ideal I would allow us to check this property since G would be given by the set $G=\{1\}$. The converse is not necessarily true: the equations defining a MU constellation may have no solutions over the real numbers but the ideal is non-empty due to the existence of complex solutions. The equation $x^2+1=0$, for example, has no *real* solutions but the ideal $I=\langle x^2+1\rangle$ is not generated by $\{1\}$. In a fortunate situation, the variety is found to be empty over the complex numbers which would then constitute a proof that a complete set of MU bases does not exist in dimension six.

3.1. Dimension two

We now confirm the validity of the algorithm based on Gröbner bases by proving (again) that no four MU bases exist in dimension two. First, we write down the system of coupled polynomial equations, the solutions to which would define the MU constellation $\{1^4\}_2$. *Four* real variables $\{x_1, x_2, y_1, y_2\}$ parameterize the candidate vectors shown in Eq. (2),

$$\begin{pmatrix} 1 \\ 0 \end{pmatrix}, \frac{1}{\sqrt{2}}\begin{pmatrix} 1 \\ 1 \end{pmatrix}, \frac{1}{\sqrt{2}}\begin{pmatrix} 1 \\ x_1 + iy_1 \end{pmatrix}, \frac{1}{\sqrt{2}}\begin{pmatrix} 1 \\ x_2 + iy_2 \end{pmatrix}, \tag{11}$$

supplemented by the constraints

$$\begin{aligned} p_1(\mathbf{x}) &\equiv x_1^2 + y_1^2 - 1 = 0\,, \\ p_2(\mathbf{x}) &\equiv x_2^2 + y_2^2 - 1 = 0\,. \end{aligned} \tag{12}$$

The constraints defining a MU constellation $\{1^4\}_2$ read explicitly

$$\begin{aligned} p_3(\mathbf{x}) &\equiv (1+x_1)^2 + y_1^2 - 2 = 0\,, \\ p_4(\mathbf{x}) &\equiv (1+x_2)^2 + y_2^2 - 2 = 0\,, \\ p_5(\mathbf{x}) &\equiv (1 + x_1x_2 + y_1y_2)^2 + (x_1y_2 - x_2y_1)^2 - 2 = 0\,. \end{aligned} \tag{13}$$

It is straightforward to derive that these equations have no solution, a fact which, for the purpose of illustration, we will confirm by showing that they have $G = \{1\}$ as a Gröbner basis.

We have constructed the Gröbner basis for Eqs. (12,13) using a package called FGb [13] implemented in Maple [14]. A standard desktop PC outputs the desired Gröbner basis in 0.016 seconds, and it is indeed equal to all polynomials over the complex numbers,

$$\langle p_1(\mathbf{x}), \ldots, p_5(\mathbf{x})\rangle = \langle 1 \rangle. \tag{14}$$

A direct confirmation of this result follows from writing down the coefficient polynomials used to construct the Gröbner basis, namely

$$\begin{aligned} r_1(\mathbf{x}) &\equiv -\frac{1}{2}(y_1y_2 + x_1 + 2)\,, \\ r_2(\mathbf{x}) &\equiv -\frac{1}{2}\left(x_1^2y_2 + y_1^2y_2 - x_1y_2 + x_2y_1 + 2y_1\right)y_1\,, \\ r_3(\mathbf{x}) &\equiv \frac{1}{2}x_1\,, \\ r_4(\mathbf{x}) &\equiv -\frac{1}{2}(x_1y_2 - x_2y_1)\,y_1\,, \\ r_5(\mathbf{x}) &\equiv \frac{1}{2}y_1y_2\,. \end{aligned} \tag{15}$$

It is not difficult to verify that the following relation holds,

$$\sum_{j=1}^{5} r_j(\mathbf{x})p_j(\mathbf{x}) = 1\,, \tag{16}$$

which establishes that $G \equiv \{1\}$. It follows that the variety $V = \{\mathbf{x} \in \mathbb{R}^4 : p_1(\mathbf{x}) = 0, \ldots, p_5(\mathbf{x}) = 0\}$ is empty implying that no four MU bases exist in the space $\mathbb{C}^2$.

Journal of Physics: Conference Series **254** (2010) 012008 doi:10.1088/1742-6596/254/1/012008

3.2. Dimension six

Unfortunately, 16GB of memory have not been sufficient to decide if the Gröbner bases corresponding to the equations generated by the constellations $\{5, 3^3\}_6$ or $\{5^2, 4, 1\}_6$ contain the element 1. The computations run out of memory *before* the algorithm terminates since the number of variables and equations is considerable, even for these smallest possible interesting constellations in dimensions six. For example, the constellation $\{5^2, 4, 1\}_6$ requires the construction of a Gröbner basis of an ideal generated by 61 equations of degree 4 in 90 real variables.

4. Using semidefinite programming

Semidefinite programming [15] is a powerful tool to obtain rigorous results aided by a computer which has been used in areas such as control theory and combinatorial optimization [16]. Recent years have seen first applications in quantum information theory to solve problems such as deciding whether a given mixed state ρ is entangled or not [17] - [20]. The success of semidefinite programming stems from the fact that the type of problems studied are *efficiently* solvable by a computer. In addition, there is a duality theorem which provides a *certificate* allowing one to directly check the result. For example, when applied to the separability problem, a semidefinite program decides whether a given mixed state is entangled—if it is, it outputs an *entanglement witness* [21], that is, a hyperplane which separates the entangled state at hand from the set of all separable states.

Another interesting application of semidefinite programming to problems about finite dimensional Hilbert spaces is to the *compatibility problem* for tripartite quantum systems. Here one asks if there exists a single state of the entire system given the states of all (proper) reduced states. Hall [22] has cast the compatibility problem in the form a semidefinite program and used it to disprove a conjecture of Butterley et al. [23]. In this case, the solution to the dual problem resulted in a certificate, called an *incompatibility witness*, which proves that a particular set of reduced states are not compatible with any multipartite state.

A *semidefinite program* (SDP) is an algorithm for solving an optimization problem of the form

$$\begin{aligned} &\text{minimise } \mathbf{c}^T\mathbf{x} \\ &\text{subject to } F(\mathbf{x}) \geq 0 \end{aligned} \tag{17}$$

where $\mathbf{c}$ is a fixed vector, and the *decision variables* $\mathbf{x}$ are constrained by the requirement that the matrix $F(\mathbf{x}) \equiv x_1 F_1 + \cdots + x_n F_n - F_0$ be positive semidefinite, with symmetric $n \times n$ matrices $F_0, F_1, \ldots, F_n$ [15]. This is a *convex* optimization problem. A linear program, where the constrains have the form $F(\mathbf{x}) = \text{diag}(\mathbf{A}\mathbf{x} - \mathbf{b})$, is an example of a SDP with important applications in economics [24] or finding optimal network flows [25].

A SDP can also be used to solve *non-linear* problems as long as they are convex. For example, the problem

$$\begin{aligned} &\text{minimise } \frac{(\mathbf{c}^T\mathbf{x})^2}{\mathbf{d}^T\mathbf{x}} \\ &\text{subject to } A\mathbf{x} \geq \mathbf{b} \end{aligned} \tag{18}$$

can be re-written as a SDP [15],

$$\begin{aligned} &\text{minimise } t \\ &\text{subject to } \begin{pmatrix} \text{diag}(\mathbf{A}\mathbf{x} - \mathbf{b}) & 0 & 0 \\ 0 & t & \mathbf{c}^T\mathbf{x} \\ 0 & \mathbf{c}^T\mathbf{x} & \mathbf{d}^T\mathbf{x} \end{pmatrix} \geq 0 \,. \end{aligned} \tag{19}$$

It is possible to cast the existence problem of MU constellations as a semidefinite program [26]. The equations $p_j(\mathbf{x}) = 0$ defining MU constellations are, unfortunately, not convex since they involve fourth-order polynomials. However, all is not lost: it is, possible to extend the applicability of this method by using tools from *non-convex optimization* [27, 28]. Lasserre has shown that, upon relaxing the non-convex constraints, one can define a *hierachy* of semidefinite programs representing ever better approximations to the original one [29], at the cost of introducing additional decision variables. At each level, this *method of relaxations* either rules out the existence of a MU constellation or is inconclusive requiring an additional step of relaxation. Each iteration inevitably leads to a computationally more difficult problem but the remarkable work of Lasserre [29] ensures that after a finite number of iterations the *exact* solution will have been obtained. In other words, the hierarchy is *asymptotically complete.* It is this type of approach which has been applied successfully to problems from quantum information [30].

Let us now show how to express the existence problem of a MU constellation as an optimization problem. To do so,

> *pick any one of the polynomials from (7), say $p_1(\mathbf{x})$, and find the minimum value of $p_1^2(\mathbf{x})$ subject to the condition that the variables satisfy the remaining constraints, $p_2(\mathbf{x}) = 0, \ldots, p_N(\mathbf{x}) = 0$.*

Effectively, we seek to minimise the value of $p_1^2(\mathbf{x})$ in the solution space of the remaining polynomials. Lasserre's method of relaxations then allows one to find a lower bound $B_L(r)$ for the function $p_1^2(\mathbf{x})$, where r is the degree of the relaxation. If for some degree of relaxation, $r = 2, 3, \ldots$ one finds a *positive* bound, $B_L(r) > 0$, then *no* MU constellation of the desired type exists. The dual formulation of the SDP would automatically provide a certificate of non-existence, offering an independent way of verifying the result. However, the original problem is transformed by the relaxation steps and so this "witness of non-existence" is less intuitively connected to the existence of an MU constellation than, for example, an entanglement witness.

Here is a sketch of the algorithm which is capable to determine whether a MU constellation $\{d-1, \lambda, \mu.\nu\}_d$ exists:

(i) write down the polynomial equations which define the MU constellation;

(ii) generate a SDP at the lowest possible level of relaxation ($r = 2$);

(iii) solve the resulting SDP—if a positive global lower bound $B_L(r)$ is found then no MU constellation $\{d-1, \lambda, \mu.\nu\}_d$ exists; otherwise, repeat Steps (ii) and (iii) at the next level of relaxation, $r := r + 1$.

If no MU constellation exists, the algorithm is guaranteed to find a positive lower bound. As r increases, the global lower bounds $B_L(r)$ converge monotonically to the exact global minimum of the function, B_L^{opt}. If the sought-after MU constellation does exist, the algorithm will find an explicit parameterisation thereof, with a high level of numerical accuracy.

4.1. Dimension two

To begin, we now spell out and run a SDP algorithm to show once more that that there are no more than three MU bases in dimension two. More explicitly, we will calculate a *strictly positive* global lower bound $B_L(r)$ for the polynomial

$$p_1^2(\mathbf{x}) = \left(x_1^2 + y_1^2 - 1\right)^2 , \tag{20}$$

subject to the requirement that the remaining polynomials in Eqs. (12) and (13) vanish (Step (i)). This leaves us with the following minimization problem,

$$\begin{array}{ll} \min & \left(x_1^2+y_1^2-1\right)^2 \\ \text{subject to} & x_2^2+y_2^2-1=0 \\ & (1+x_1)^2+y_1^2-2=0 \\ & (1+x_2)^2+y_2^2-2=0 \\ & (1+x_1x_2+y_1y_2)^2+(x_1y_2-x_2y_1)^2-2=0 \end{array} \tag{21}$$

with four decision variables $\mathbf{x}=(x_1,\ldots,y_2)$. The MU constellation $\{1^4\}_2$ exists if and only if no global lower bound of $p_1^2(\mathbf{x})$ exceeds the value zero, or $B_L(r)\leq 0$ for all r. Any *positive* bound implies the non-existence of a MU constellation $\{1^4\}_2$.

Using the Matlab package **gloptipoly3** [31] which is based on the theory presented in [32], we are able to convert the problem (21) into a semidefinite program, thus completing Step (ii). The resulting SDP has been solved using the **SeDuMi** MatLab package developed by Strum et al. [33], corresponding to Step (iii). The results of the computations at three levels of relaxation, $r=2,3,4$, are presented in Table 1. We find that already the first level of relaxation, $r=2$, provides a positive lower bound for the polynomial $p_1^2(\mathbf{x})$, $B_L(2)>0$. This result confirms that semidefinite programming is capable to positively identify non-existing MU constellations.

The third and fourth columns of Table 1 show how the size of the SDP grows as we increase the level of relaxation. The number of decision variables, N_d, jumps from 69 to 494. Similarly, the dimension of matrices defining the semidefinite constraint $F(\mathbf{x})\geq 0$ is almost five times larger at $r=4$ than at the lowest level of relaxation, $r=2$. The algorithm proves to be very efficient taking only 0.11 seconds on a desktop PC to convert the original problem and solve the SDP for $r=2$. This time rises to 1.67 seconds when $r=4$.

r	B_L	N_d	F
2	1.4038×10^{-8}	69	15×15
3	0.5359	209	35×35
4	0.5359	494	70×70

Table 1. Lower bounds of the minimization problem defined in Eq. (21), with levels of relaxation denoted by r, while N_d and F give the number of decision variables and the size of the semidefinite inequalities in the resulting SDP, respectively.

The high level of numerical accuracy achieved by the optimization program **SeDuMi** allows us, in fact, to find an analytic expression for the lower bound, namely $B_L^{opt}=(1-\sqrt{3})^2$. At the third and fourth levels of relaxation, the lower bound is optimal in that the function $p_1^2(\mathbf{x})$ reaches the value B_L. It is possible to output the parameter values $\mathbf{x}$ which achieve this lower bound. There are two sets of vectors corresponding to the global minimum value $B_L^{opt}=(1-\sqrt{3})^2$ given by

$$V_\pm=\left\{\begin{pmatrix}1\\0\end{pmatrix},\frac{1}{\sqrt{2}}\begin{pmatrix}1\\1\end{pmatrix},\frac{1}{\sqrt{2}}\begin{pmatrix}1\\\pm i\end{pmatrix},\frac{1}{\sqrt{2}}\begin{pmatrix}1\\\alpha(1\mp i)\end{pmatrix}\right\}, \tag{22}$$

where $\alpha=(\sqrt{3}-1)/2$. Interestingly, the sets V_+ and V_- both contain three MU bases plus one additional vector.

4.2. Dimension six

According to Eq. (8) the number of variables parameterising candidates for MU constellation increases quadratically with the number of undetermined vectors. Thus, a successful application of semidefinite programming in dimension two does not guarantee that a similar algorithm will terminate in dimension six. As for Gröbner bases, finite computational resources may limit the type of problems which can be solved. We will consider two MU constellations in $\mathbb{C}^6$ with small and large numbers of variables, respectively.

The columns of the *spectral matrix* [34]

$$S = \frac{1}{\sqrt{6}} \begin{pmatrix} 1 & 1 & 1 & 1 & 1 & 1 \\ 1 & 1 & \omega & \omega & \omega^2 & \omega^2 \\ 1 & \omega & 1 & \omega^2 & \omega^2 & \omega \\ 1 & \omega & \omega^2 & 1 & \omega & \omega^2 \\ 1 & \omega^2 & \omega^2 & \omega & 1 & \omega \\ 1 & \omega^2 & \omega & \omega^2 & \omega & 1 \end{pmatrix}, \tag{23}$$

define an orthonormal basis of $\mathbb{C}^6$ since S is unitary. It has been shown that the pair $\{I, S\}$ cannot be extended to a triple of MU bases [12] (the identity I is associated with the standard basis of $\mathbb{C}^6$). Furthermore, there is no pair of orthogonal vectors $\{|u\rangle, |v\rangle\}$ MU to the column vectors of I and S. The two undetermined vectors $\{|u\rangle, |v\rangle\}$ depend on 20 real variables which must satisfy 21 constraints given by 4th order polynomials. Due to the relatively small number of variables involved, one may expect that this property can be checked using semidefinite programming. Upon transforming the original problem, the equivalent SDP has been found to require $10,625$ decision variables satisfying $4,851$ linear constraints while the semidefinite inequalities involve matrices of dimension 231×231—at the lowest level of relaxation, $r = 2$. It took approximately 210 minutes and 5.4G of memory for the SDP to terminate, resulting in a global lower bound for the square of one of the polynomials defined by the MU conditions,

$$B_L(2) = 2.28 \times 10^{-8}\,. \tag{24}$$

Since $B_L(2)$ is *positive*, the SDP algorithm confirms the result of [12]: no two orthonormal vectors can be found which are mutually unbiased to the column vectors of the pair $\{I, S\}$. At the next level of relaxation, $r = 3$, the resulting SDP is already too large to terminate: there are $230,229$ variables while the constraints involve semidefinite matrices of dimension 1771×1771. Consequently, we have not been able to improve the lower bound (24).

It should not come as a surprise our attempts to construct the SDP for the constellation $\{5^3, 1\}_6$ have not been successful, even at the lowest level of relaxation. Judging by the increase in the number of decision variables and the size of the semidefinite constraints seen in the previous examples, it is likely to be very large. Solving the resulting SDP would seem optimistic. However, it is possible that the resulting SDP has some specific structure which would allow one to use appropriate techniques.

5. Summary and outlook

We have reviewed two known computer-aided methods suitable to obtain *rigorous* bounds on the number of MU bases existing in the space $\mathbb{C}^d$, based on a discretization of parameter space and on Gröbner bases. Both methods are currently too costly from a computational point of view to decide on the (non-) existence of more than three MU bases in dimension six.

In addition, we have rephrased the problem in terms of semidefinite programming which allows one to use rigorous methods of *global optimization*. This approach has been shown to reproduce two known results, the non-existence of four MU bases in dimension two, and the fact that the spectral matrix S cannot be part of a complete set of seven mutually unbiased bases in dimension six.

Seemingly innocent questions about sets of mutually unbiased vectors in low-dimensional Hilbert spaces such as $\mathbb{C}^6$ continue to resist a variety of rigorous attempts of solution, confirming once more the dictum [35]

Hilbert space is a big place.

Interestingly, some MU constellations in dimension six are just within reach of *numerical* methods [5] which, however, only make their non-existence plausible. It remains to be seen whether a proof of the (non-) existence of a MU constellation such as $\{5^3, 1\}_6$ will be found first by a computer-aided approach or by a purely analytic reasoning. It is safe to say though that obtaining a proof will be a matter of time only: the algorithmic methods presented here ascertain that we do *not* deal with an *undecidable* question when searching for complete sets of MU bases in composite dimensions.

References

[1] Zauner G 1999 *Quantendesigns. Grundzüge einer nichtkommutativen Designtheorie* PhD thesis (Univ. of Wien)
[2] Durt T, Englert B-G, Bengtsson I and Życzkowski K 2010 On mutually unbiased bases *Int. J. Quant. Info.* **8** 535 (*Preprint* arXiv:1004.3348)
[3] Jaming P, Matolcsi M, Mora P, Szöllősi F, and Weiner M 2009 A generalized Pauli problem and an infinite family of MUB-triplets in dimension 6 *J. Phys. A* **42** 245305 (*Preprint* arXiv:0902.0882)
[4] Brierley S, Weigert S and Bengtsson I 2009 All mutually unbiased bases in dimensions two to five *Int. J. Quant. Info.* **10** 803 (*Preprint* arXiv:0907.4097)
[5] Brierley S and Weigert S 2008 Maximal sets of mutually unbiased quantum states in dimension six *Phys. Rev. A* **78** 042312 (*Preprint* arXiv:0808.1614)
[6] Haargerup U 1996 Orthogonal maximal abelian *-subalgebras of the $n \times n$ matrices and cyclic n-roots *Proc.: Operator Algebras and Quantum Field Theory* (Rome) ed S Doplicher 296
[7] Matolcsi M 2009 private communication
[8] Lam C, Thiel L and Swiercz S 1989 The non-existence of finite projective planes of order 10 *Can. J. Math.* **41** 1117
[9] Saniga M, Planat M and Rosu H 2004 Mutually unbiased bases and finite projective planes *J. Opt. B* **6** L19 (*Preprint* arXiv:math-ph/0403057)
[10] Buchberger B 1965 PhD thesis (Univ. of Innsbruck); English translation: 2006 *J. Symb. Comput.* **41** 471
[11] Grassl M 2004, On SIC-POVMs and MUBs in dimension 6 *Proc.: ERATO Conference on Quantum Information Science* ed J. Gruska (Tokyo) p 60 (*Preprint* arXiv:quant-ph/0406175)
[12] Brierley S and Weigert S 2009 Constructing mutually unbiased bases in dimension six *Phys. Rev. A* **79** 052316 (*Preprint* arXiv:0901.4051)
[13] Faugère J-C 1999 A new efficient algorithm for computing Gröbner bases (f4) *J.Pure Appl. Algebra* **139** 61
[14] Maple 11 Waterloo Maple Inc. Waterloo, Ontario, Canada.
[15] Vandenberghe L and Boyd S 1996 Semidefinite programming *SIAM Review 38* **1** 49
[16] Vandenberghe L and Boyd S 1999 Applications of semidefinite programming *Appl. Num. Math.* **29** 283
[17] Brandão F and Vianna R 2004 A robust semidefinite programming approach to the separability problem *Phys. Rev. A* **70** 062309 (*Preprint* arXiv:quant-ph/0405008)
[18] Brandão F and Vianna R2004 Separable multipartite mixed states - operational asymptotically necessary and sufficient conditions *Phys. Rev. Lett.* **93** 220503 (*Preprint* arXiv:quant-ph/0405063)
[19] Doherty A, Parrilo P and Spedalieri F 2002 Distinguishing separable and entangled states *Phys. Rev. Lett.* **88** 187904 (*Preprint* arXiv:quant-ph/0112007)
[20] Doherty A, Parrilo P and Spedalieri F 2004 Complete family of separability criteria *Phys. Rev. A* **69** 022308 (*Preprint* arXiv:quant-ph/0308032)
[21] Horodecki R, Horodecki P, Horodecki M and Horodecki K 2009 Quantum entanglement *Rev. Mod. Phys.* **81** 865 (*Preprint* arXiv:quant-ph/0702225)
[22] Hall W 2007 Compatibility of subsystem states and convex geometry *Phys. Rev. A* **75** 032102 (*Preprint* arXiv:quant-ph/0610031)
[23] Butterley P, Sudbery A and Szulc J 2006 Compatibility of subsystem states *Found. Phys.* **36** 83 (*Preprint* arXiv:quant-ph/0407227)

[24] Mathiesen L 1985 Computation of economic equilibria by a sequence of linear complementarity problems *Math. Programming Stud.* **23** 144

[25] Bertsimas D and Sim M 2003 Robust discrete optimization and network flows *Math. Programming* **98** 49

[26] Brierley S 2009 *Mutually unbiased bases in low dimensions* PhD thesis (Univ. of York)

[27] Kojima M and Tunçel L 2000 Cones of matrices and successive convex relaxations of nonconvex sets *SIAM Journal on Optimization* **10** 750

[28] Shor N 1987 Quadratic optimization problems *Sov. J. Circuits Systems Sciences* **25** 1

[29] Lasserre J 2001 Global optimization with polynomials and the problem of moments *SIAM Journal on Optimization* **11** 796

[30] Eisert J, Hyllus P, Gühne O and Curty M 2004 Complete hierarchies of efficient approximations to problems in entanglement theory *Phys. Rev. A* **70** 062317 (*Preprint* arXiv:quant-ph/0407135)

[31] Henrion D, Lasserre J and Loefberg J 2008 *GloptiPoly 3: Moments, optimization and semidefinite programming v3.4* available at `http://www.laas.fr/~henrion/software/gloptipoly3/`

[32] Lasserre J 2008 A semidefinite programming approach to the generalized problem of moments *Math. Programming* **112** 65

[33] Strum J 1999 Using SeDuMi: A matlab toolbox for optimization over symmetric cones *Optimization Methods and Software* **11** 625 (available at `www.sedumi.ie.lehigh.edu/`)

[34] Tao T 2004 Fuglede's conjecture is false in 5 and higher dimensions *Math. Res. Lett.* **11** 251 (*Preprint* arXiv:math/0306134)

[35] Attributed to Carlton Caves; a paper by Caves C and Fuchs C (1996 The Dilemma of Einstein, Podolsky, and Rosen - 60 Years Later ed by Mann A and Revzen M *Ann. Israel Phys. Soc.* **12** 226) only has *Hilbert space is gratuitously big.*

Protecting subspaces by acting on the outside

Jonathan Busch and Almut Beige

The School of Physics and Astronomy, University of Leeds, Leeds, LS2 9JT, United Kingdom

E-mail: a.beige@leeds.ac.uk

Abstract. Many quantum control tasks aim at manipulating the state of a quantum mechanical system within a finite subspace of states. However, couplings to the outside are often inevitable. Here we discuss strategies which keep the system in the controlled subspace by applying strong interactions onto the outside. This is done by drawing analogies to simple toy models and to the quantum Zeno effect. Special attention is paid to the constructive use of dissipation in the protection of subspaces.

Received: 18 February 2010

1. Introduction

Quantum control techniques like Hamiltonian engineering are very successful when manipulating finite-dimensional Hilbert spaces [1, 2, 3]. However, quantum control tasks often require the control of an infinite-dimensional state space. This applies for example, when the quantum system of interest couples to an infinitely large reservoir or when it contains bosonic modes with infinitely many states. In such a situation one should take advantage of mechanisms which restrict the time evolution of the system effectively onto a finite-dimensional space. The corresponding evolution can then be used to engineer appropriate control sequences [4, 5].

The protection of finite-dimensional subspaces against the leakage of population (cf. Fig. 1) can be achieved in many different ways. One approach is to use Hamiltonians which act only on small subsets of states and naturally restrict the time evolution of the system onto a finite-dimensional state space. This technique is used for example in ion trap quantum computing, where gate operations are realised by applying the required interactions in successive steps [6].

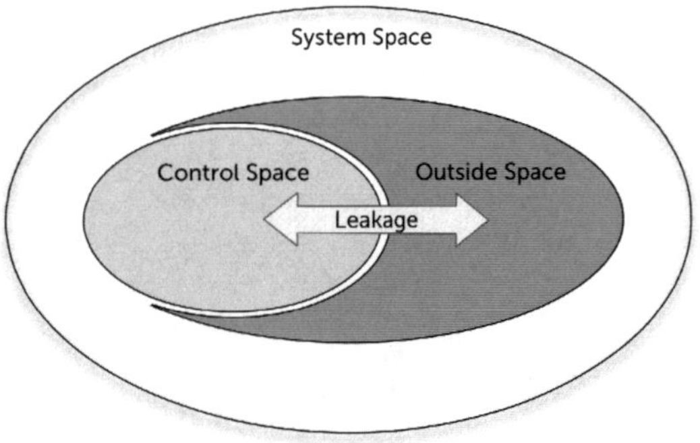

Figure 1. Illustration of the control problem. We wish to control the system evolution within a subspace (yellow) of the total space (white) whilst there is coupling to an external subspace (blue).

During each time step, the number of phonons in the system increases at most by one. The populated state space remains finite and the excitation of coherent phonon states has been avoided. At the end of every control sequence, the ions return into a state with zero phonon excitation [7, 8].

Another very efficient tool for the protection of subspaces against leakage errors are stimulated Raman adiabatic passages (STIRAP) [9, 10]. These employ the adiabatic theorem to induce transitions between states with no direct coupling between them. In composite quantum systems, like atoms which move slowly through an optical cavity, STIRAP can create ground state entanglement without populating excited electronic states and without creating photons inside the resonator [11]. An alternative approach for protecting subspaces against leakage errors is to use especially designed impulsive pulse sequences. This technique has initially been developed to minimise radiation damage when exciting specific ground-state vibrational modes of molecules [12]. A generalisation of both strategies is to simply use numerical simulations which impose state dependent constraints to design optimal control sequences [13].

In this paper we discuss a very different strategy for protecting the controlled subspace against leakage. Instead of imposing a well designed dynamics onto its states, we consider strong interactions which act *only* on the outside. If these interactions introduce a time scale into the system which is much shorter than the time scale on which the leakage of population out of the controlled subspace would occur in the unprotected case, one can show that these unwanted transitions become strongly inhibited [14, 15, 16]. This approach has many similarities with bang-bang and its generalisation dynamical decoupling [17, 18, 19] which also interrupt a relatively slow evolution with strong interactions.

The second half of the paper pays special attention to the constructive use of dissipation in the protection of subspaces against population leakage and assumes non-zero spontaneous decay rates of the outside states. We are interested in cases, where the time evolution of the system can be understood in terms of rapidly-repeated measurements on whether the system remains in the controlled subspace or not [11, 20, 21, 22, 23, 24, 25]. These force the system to remain there much longer than in the unprotected case which can be understood in terms of the quantum Zeno effect [26, 27, 28]. Notice that this approach provides a build-in error detection mechanism, when it is possible to register and act upon unwanted measurement outcomes. Moreover, as we shall see below, dissipation can protect subspaces against leakage errors even in situations, where other methods would simply fail.

Refs. [29, 30] discuss similarities between bang-bang, dynamical decoupling, the protection of subspaces with strong interactions, and the protection of subspaces using dissipation. The authors conclude that all these approaches are essentially equivalent, since all of them can be understood in terms of the quantum Zeno effect [26]. The purpose of the present manuscript is to give more insight into the underlying processes. This is done by analysing relatively simple toy models which allow us to compare the above mentioned methods qualitatively as well as quantitatively.

This paper is organized in five sections. In the next section, we consider a two-level system with resonant coherent coupling to obtain information about the expected leakage rates in a simple unprotected subspace scenario. In Section 3, we extend the outside space and show how strong interactions acting on the outside space can be used (or how they should not be used) to protect the controlled subspace against leakage error. Section 4 analyses closely related level schemes but with non-zero spontaneous decay rates. An example is given where dissipation results in the protection of a subspace which would not be there otherwise. We finally summarise our findings in Section 5.

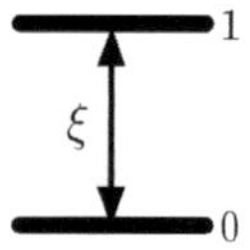

Figure 2. Toy model illustrating the leakage of population from an unprotected controlled subspace (represented by $|0\rangle$) with coupling strength ξ into an outside space (represented by $|1\rangle$).

2. An unprotected subspace

Let us first consider a case where no effort is made to protect a controlled subspace from leaking population into outside states. For simplicity, we assume that the controlled subspace contains only a single state $|0\rangle$. As shown in Fig. 2, there is moreover only one relevant state outside the controlled subspace which we denote by $|1\rangle$. The leakage of population from level 0 into level 1 could be due to resonant interactions (like a laser field). Although this is an almost trivial case, the analysis of the time evolution of this level scheme introduces the relevant time scales of the system. This will enable us later to characterise and to compare the effectiveness of different strategies for the protection of controlled subspaces against leakage errors.

In the following, we assume that the laser is in resonance with the 0–1 transition and denote its (real) Rabi frequency by ξ. Moreover, $\hbar\omega_i$ denotes the energy of states $|i\rangle$. The system Hamiltonian in the usual rotating wave and dipole approximation can be written as

$$H = \hbar\xi \, e^{i\omega_\xi t} \, |0\rangle\langle 1| + \text{h.c.} + \sum_{i=0}^{1} \hbar\omega_i \, |i\rangle\langle i| \tag{1}$$

with $\omega_\xi \equiv \omega_1 - \omega_0$. To solve the corresponding time evolution, we first change into an interaction picture with respect to $H_0 = \sum_{i=0}^{1} \hbar\omega_i \, |i\rangle\langle i|$. This transfers the Hamiltonian (1) into the interaction Hamiltonian

$$H_{\rm I} = \hbar\xi \, |0\rangle\langle 1| + \text{h.c.} \tag{2}$$

To estimate the leakage rate of the controlled subspace in this case, we now calculate the population P_0 in $|0\rangle$ at time t, given that the system was initially in $|0\rangle$.

One way of doing this is to consider the usual Pauli operators σ_2 and σ_3,

$$\sigma_2 = -i\,(\,|0\rangle\langle 1| - |1\rangle\langle 0|\,) \text{ and } \sigma_3 = |0\rangle\langle 0| - |1\rangle\langle 1| \, , \tag{3}$$

and to use the relation

$$\langle \dot{A} \rangle = -\frac{i}{\hbar} \, \langle [A, H_{\rm I}] \rangle \tag{4}$$

for the time evolution of the expectation value of an operator A in the interaction picture to obtain a closed set of rate equations. This yields the differential equations

$$\begin{pmatrix} \langle \dot{\sigma}_2 \rangle \\ \langle \dot{\sigma}_3 \rangle \end{pmatrix} = 2\xi \begin{pmatrix} 0 & -1 \\ 1 & 0 \end{pmatrix} \begin{pmatrix} \langle \sigma_2 \rangle \\ \langle \sigma_3 \rangle \end{pmatrix} \tag{5}$$

which can be solved easily analytically.

A more straightforward way of solving the time evolution of the system is to write its state vector as $|\psi\rangle = \sum_{i=0,1} c_i \, |i\rangle$ and to use the Schrödinger equation to obtain differential equations for the complex coefficients c_i. However, the above approach of deriving rate equations for expectation values is more efficient, since we are only interested in the leakage of population

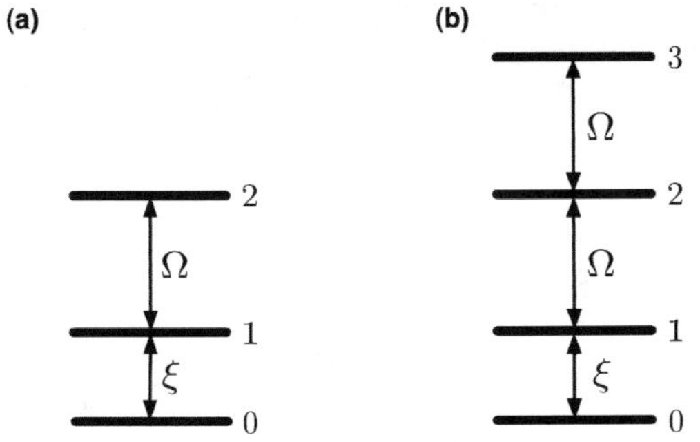

Figure 3. Toy models to illustrate the possible protection of a controlled subspace (represented by $|0\rangle$) with strong interactions with coupling strength Ω in the outside space. Here the outside space contains either the two states $|1\rangle$ and $|2\rangle$ (a) or the three states $|1\rangle$, $|2\rangle$, and $|3\rangle$ (b).

out of the controlled subspace. Moreover, this approach can be extended easily to include more complex level schemes as well as the effect of spontaneous photon emission.

Since σ_3 commutes with H_0, we can calculate P_0 using the relation

$$P_0(t) = \frac{1}{2}(1 + \langle \sigma_3(t) \rangle) \,. \tag{6}$$

Solving Eq. (5) for time-independent coupling constants ξ and for the case where the system is initially in $|0\rangle$, we find that the population in the initial state changes according to

$$P_0(t) = \frac{1}{2}\left(1 + \cos(2\xi t)\right) = \cos^2(\xi t) \,. \tag{7}$$

This means, in the absence of any protection, the system remains inside the controlled subspace only on a time scale which is short compared to $1/\xi$.

3. Protecting a subspace with strong interactions

One way to protect the controlled subspace against errors is to involve the relevant outside states into a relatively fast time evolution. Indeed it has been found that strong interactions can have the same effect as rapidly repeated measurements whether the system remains in its initial subspace or not [29, 30]. In good agreement with the predictions of the quantum Zeno effect [26], these measurements strongly inhibit transitions out of the controlled subspace. In the following, we illustrate this approach with the help of the two toy models shown in Fig. 3. The purpose of the interactions with amplitude Ω is to induce fast oscillations of the amplitude of the state $|1\rangle$. These cause $\langle \sigma_2 \rangle$ in Eq. (3) to oscillate rapidly in time, such that $\langle \dot{\sigma}_3 \rangle$ in Eq. (5) becomes zero on average and the system remains approximately in $|0\rangle$.

As we shall see below, this strategy works well for the level scheme in Fig. 3(a). However, strong interactions acting on the outside space do not always protect the controlled subspace against leakage errors. Problems arise for example in the level scheme in Fig. 3(b). There the interactions in the outside space are more complex than the interactions which cause the leakage. The result is that the generation of approximate dark states in the outside space. These are zero eigenvectors of the fast system dynamics. Transitions between dark states and the controlled subspace are hence not protected by time scale separation, even when Ω becomes very large.

3.1. Single-coupling case

We begin with an analysis of the three-level system shown in Fig. 3(a). Again, the controlled subspace contains only a single state, $|0\rangle$, while the outside subspace contains the two states $|1\rangle$ and $|2\rangle$. In order to maximise the effect of the applied interactions, we assume resonant couplings. As before, ξ is the coupling constant for the 0–1 transition, while Ω denotes the coupling constant for the 1–2 transition. Here we are especially interested in the case, where $\xi \ll \Omega$. Again we have a closer look at the time evolution of the population P_0 in the controlled subspace.

As in Section 2, we denote the energy of level i by $\hbar\omega_i$. The Hamiltonian for the level configuration in Fig. 3(a) can then be written as

$$H = \hbar\xi\, e^{i\omega_\xi t}\, |0\rangle\langle 1| + \hbar\Omega\, e^{i\omega_\Omega t}\, |1\rangle\langle 2| + \text{h.c.} + \sum_{i=0}^{2} \hbar\omega_i\, |i\rangle\langle i| \tag{8}$$

with $\omega_\xi \equiv \omega_1 - \omega_0$ and $\omega_\Omega \equiv \omega_2 - \omega_1$. Transforming this Hamiltonian into the interaction picture with respect to $H_0 = \sum_{i=0}^{2} \hbar\omega_i\, |i\rangle\langle i|$, we obtain

$$H_{\rm I} = \hbar\xi\, |0\rangle\langle 1| + \hbar\Omega\, |1\rangle\langle 2| + \text{h.c.} \tag{9}$$

This interaction Hamiltonian is time independent and contains only the weak coupling between $|0\rangle$ and $|1\rangle$ and the strong coupling between $|1\rangle$ and $|2\rangle$.

In order to obtain a closed system of rate equations, we now consider the expectation values of the Gell-Mann matrices [31]

$$\sigma_2 = -i\,(|0\rangle\langle 1| - |1\rangle\langle 0|)\,, \quad \sigma_7 = -i\,(|1\rangle\langle 2| - |2\rangle\langle 1|)\,, \quad \sigma_4 = |0\rangle\langle 2| + |2\rangle\langle 0|\,,$$
$$\sigma_3 = |0\rangle\langle 0| - |1\rangle\langle 1|\,, \quad \sigma_8 = \frac{1}{\sqrt{3}}(|0\rangle\langle 0| + |1\rangle\langle 1| - 2\,|2\rangle\langle 2|)\,. \tag{10}$$

These are generalisations of the Pauli operators used in Section 2. Overall there are eight Gell-Mann matrices which can be used to model the time evolution of coupled three-level systems in a convenient way. However, due to the specific form of the interactions in the level scheme in Fig. 3(a), we need to consider only five of them. Using relation (4), we find the following closed system of differential equations

$$\begin{pmatrix} \langle\dot\sigma_2\rangle \\ \langle\dot\sigma_3\rangle \\ \langle\dot\sigma_4\rangle \\ \langle\dot\sigma_7\rangle \\ \langle\dot\sigma_8\rangle \end{pmatrix} = \begin{pmatrix} 0 & -2\xi & -\Omega & 0 & 0 \\ 2\xi & 0 & 0 & -\Omega & 0 \\ \Omega & 0 & 0 & -\xi & 0 \\ 0 & \Omega & \xi & 0 & -\sqrt{3}\Omega \\ 0 & 0 & 0 & \sqrt{3}\Omega & 0 \end{pmatrix} \begin{pmatrix} \langle\sigma_2\rangle \\ \langle\sigma_3\rangle \\ \langle\sigma_4\rangle \\ \langle\sigma_7\rangle \\ \langle\sigma_8\rangle \end{pmatrix}. \tag{11}$$

These differential equations can be solved for example by calculating analytical expressions for the eigenvalues of this matrix or by simply using *Mathematica*.

From Eq. (10) we see that the population in the controlled subspace equals

$$P_0 = \frac{1}{3} + \frac{1}{2}\,\langle\sigma_3\rangle + \frac{1}{2\sqrt{3}}\,\langle\sigma_8\rangle\,. \tag{12}$$

Substituting the solution of the above rate equations into this equation, we find that the population in $|0\rangle$ at time t evolves according to

$$P_0(t) = \frac{2\Omega^4 + \xi^4}{2\mu^4} + \frac{2\Omega^2\xi^2}{\mu^4}\cos(\mu t) + \frac{\xi^4}{2\mu^4}\cos(2\mu t) \tag{13}$$

with $\mu^2 \equiv \Omega^2 + \xi^2$, if the system was initially in $|0\rangle$. For $\xi \ll \Omega$, Eq. (13) simplifies to

$$P_0(t) = 1 - \frac{2\xi^2}{\Omega^2}\,[1 - \cos(\Omega t)] \tag{14}$$

which holds up to first order in ξ^2/Ω^2. As shown in Fig. 4, the system remains to a very good approximation, i.e. up to variations with an amplitude proportional to ξ^2/Ω^2, in $|0\rangle$. This means, for $\xi^2 \ll \Omega^2$, the controlled subspace is effectively protected against leakage errors.

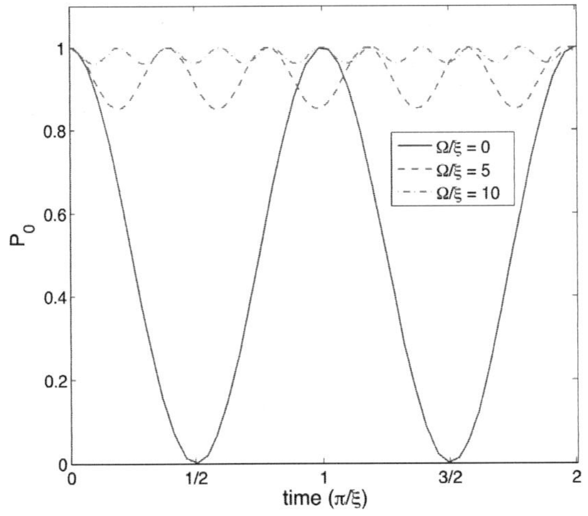

Figure 4. Time dependence of P_0 for the level scheme shown in Fig. 3(a) for different ratios of Ω/ξ. The system is initially in $|0\rangle$. For $\Omega = 0$, the system leaves its initial state space on a time scale given by $1/\xi$. For $\Omega > 10\,\xi$, the system remains there with a fidelity above 95 % which constitutes an effective protection of the initial state space.

3.2. Double-coupling case

Using the same notation as in the previous subsection, the Hamiltonian for the level configuration in Fig. 3(b) in the Schrödinger picture equals

$$H = \hbar\xi\, \mathrm{e}^{\mathrm{i}\omega_\xi t}|0\rangle\langle 1| + \hbar\Omega\, \mathrm{e}^{\mathrm{i}\omega_\Omega t}\,(|1\rangle\langle 2| + |2\rangle\langle 3|) + \mathrm{h.c.} + \sum_{i=0}^{3} \hbar\omega_i\, |i\rangle\langle i|\,. \tag{15}$$

Again we first simplify this Hamiltonian by changing into the interaction picture with respect to the free Hamiltonian $H_0 = \sum_{i=0}^{3} \hbar\omega_i\, |i\rangle\langle i|$. This yields

$$H_{\mathrm{I}} = \hbar\xi\, |0\rangle\langle 1| + \hbar\Omega\, (|1\rangle\langle 2| + |2\rangle\langle 3|) + \mathrm{h.c.} \tag{16}$$

Instead of solving the corresponding Schrödinger equation, we apply again Eq. (4) to obtain a closed set of rate equations.

To predict the time evolution of the population P_0 in the controlled subspace, we now have to consider nine generalised Gell-Mann matrices [31]. These are

$$\begin{aligned}
&\sigma_2 = -\mathrm{i}\,(|0\rangle\langle 1| - |1\rangle\langle 0|)\,, \quad \sigma_7 = -\mathrm{i}\,(|1\rangle\langle 2| - |2\rangle\langle 1|)\,, \\
&\sigma_{10} = -\mathrm{i}\,(|0\rangle\langle 3| - |3\rangle\langle 0|)\,, \quad \sigma_{14} = -\mathrm{i}\,(|2\rangle\langle 3| - |3\rangle\langle 2|)\,, \\
&\sigma_4 = |0\rangle\langle 2| + |2\rangle\langle 0|\,, \quad \sigma_{11} = |1\rangle\langle 3| + |3\rangle\langle 1|\,, \\
&\sigma_3 = |0\rangle\langle 0| - |1\rangle\langle 1|\,, \quad \sigma_8 = \frac{1}{\sqrt{3}}\,(|0\rangle\langle 0| + |1\rangle\langle 1| - 2\,|2\rangle\langle 2|)\,, \\
&\sigma_{15} = \frac{1}{\sqrt{6}}\,(|0\rangle\langle 0| + |1\rangle\langle 1| + |2\rangle\langle 2| - 3\,|3\rangle\langle 3|)\,.
\end{aligned} \tag{17}$$

Moreover, we notice that the interaction Hamiltonian (16) can be written as

$$H_{\mathrm{I}} = \hbar\xi\,\sigma_1 + \hbar\Omega\,(\sigma_6 + \sigma_{13}) \tag{18}$$

with

$$\sigma_1 = |0\rangle\langle 1| + |1\rangle\langle 0|\,, \quad \sigma_6 = |1\rangle\langle 2| + |2\rangle\langle 1|\,, \quad \sigma_{13} = |2\rangle\langle 3| + |3\rangle\langle 2|\,. \tag{19}$$

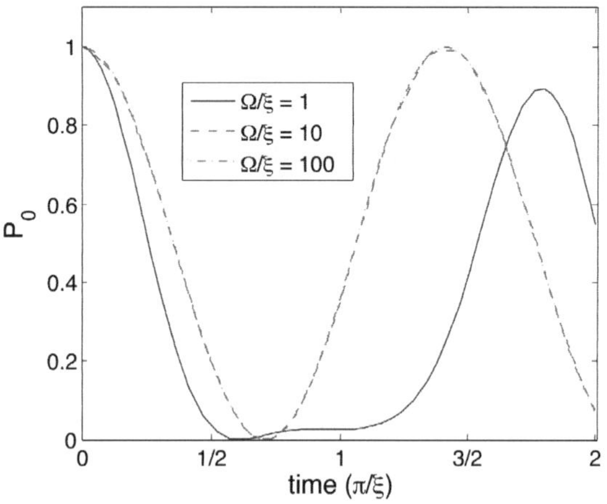

Figure 5. Time dependence of P_0 for the level scheme shown in Fig. 3(b) for different ratios of Ω/ξ. Here the controlled subspace is no longer protected against leakage, even when Ω becomes as large as $100\,\xi$. The reason is the leakage of population into the dark state $|\lambda_0\rangle$ which is illustrated in Fig. 6.

Substituting Eqs. (17)–(19) into Eq. (4) and evaluating the relevant commutators, we see that the expectation of the operators in Eq. (17) evolve according to

$$\begin{pmatrix} \langle\dot\sigma_2\rangle \\ \langle\dot\sigma_3\rangle \\ \langle\dot\sigma_4\rangle \\ \langle\dot\sigma_7\rangle \\ \langle\dot\sigma_8\rangle \\ \langle\dot\sigma_{10}\rangle \\ \langle\dot\sigma_{11}\rangle \\ \langle\dot\sigma_{14}\rangle \\ \langle\dot\sigma_{15}\rangle \end{pmatrix} = \begin{pmatrix} 0 & -2\xi & -\Omega & 0 & 0 & 0 & 0 & 0 & 0 \\ 2\xi & 0 & 0 & -\Omega & 0 & 0 & 0 & 0 & 0 \\ \Omega & 0 & 0 & -\xi & 0 & \Omega & 0 & 0 & 0 \\ 0 & \Omega & \xi & 0 & -\sqrt{3}\Omega & 0 & -\Omega & 0 & 0 \\ 0 & 0 & 0 & \sqrt{3}\Omega & 0 & 0 & 0 & -\frac{2}{\sqrt{3}}\Omega & 0 \\ 0 & 0 & -\Omega & 0 & 0 & 0 & \xi & 0 & 0 \\ 0 & 0 & 0 & \Omega & 0 & -\xi & 0 & -\Omega & 0 \\ 0 & 0 & 0 & 0 & \frac{2}{\sqrt{3}}\Omega & 0 & \Omega & 0 & -\frac{2\sqrt{2}}{\sqrt{3}}\Omega \\ 0 & 0 & 0 & 0 & 0 & 0 & 0 & \frac{2\sqrt{2}}{\sqrt{3}}\Omega & 0 \end{pmatrix} \begin{pmatrix} \langle\sigma_2\rangle \\ \langle\sigma_3\rangle \\ \langle\sigma_4\rangle \\ \langle\sigma_7\rangle \\ \langle\sigma_8\rangle \\ \langle\sigma_{10}\rangle \\ \langle\sigma_{11}\rangle \\ \langle\sigma_{14}\rangle \\ \langle\sigma_{15}\rangle \end{pmatrix}. \tag{20}$$

This system of linear differential equations can, in principle, be solved analytically. However, for simplicity, we restrict ourselves to the presentation of a numerical solution.

Using the Gell Mann matrices defined in Eq. (17), the population in the initial state P_0 can now be written as

$$P_0 = \frac{1}{4} + \frac{1}{2}\langle\sigma_3\rangle + \frac{1}{2\sqrt{3}}\langle\sigma_8\rangle + \frac{1}{2\sqrt{6}}\langle\sigma_{15}\rangle. \tag{21}$$

The time evolution of P_0 obtained from substituting the numerical solution of the differential equation (20) into Eq. (21) is shown in Fig. 5. Comparing the result for different values of Ω/ξ with the time evolution in the $\Omega = 0$ case, we see that the controlled subspace is not protected, even when Ω is much larger than ξ. Leakage of population out of the controlled subspace happens on the same time scale as in the unprotected case.

Why does the protection of the controlled subspace work in the level scheme shown in Fig. 3(a) but not in the very similar level scheme shown in Fig. 3(b)? The reason for this becomes clear when we rewrite the Hamiltonian in Eq. (16) in terms of the states $|0\rangle$,

$$|\lambda_0\rangle \equiv \frac{1}{\sqrt{2}}(|1\rangle - |3\rangle)\,, \quad |\lambda_1\rangle \equiv \frac{1}{\sqrt{2}}(|1\rangle + |3\rangle) \tag{22}$$

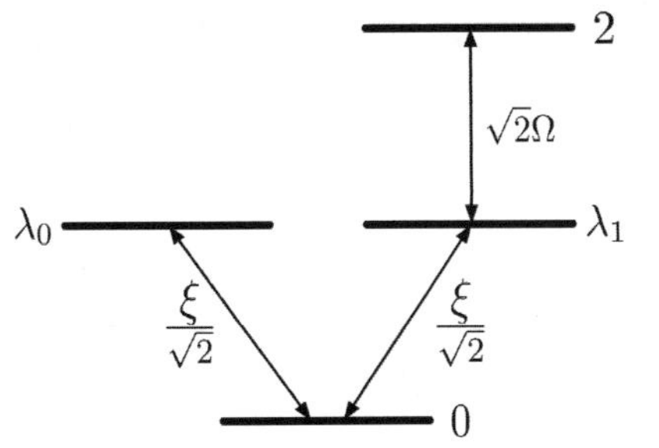

Figure 6. Illustration of the effect of the Hamiltonian (23) onto the states involved in the time evolution of the system. This level scheme is identical to the one shown in Fig. 3(b) but now we clearly see why the initial state $|0\rangle$ is no longer protected against leakage errors.

and $|2\rangle$. Using this notation, $H_{\rm I}$ becomes

$$H_{\rm I} \;=\; \frac{1}{\sqrt{2}}\hbar\xi\,|0\rangle\langle\lambda_0| + \frac{1}{\sqrt{2}}\hbar\xi\,|0\rangle\langle\lambda_1| + \sqrt{2}\hbar\Omega\,|\lambda_1\rangle\langle 2| + \text{h.c.} \tag{23}$$

The effect of this Hamiltonian is illustrated in Fig. 6. It shows that the system is only protected against leakage into the $|\lambda_1\rangle$ state, since this state experiences a strong interaction. However, the system is not protected against leakage into $|\lambda_0\rangle$, since $|\lambda_0\rangle$ is a zero eigenstate of the Ω terms in Eq. (23). This means, $|\lambda_0\rangle$ is not involved in a fast evolution and the transfer from $|0\rangle$ to $|\lambda_0\rangle$ occurs on the same time scale as in the unprotected case. In the final section, we show that dissipation is able to remove such dark states from the system so that the controlled subspace becomes protected again.

4. Protecting a subspace with dissipation

In this section we analyse three examples (cf. Fig. 7) which illustrate the possible protection of the controlled subspace *using dissipation*. The controlled subspace contains again only the $|0\rangle$ state, while the outside space contains one, two or three states. The only difference to the examples discussed in Sections 2 and 3 is the presence of a non-zero spontaneous decay rate Γ. As we shall see below, the controlled subspace is well protected against leakage in all three scenarios, when the interactions in the outside space described by Ω and the spontaneous decay rate Γ are sufficiently larger than ξ.

4.1. Single-coupling case with dissipation

Let us first have a look at the three-level system shown in Fig. 7(b). To describe its time evolution, we go again into the interaction picture with respect to the free evolution and analyse the master equation

$$\dot{\rho} = -\frac{\rm i}{\hbar}[\,H_{\rm I},\rho\,] + \frac{\Gamma}{2}\Big[\,2\,|1\rangle\langle 2|\,\rho\,|2\rangle\langle 1| - \rho\,|2\rangle\langle 2| - |2\rangle\langle 2|\,\rho\,\Big]\,. \tag{24}$$

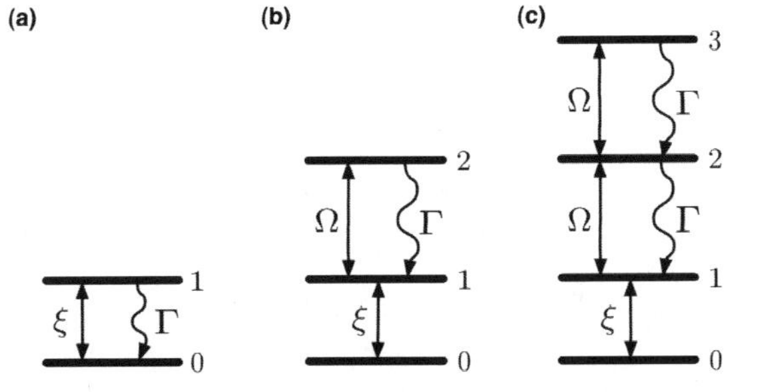

Figure 7. Toy models to illustrate the possible protection of the controlled subspace (represented by $|0\rangle$) with a non-zero spontaneous decay rate Γ and strong interactions with coupling strength Ω in the outside space.

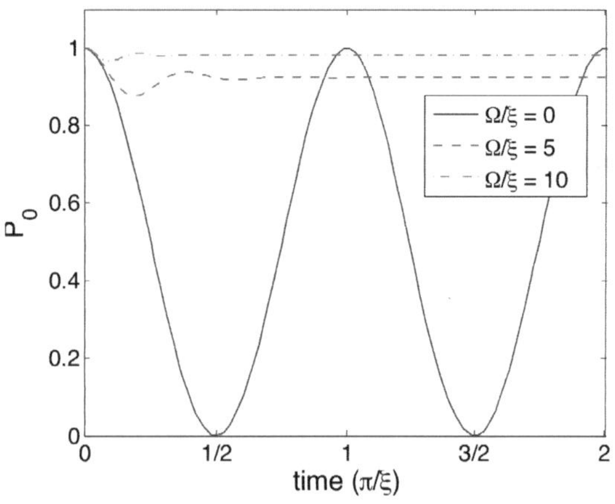

Figure 8. Time dependence of P_0 for the level scheme shown in Fig. 7(b) for $\Gamma = \Omega$ and different ratios of Ω/ξ. On average, the protection of the controlled subspace is more or less the same as in Fig. 4 which corresponds to the same level scheme but with $\Gamma = 0$ (cf. Fig. 3(a)).

The interaction Hamiltonian $H_{\rm I}$ is the same as in Eq. (9). In order to predict the time evolution of the population in the controlled subspace, we derive again a closed system of rate equations. The time derivative of the expectation value of an operator A is now given by

$$\langle \dot{A} \rangle = \mathrm{Tr}\,(A\dot{\rho}) \ . \tag{25}$$

Taking this into account, we find that the Gell Mann matrices in Eq. (10) evolve according to

$$\begin{pmatrix} \langle \dot{\sigma}_2 \rangle \\ \langle \dot{\sigma}_3 \rangle \\ \langle \dot{\sigma}_4 \rangle \\ \langle \dot{\sigma}_7 \rangle \\ \langle \dot{\sigma}_8 \rangle \end{pmatrix} = \begin{pmatrix} 0 & -2\xi & -\Omega & 0 & 0 \\ 2\xi & 0 & 0 & -\Omega & \frac{1}{\sqrt{3}}\Gamma \\ \Omega & 0 & -\frac{1}{2}\Gamma & -\xi & 0 \\ 0 & \Omega & \xi & -\frac{1}{2}\Gamma & -\sqrt{3}\Omega \\ 0 & 0 & 0 & \sqrt{3}\Omega & -3\Gamma \end{pmatrix} \begin{pmatrix} \langle \sigma_2 \rangle \\ \langle \sigma_3 \rangle \\ \langle \sigma_4 \rangle \\ \langle \sigma_7 \rangle \\ \langle \sigma_8 \rangle \end{pmatrix} + \begin{pmatrix} 0 \\ -\frac{1}{3}\Gamma \\ 0 \\ 0 \\ \sqrt{3}\Gamma \end{pmatrix} . \tag{26}$$

These equations resemble the ones shown in Eq. (11). The additional Γ terms take the effect of dissipation into account.

The population in the controlled subspace can be obtained by substituting for example the numerical solution of these equations into Eq. (12). The result is shown in Fig. 8. For simplicity we assumed $\Gamma = \Omega$. For $\Omega = 0$, we see again the Rabi oscillations of the unprotected case. However, when Ω becomes sufficiently larger than ξ, then the system remains to a very good approximation in its initial state. On average, the protection of the controlled subspace is more or less the same as in Section 3.1, where we had $\Gamma = 0$ (cf. Fig. 4). There seems to be no advantage of having a non-zero spontaneous decay rate in the system! Notice that having $\Gamma \neq 0$ in the level scheme in Fig. 7(b) is only advantageous when someone actually observes whether the system emits photons or not. Indeed, one can show that the system remains in its initial state $|0\rangle$ with a very high fidelity under the condition of *no* photon emission [32]. If a photon emission is detected, then the system has left the controlled subspace and the anticipated control experiment needs to be restarted.

Comparing the level scheme in Fig. 7(b) with the level scheme analysed in Refs. [32, 33, 34], we see that its dynamics exhibits so-called macroscopic light and dark periods. Indeed, for ξ much smaller than Ω and Γ, the initial state $|0\rangle$ is an approximate zero eigenstate of the system dynamics. The absence of photon emissions hence confirms that the system is in this state. As a consequence of the quantum Zeno effect, it therefore remains there for a relatively long time. On average, this time equals $\Omega^2/\Gamma\xi^2$ which is much larger than $1/\xi$ [32]. In other words, the system exhibits a macroscopic dark period. The system may eventually drop out

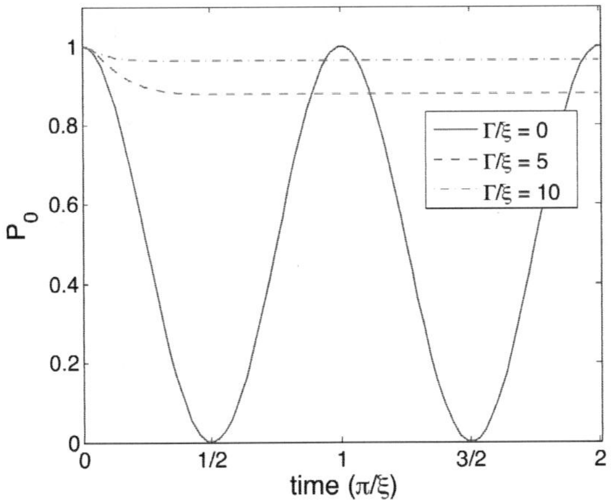

Figure 9. Time dependence of P_0 for the level scheme shown in Fig. 7(a) for different ratios of Γ/ξ. For $\Gamma \gg \xi$, the system remains in the controlled subspace with a very high fidelity.

of the controlled subspace, thereby entering a so-called macroscopic light period and causing fluorescence at a rate which depends on Ω and Γ. This behaviour is not reflected in Fig. 8, since the density matrix description used in this paper does not allow us to distinguish the different trajectories of the system.

4.2. Single-state outside

Let us now have a look at the simple level configuration in Fig. 7(a). Its time evolution is given by the master equation

$$\dot{\rho} = -\frac{\mathrm{i}}{\hbar}[H_\mathrm{I}, \rho] + \frac{\Gamma}{2}\Big[2\,|0\rangle\langle 1|\,\rho\,|1\rangle\langle 0| - \rho\,|1\rangle\langle 1| - |1\rangle\langle 1|\,\rho\Big]. \tag{27}$$

In the interaction picture with respect to the free evolution, the interaction Hamiltonian H_I is the same as in Eq. (2). To predict the time evolution of P_0 we proceed as in Section 2 and consider again the Pauli operators in Eq. (3). Their expectation values evolve now according to

$$\begin{pmatrix} \langle\dot{\sigma}_2\rangle \\ \langle\dot{\sigma}_3\rangle \end{pmatrix} = \begin{pmatrix} -\frac{1}{2}\Gamma & -2\xi \\ 2\xi & -\Gamma \end{pmatrix} \begin{pmatrix} \langle\sigma_2\rangle \\ \langle\sigma_3\rangle \end{pmatrix} + \begin{pmatrix} 0 \\ \Gamma \end{pmatrix}. \tag{28}$$

Combining the result of this equation with Eq. (6) yields the time dependence of the population P_0 in the controlled subspace.

Fig. 9 shows a numerical solution of the time dependence of P_0 for different ratios Γ/ξ. For $\Gamma = 0$, we observe the Rabi oscillations in and out of the initial subspace which occur in the unprotected case. For $\Gamma \gg \xi$, the state vector becomes $|0\rangle$ with a very high fidelity. But even for relatively modest values for Γ/ξ, the density matrix ρ settles quickly into a steady state with the system predominantly in $|0\rangle$. The reason for this very strong protection of the controlled subspace is that, even when it leaves, the system returns very rapidly via the spontaneous emission of a photon.

4.3. Double-coupling case with dissipation

This final subsection analyses the time evolution of the four-level system shown in Fig. 7(c). The only difference to the level scheme in Fig. 3(b) is the presence of the non-zero spontaneous decay rate Γ. To calculate the time evolution of the population in the controlled subspace, we

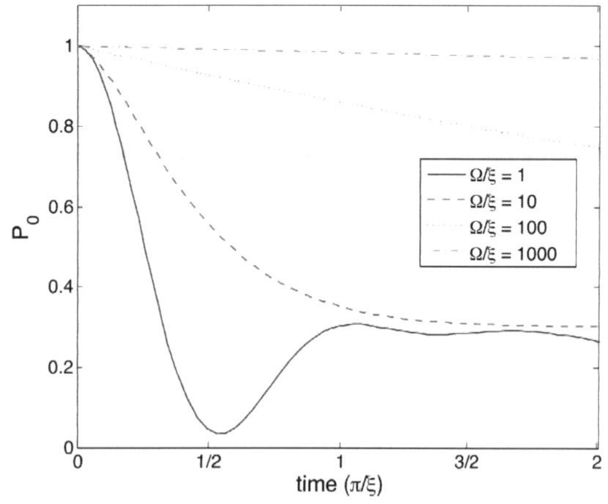

Figure 10. Time dependence of P_0 for the level scheme shown in Fig. 7(c) for $\Gamma = \Omega$ and different ratios of Ω/ξ. Compared to Fig. 5, we now observe an increasing effectiveness of protection of the controlled subspace with increasing values of Ω.

now consider the master equation

$$\begin{aligned}\dot{\rho} &= -\frac{\mathrm{i}}{\hbar}[H_{\mathrm{I}},\rho] + \frac{\Gamma}{2}\Big[2\,|1\rangle\langle 2|\,\rho\,|2\rangle\langle 1| - \rho\,|2\rangle\langle 2| - |2\rangle\langle 2|\,\rho\Big] \\ &\quad + \frac{\Gamma}{2}\Big[2\,|2\rangle\langle 3|\,\rho\,|3\rangle\langle 2| - \rho\,|3\rangle\langle 3| - |3\rangle\langle 3|\,\rho\Big], \end{aligned} \quad (29)$$

whose interaction Hamiltonian H_{I} can be found in Eq. (16). Proceeding as above, we find that the time evolution of the Gell Mann matrices (17) is now given by the differential equations

$$\begin{aligned}&(\langle\dot{\sigma}_2\rangle, \langle\dot{\sigma}_3\rangle, \langle\dot{\sigma}_4\rangle, \langle\dot{\sigma}_7\rangle, \langle\dot{\sigma}_8\rangle, \langle\dot{\sigma}_{10}\rangle, \langle\dot{\sigma}_{11}\rangle, \langle\dot{\sigma}_{14}\rangle, \langle\dot{\sigma}_{15}\rangle)^{\mathrm{T}} \\ &\quad = M\,(\langle\sigma_2\rangle, \langle\sigma_3\rangle, \langle\sigma_4\rangle, \langle\sigma_7\rangle, \langle\sigma_8\rangle, \langle\sigma_{10}\rangle, \langle\sigma_{11}\rangle, \langle\sigma_{14}\rangle, \langle\sigma_{15}\rangle)^{\mathrm{T}} \\ &\quad\quad + \left(0, -\frac{1}{4}\Gamma, 0, 0, \frac{1}{4\sqrt{3}}\Gamma, 0, 0, 0, \frac{1}{\sqrt{6}}\Gamma\right)^{\mathrm{T}} \end{aligned} \quad (30)$$

with

$$M = \begin{pmatrix} 0 & -2\xi & -\Omega & 0 & 0 & 0 & 0 & 0 & 0 \\ 2\xi & 0 & 0 & -\Omega & \frac{1}{\sqrt{3}}\Gamma & 0 & 0 & 0 & -\frac{1}{2\sqrt{6}}\Gamma \\ \Omega & 0 & -\frac{1}{2}\Gamma & -\xi & 0 & \Omega & 0 & 0 & 0 \\ 0 & \Omega & \xi & -\frac{1}{2}\Gamma & -\sqrt{3}\Omega & 0 & -\Omega & 0 & 0 \\ 0 & 0 & 0 & \sqrt{3}\Omega & -\Gamma & 0 & 0 & -\frac{2}{\sqrt{3}}\Omega & \frac{3}{2\sqrt{2}}\Gamma \\ 0 & 0 & -\Omega & 0 & 0 & -\frac{1}{2}\Gamma & \xi & 0 & 0 \\ 0 & 0 & 0 & \Omega & 0 & -\xi & -\frac{1}{2}\Gamma & -\Omega & 0 \\ 0 & 0 & 0 & 0 & \frac{2}{\sqrt{3}}\Omega & 0 & \Omega & -\Gamma & -\frac{2\sqrt{2}}{\sqrt{3}}\Omega \\ 0 & 0 & 0 & 0 & 0 & 0 & 0 & \frac{2\sqrt{2}}{\sqrt{3}}\Omega & -\Gamma \end{pmatrix}. \quad (31)$$

Fig. 10 shows the time dependence of P_0 for the case where the system is initially in the controlled subspace and has been obtained by substituting the numerical solution of these equations into Eq. (21).

Comparing Figs. 5 and 10, we see that the presence of a sufficiently large spontaneous decay rate Γ combined with the presence of a relatively large coupling constant Ω now results in an effective protection of the controlled subspace against leakage errors. There are different ways of seeing how this protection (which was not there before) has been achieved. One way is to have a closer look at the above master equation and to notice that the state $|\lambda_0\rangle$ is no longer a zero eigenstate of the system dynamics. Whenever, population accumulates in this state, the system returns (either via the emission of a photon or as a result of its no-photon evolution) on the time scale given by Γ into $|1\rangle$, where it experiences fast driving with Ω. This example confirms that dissipation can provide a very efficient tool for restricting the time evolution of a system onto a controlled subspace.

Another way to gain an intuition into the behaviour of the level scheme in Fig. 7(c) is to compare it to the level scheme in Fig. 7(b) which we analysed in Section 4.1. Observing whether the system emits photons or not, one would notice again two very distinct phases of operation. The system either emits photons at a high rate or it remains dark for a relatively long time. A macroscopic light period, on one hand, indicates that the state vector lies entirely outside the controlled subspace. A macroscopic dark period, on the other hand, indicates that the system is in $|0\rangle$. In other words, if the system is initially in the controlled subspace, it remains there on average much longer than in the unprotected case. The result is the protection of the system against leakage errors which, when they occur, are heralded by an easy-to-detect fluorescence signal.

5. Conclusions

This paper illustrates two methods to protect a controlled subspace against the leakage of population into the outside space: one using strong interactions in the outside space and one using dissipation. This is done with the help of relatively simple toy models whose time evolution can be analysed relatively easily. For simplicity, we assume that the controlled subspace consists only of one state, namely $|0\rangle$. The outside space contains either one, two or three states denoted $|1\rangle$, $|2\rangle$, and $|3\rangle$ (cf. Figs. 3 and 7). Section 2 discusses the unprotected case and shows that unwanted transitions from $|0\rangle$ to $|1\rangle$ (due to resonant coupling) occur on a time scale given by a relatively small parameter ξ (cf. Fig. 2).

In Section 3, the decoherence time of the system is increased to one which scales as ξ^2 by applying relatively fast interactions with coupling strength Ω to the outside space. However, these strong interactions are not always sufficient for protecting a controlled subspace against leakage errors. While it works well for the level scheme shown in Fig. 3(a), no protection occurs for the level scheme shown in Fig. 3(b). The reason is the existence of an approximate zero eigenstate outside the controlled subspace. This state does not experience fast driving and therefore behaves as the state $|1\rangle$ in the unprotected case.

Section 4 considers three scenarios where a spontaneous decay rate Γ has been added to the level schemes analysed in Sections 2 and 3. All the level schemes shown in Fig. 7 exhibit a strong protection of the controlled subspace. One way to understand the mechanism which inhibits the population transfer out of the controlled subspace is to interpret the behaviour of the system in terms of the quantum Zeno effect [26, 27, 28]. Suppose being outside the controlled subspace results necessarily in the spontaneous emission of a photon. Then, observing whether a photon emission takes place or not is equivalent to performing a measurement on whether the system is in the controlled subspace or not. If these measurements occur on a sufficiently short time scale, then a system initially in the controlled subspace remains there much longer than in an unobserved case. A similar interpretation applies to the protection of the controlled subspace with a strong interaction illustrated in Fig. 3(a) [29, 30].

Finally, let us remark that the absence of decoherence within the controlled subspace results in general in an effective time evolution which can be described by the effective Hamiltonian [14, 17, 22, 35]

$$H_{\rm eff} = \mathbf{P}_{\rm CS} \, H \, \mathbf{P}_{\rm CS} \, , \tag{32}$$

where $\mathbf{P}_{\rm CS}$ denotes the projector onto the controlled subspace and H is the total system Hamiltonian. Once the protection is in place, the interactions described by $H_{\rm eff}$ can be designed as required by the control task at hand. Applications of such control tasks can be found for example in quantum information processing, i.e. in the realisation of gate operations and the preparation of highly entangled states. Recent ideas for achieving these tasks use dissipation in an even more constructive way, for example, by heralding successful state preparations with macroscopic fluorescence signals [36, 37, 38, 39] or by letting non-unitary evolutions guide the system into the desired target states [40, 41, 42, 43, 44, 45].

Acknowledgments

A. B. thanks Christiane Koch, Tony Sudbury, and Lorenza Viola for very interesting and valuable discussions. She also acknowledges a James Ellis University Research Fellowship from the Royal Society and the GCHQ. This work was moreover supported by the UK Research Council EPSRC through the QIP IRC and by the EU Research and Training Network EMALI.

References

[1] Huang G M, Tarn T J and Clark J W 1983 *J. Math. Phys.* **24** 2608
[2] Ramakrishna V, Salapaka M V, Dahlem M, Rabitz H and Peirce A 1995 *Phys. Rev. A* **51** 960
[3] Schirmer S G, Fu H and Solomon A I 2001 *Phys. Rev. A* **63** 063410
[4] Rabitz H, de Vivie-Riedle R, Motzkus M and Kompa K 2000 *Science* **288** 824
[5] Khaneja N, Brockett R and Glaser S J 2001 *Phys. Rev. A* **63** 032308
[6] Rangan C, Bloch A M, Monroe C and Bucksbaum P H 2004 *Phys. Rev. Lett.* **92** 113004
[7] Schmidt-Kaler F, Häffner H, Riebe M, Gulde S, Lancaster G P T, Deuschle T, Becher C, Roos C F, Eschner J and Blatt R 2003 *Nature* **422** 408
[8] Leibfried D, DeMarco B, Meyer V, Lucas D, Barrett M, Britton J,Itano W M, Jelenkovic B, Langer C, Rosenband T and Wineland D J 2003 *Nature* **422** 412
[9] Oreg J, Hioe F T and Eberly J H 1984 *Phys. Rev. A* **29** 690
[10] Bergmann K, Theuer H and Shore B W 1998 *Rev. Mod. Phys.* **70** 1003
[11] Marr C, Beige A and Rempe G 2003 *Phys. Rev. A* **68** 033817
[12] Kosloff R, Hammerich A D and Tannor D 1992 *Phys. Rev. Lett.* **69** 2172
[13] Palao J P, Kosloff R and Koch C P 2008 *Phys. Rev. A* **77** 063412
[14] Shore B W 1981 *Phys. Rev. A* **24** 1413
[15] Peres A 1980 *Am. J. Phys.* **48** 931
[16] Facchi P and Pascazio S 2002 *Phys. Rev. Lett.* **89** 080401
[17] Viola L and Lloyd S 1998 *Phys. Rev. A* **58** 2733
[18] Viola L, Knill E and Lloyd S 1999 *Phys. Rev. Lett.* **82** 2417
[19] Zanardi P 1999 *Phys. Lett. A* **258** 77
[20] Zurek W H 1984 *Phys. Rev. Lett.* **53** 391
[21] Dugi'c M 2000 *Quantum Computers and Computing* **1** 102
[22] Beige A, Braun D, Tregenna B and Knight P L 2000 *Phys. Rev. Lett.* **85** 1762
[23] Pachos J and Walther H 2002 *Phys. Rev. Lett.* **89** 187903
[24] Beige A 2003 *Phys. Rev. A* **67** 020301(R)
[25] Pechen A and Rabitz H 2006 *Phys. Rev. A* **73** 062102
[26] Misra B and Sudarshan E C G 1977 *J. Math. Phys.* **18** 756
[27] Sudbery A 1984 *Ann. Phys.* **157** 512
[28] Beige A and Hegerfeldt G C 1996 *Phys. Rev. A* **53** 53
[29] Facchi P, Lidar D A and Pascazio S 2004 *Phys. Rev. A* **69** 032314
[30] Facchi P, Tasaki S, Pascazio S, Nakazato H, Tokuse A and Lidar D A 2005 *Phys. Rev. A* **71** 022302
[31] Georgi H1999 *Lie algebras in particle physics*, Frontiers in Physics, Vol. **54** (Perseus Books)
[32] Beige A and Hegerfeldt G C 1997 *J. Phys. A* **30** 1323
[33] Dehmelt H G 1975 *Bull. Am. Phys. Soc.* **20** 60
[34] Cook R J and Kimble H J 1985 *Phys. Rev. Lett.* **54** 1023
[35] Facchi P and Pascazio S 2008 *J. Phys. A* **41** 493001
[36] Metz J, Trupke M and Beige A 2006 *Phys. Rev. Lett.* **97** 040503
[37] Metz J and Beige A 2007 *Phys. Rev. A* **76** 022331
[38] Busch J, Kyoseva E S, Trupke M and Beige A 2008 *Phys. Rev. A* **78** 040301(R)

[39] Khudaverdyan M, Alt W, Kampschulte T, Reick S, Thobe A, Widera A and Meschede D 2009 *Phys. Rev. Lett.* **103** 123006

[40] Sklarz S E, Tannor D J and Khaneja N 2004 *Phys. Rev. A* **69** 053408

[41] Vacanti G and Beige A 2009 *New J. Phys.* **11** 083008

[42] Wang X T and Schirmer S G 2009 *Phys. Rev. A* **80** 042305

[43] Verstraete F, Wolf M M and Cirac J I 2008 Quantum computation, quantum state engineering, and quantum phase transitions driven by dissipation *Preprint* arXiv:0804.1888

[44] Kraus B, Büchler H P, Diehl S, Kantian A, Micheli A and Zoller P 2008 *Phys. Rev. A* **78** 042397

[45] Ticozzi F and Viola L 2008 Analysis and synthesis of attractive quantum Markovian dynamics *Preprint* arXiv:0809.0613

Journal of Physics: Conference Series **254** (2010) 012010 doi:10.1088/1742-6596/254/1/012010

The entanglement of few-particle systems when using the local-density approximation

J P Coe and I D'Amico

Department of Physics, University of York, York YO10 5DD, United Kingdom

E-mail: jpc503@york.ac.uk, ida500@york.ac.uk

Abstract. In this chapter we discuss methods to calculate the entanglement of a system using density-functional theory. We firstly introduce density-functional theory and the local-density approximation (LDA). We then discuss the concept of the 'interacting LDA system'. This is characterised by an interacting many-body Hamiltonian which reproduces, uniquely and exactly, the ground state density obtained from the single-particle Kohn-Sham equations of density-functional theory when the local-density approximation is used. We motivate why this idea can be useful for appraising the local-density approximation in many-body physics particularly with regards to entanglement and related quantum information applications. Using an iterative scheme, we find the Hamiltonian characterising the interacting LDA system in relation to the test systems of Hooke's atom and helium-like atoms. The interacting LDA system ground state wavefunction is then used to calculate the spatial entanglement and the results are compared and contrasted with the exact entanglement for the two test systems. For Hooke's atom we also compare the entanglement to our previous estimates of an LDA entanglement. These were obtained using a combination of evolutionary algorithm and gradient descent, and using an LDA-based perturbative approach. We finally discuss if the position-space information entropy of the density—which can be obtained directly from the system density and hence easily from density-functional theory methods—can be considered as a proxy measure for the spatial entanglement for the test systems.

Received: 19 February 2010

1. Introduction

The mapping of an interacting many-body system to a collection of effective single particle equations is a ubiquitous and powerful approach in theoretical physics. This is perhaps most notably embodied in density-functional theory (DFT) where a non-interacting many-body system, the Kohn-Sham system [1], is proven to have exactly the same ground state density as the original interacting system. By solving a set of single-particle Kohn-Sham (KS) equations, the ground state density can be calculated and from this in principle all the ground-state properties of the interacting many-body system [2]. The KS equations would give the exact density due to the inclusion in their effective single-particle potential of the so-called "exchange-correlation" potential (v_{xc}). In practice v_{xc} is an unknown functional of the density so approximations for v_{xc} have been designed. The most popular and often effective method is the local-density approximation (LDA). DFT has been so far one of the most powerful methods to calculate many-body properties of complex systems, and its efficiency lies in the observation that modelling many-particle systems exactly is generally computationally intractable as, for example, one is dealing with a many-body wavefunction whose storage requires excessive amounts of memory. Solving a system of single-particle equations or dealing with the density, which is a function of only one variable, represents indeed a computationally attractive alternative. However we do not know in general how to express some

Journal of Physics: Conference Series **254** (2010) 012010 doi:10.1088/1742-6596/254/1/012010

ground state properties, such as the entanglement, in terms of the ground state particle density, and this implies that common approximation schemes used within DFT do not have a straightforward extension to the calculations of these properties. Here we present a scheme to calculate the entanglement associated with the LDA approximation and investigate how well it compares with the exact entanglement. Our test-bed systems are helium-like atoms and Hooke's atoms. This information could be useful when considering which approximation to invoke to model the suitability of systems as a component of a quantum information processing device. As the entanglement is considered as one of the main resources for these devices, the knowledge of how much entanglement could be created in a certain system appears essential for quantum information purposes. In addition, comparison of the entanglement associated with a DFT approximation to the exact entanglement offers another way to appraise the accuracy of the approximation. The latter use is illustrative of how the entanglement and theoretical tools developed for quantum information may now be used for further understanding of phenomena and modelling techniques in many-body physics, even in systems for which directly extracting the entanglement to exploit as a resource in quantum information schemes may be difficult or not possible. For example, the behaviour of the entanglement has been used to identify quantum phase transitions (see, e.g., [3, 4]), and to indicate how to increase efficiency when using the density matrix renormalisation group algorithm (DMRG) to model quantum lattice systems [5].

In this chapter we firstly introduce DFT and the LDA. We explain the concept of the 'interacting LDA system' (i-LDA) and how this may be used to calculate the entanglement corresponding to the use of the LDA. We then introduce our test-bed systems, Hooke's atom and the helium-like atoms. Next we consider how these systems can be solved using numerically 'exact' methods and also approximately using the LDA within DFT. The exact and i-LDA spatial entanglement results are then displayed and compared. We contrast these results with the position space information entropy which can easily be expressed as a functional of the density—and hence calculated using traditional DFT schemes—and investigate its usefulness as a proxy measure for the entanglement. Finally, we compare the i-LDA entanglement to our previous best approximations to an LDA entanglement, found using a combination of evolutionary algorithm and gradient descent, and an LDA-based perturbative approach [6].

2. Density-functional theory

Directly solving the many-body Schrödinger equation is a formidable, arduous task and doing so accurately for systems of more than a few electrons remains outside the realms of contemporary computing. This can be partly illuminated by considering the wavefunction on a mesh of 1000^3 points for one electron; for the same accuracy with N electrons one would require 1000^{3N} points, clearly an excessive amount of memory as N becomes large. Yet nature, in a sense performs similar calculations all the time, but these are too complex to be reproduced by 'classical' computation. Quantum computing offers some hope in the simulation of quantum systems (see for example Ref. [7] and Ref. [8]). However, unless quantum computers of a substantial number of qubits are realised, simulating large systems may not be feasible. There is though a powerful and important method that allows classical computing to efficiently model many-body electron systems through the density: density-functional theory. The density depends upon only one co-ordinate (so only 1000^3 mesh points in our example) therefore by using this, rather than the many-body wavefunction, solving many-body systems can be made computationally more tractable. In its simplest formulation density-functional theory expresses observables of the ground-state of an interacting electron system as functionals of the ground-state density,

$$n(\boldsymbol{r_1}) = N \int |\Psi(\boldsymbol{r_1}...\boldsymbol{r_N})|^2 \boldsymbol{dr}_2...\boldsymbol{dr}_N \tag{1}$$

and is exact in principle due to a theorem by Hohenberg and Kohn [2] which we present next.

2.1. The Hohenberg-Kohn Theorem

In an electron system with Hamiltonian

$$\hat{H} = \hat{T} + \hat{V}_{ee} + \sum_i v_{ext}(\boldsymbol{r}_i) \tag{2}$$

with a non-degenerate ground state, the ground state density uniquely determines the ground state many-body wavefunction and therefore the external potential. This is known as the Hohenberg-Kohn theorem [2]. The usual proof shows how the density uniquely determines the wavefunction [9]. Assume there exist two Hamiltonians H_1 and H_2 with different ground-state wavefunctions, $|\Psi_1\rangle$ and $|\Psi_2\rangle$ respectively, that give rise to the same density. Then

$$\begin{aligned}\langle\Psi_1| H_2 |\Psi_1\rangle &= \langle\Psi_1| H_2 - H_1 |\Psi_1\rangle + \langle\Psi_1| H_1 |\Psi_1\rangle \\ &= \langle\Psi_1| \sum_i (v_{2,ext}(\boldsymbol{r}_i) - v_{1,ext}(\boldsymbol{r}_i)) |\Psi_1\rangle + E_1 \\ &= \int (v_{2,ext}(\boldsymbol{r}) - v_{1,ext}(\boldsymbol{r}))\, n(\boldsymbol{r}) d\boldsymbol{r} + E_1.\end{aligned} \tag{3}$$

Then by the variational principle

$$\int (v_{2,ext}(\boldsymbol{r}) - v_{1,ext}(\boldsymbol{r}))\, n(\boldsymbol{r}) d\boldsymbol{r} + E_1 > E_2, \tag{4}$$

with E_2 the ground state related to H_2.

This can be repeated but with 1 and 2 exchanged to give

$$\begin{aligned}\langle\Psi_2| H_1 |\Psi_2\rangle &= \langle\Psi_2| H_1 - H_2 |\Psi_2\rangle + \langle\Psi_2| H_2 |\Psi_2\rangle \\ &= \langle\Psi_2| \sum_i (v_{1,ext}(\boldsymbol{r}_i) - v_{2,ext}(\boldsymbol{r}_i)) |\Psi_2\rangle + E_2 \\ &= \int (v_{1,ext}(\boldsymbol{r}) - v_{2,ext}(\boldsymbol{r}))\, n(\boldsymbol{r}) d\boldsymbol{r} + E_2.\end{aligned} \tag{5}$$

By the variational principle

$$\int (v_{1,ext}(\boldsymbol{r}) - v_{2,ext}(\boldsymbol{r}))\, n(\boldsymbol{r}) d\boldsymbol{r} + E_2 > E_1. \tag{6}$$

Adding the inequalities gives

$$E_1 + E_2 > E_1 + E_2. \tag{7}$$

This inequality is a fallacy so the initial assumption must have been incorrect: there do not exist two ground state wavefunctions that give rise to the same density. Hence by reductio ad absurdum the density uniquely determines the wavefunction.

As we may express the external potential as

$$-\frac{\hat{T}\psi}{\psi} - \frac{\hat{V}_{ee}\psi}{\psi} + E = \sum_i v_{ext}(\boldsymbol{r}_i) \tag{8}$$

then the density by determining the wavefunction also determines the external potential (up to an additive constant), and therefore all of the properties of the system.

Journal of Physics: Conference Series **254** (2010) 012010 doi:10.1088/1742-6596/254/1/012010

Although the Hohenberg-Kohn theorem proves the existence of the one-to-one relationship between ground state densities and wavefunctions it does not suggest how to construct the density nor find the ground-state properties of the system from the density. To achieve this the Kohn-Sham equations [1] are used.

2.2. *The Kohn-Sham equations*

The Kohn-Sham (KS) equations allow the ground-state density to be found efficiently and the system energy to be expressed using the density. The KS equations are derived by considering the system energy as a functional of the density $E[n]$, this is minimised subject to a constraint incorporating N, the number of electrons,

$$\int n(\boldsymbol{r})d\boldsymbol{r} = N. \tag{9}$$

Using Lagrange multipliers with the functional derivative

$$\frac{\delta F[n(y)]}{\delta n(x)} = \lim_{\epsilon \to 0} \frac{F[n(y) + \epsilon\delta(y-x)] - F[n(y)]}{\epsilon} \tag{10}$$

gives

$$\frac{\delta E[n]}{\delta n(\boldsymbol{r})} - \mu = 0. \tag{11}$$

As the expression of the kinetic energy as a functional of the density for an interacting electron system is unknown, Kohn and Sham re-wrote the exact energy functional in terms of the kinetic energy of non-interacting electrons $T_{NI}[n]$, the Hartree energy $U[n]$, and the potential energy from the electron density. This requires the introduction of a functional correctional term, the exchange-correlation energy $E_{xc}[n]$. The expression for $E[n]$ then becomes

$$E[n] = T_{NI}[n] + U[n] + \int v_{ext}(\boldsymbol{r})n(\boldsymbol{r})d\boldsymbol{r} + E_{xc}[n] \tag{12}$$

where

$$U[n] = \frac{1}{2}\int\int \frac{n(\boldsymbol{r})n(\boldsymbol{r}')}{|\boldsymbol{r}-\boldsymbol{r}'|}d\boldsymbol{r}d\boldsymbol{r}'. \tag{13}$$

Using Lagrange multipliers now gives

$$\frac{\delta T_{NI}[n]}{\delta n(\boldsymbol{r})} + v_{eff}(\boldsymbol{r}) - \mu = 0. \tag{14}$$

Here

$$v_{eff}(\boldsymbol{r}) = v_{\mathrm{H}}(\boldsymbol{r};[n]) + v_{ext}(\boldsymbol{r}) + v_{\mathrm{xc}}(\boldsymbol{r};[n]) \tag{15}$$

where they define the exchange-correlation potential

$$v_{\mathrm{xc}}(\boldsymbol{r};[n]) = \frac{\delta E_{xc}[n]}{\delta n(\boldsymbol{r})} \tag{16}$$

and the Hartree potential

$$v_{\mathrm{H}}(\boldsymbol{r};[n]) = \int \frac{n(\boldsymbol{r}')}{|\boldsymbol{r}-\boldsymbol{r}'|}d\boldsymbol{r}'. \tag{17}$$

Now Eq. 14 can be read as a non-interacting system—the KS system—with a potential v_{eff}. So to find the density and hence the energy the KS equations must be solved self-consistently:

$$\left(-\frac{1}{2}\nabla^2 + v_{eff}(\boldsymbol{r})\right)\phi_i(\boldsymbol{r}) = \epsilon_i\phi_i(\boldsymbol{r}) \tag{18}$$

and

$$n(\boldsymbol{r}) = \sum_{i=1}^{N} |\phi_i(\boldsymbol{r})|^2. \tag{19}$$

As

$$T_{NI}[n] = \sum_{i=1}^{N}\int \phi_i^*(\boldsymbol{r})\left(-\frac{1}{2}\nabla^2\phi_i(\boldsymbol{r})\right)\boldsymbol{dr} \tag{20}$$

then the energy $E[n]$ may be calculated from the density using E_{xc}.

This scheme would be exact but the general functional form of E_{xc} is unknown hence it has to be approximated. One of the simplest yet often effective approximations is using the fact that, for systems with slowly varying density, the exchange-correlation density can be locally approximated by that of a uniform electron gas. This is known as the local density approximation.

2.3. The local-density approximation

The local-density approximation for the exchange-correlation energy depends only upon the density as opposed to generalised gradient approximations which include first and higher derivatives. It may be written using the form

$$E_{xc} = \int \boldsymbol{dr} e_{xc}(n(\boldsymbol{r})). \tag{21}$$

The exchange-correlation energy density may be separated into its exchange and correlation parts, $e_{xc} = e_x + e_c$. Then, for a three-dimensional homogeneous electron gas, e_x is known exactly [10]

$$e_x = -\frac{3}{4}\left(\frac{3}{\pi}\right)^{1/3} n^{\frac{4}{3}}, \tag{22}$$

while e_c has to be approximated. We use e_c as calculated numerically by Perdew-Wang [11] by fitting the parameters A, α, $\beta_1 \ldots \beta_4$ to the results of Monte-Carlo simulations of the homogeneous electron gas [12]. For zero spin polarisation this has the form

$$e_c = -2An(1+\alpha r_s)\log\left(1 + \frac{1}{2A(\beta_1 r_s^{1/2} + \beta_2 r_s + \beta_3 r_s^{3/2} + \beta_4 r_s^2)}\right), \tag{23}$$

where r_s is the radius of the sphere whose volume is equal to the volume per electron in a homogeneous electron gas,

$$r_s = \left(\frac{3}{4\pi n(r)}\right)^{\frac{1}{3}}. \tag{24}$$

The above approximation is exact for the homogeneous electron gas so would be expected to be more accurate for systems close to this limit.

3. The interacting LDA system

We wish to compare the entanglement calculated using a local-density approximation to the exact entanglement, and although in theory the density determines all the ground-state properties of the system it is unknown how to use the density to represent certain observables, including the entanglement. One method to calculate the entanglement within LDA is to construct that many-body interacting system whose ground state wavefunction reproduces exactly the LDA density. We named this 'the interacting LDA system' (i-LDA) [6]. By the Hohenberg-Kohn theorem [2] this system is unique.

Given a test-bed interacting many-body system for which ground-state properties, such as the entanglement, can be calculated exactly, by comparing their values with the ones obtained using the corresponding i-LDA wavefunction, we can infer how precise the LDA can be as an approximation to those quantities. Notice that the external potentials characterising the initial many-body system and the i-LDA system are different, and that *the LDA density is at the same time an approximation to the density of the test-bed system and the exact ground state density of the i-LDA system.*

Such a route to assess system properties within the LDA is computationally intensive, but, for small test systems, allows a way to investigate the accuracy of the LDA for quantities for which functional expressions in terms of the density are currently unknown. In addition this procedure offers a novel way to appraise the LDA by considering the differences between the exact and i-LDA external potentials, thereby furnishing researchers with spatially resolved information about an approximation's accuracy. A similar procedure to the one described can be applied to other approximations within DFT.

We introduced a method to approximate the i-LDA wavefunction in Ref. [6], then in Ref. [13] we developed an iterative scheme to find the Hamiltonian of the i-LDA system 'exactly' (within numerical accuracy). Then the 'exact' i-LDA wavefunction could then be derived. This is the scheme we use in this chapter to allow the entanglement corresponding to the LDA to be ascertained.

4. The test-bed systems

The helium atom comprises two electrons held by a nuclear charge arising from two protons in the nucleus. By using the Born-Oppenheimer approximation the motion of the nucleus may be ignored. The system generalises to helium-like atoms by considering a nuclear charge of Z giving rise to the following non-relativistic Hamiltonian in atomic units ($e = \hbar = m = 4\pi\epsilon_0 = 1$)

$$H = -\frac{1}{2}\nabla_1^2 - \frac{1}{2}\nabla_2^2 - \frac{Z}{r_1} - \frac{Z}{r_2} + \frac{1}{|\boldsymbol{r_1} - \boldsymbol{r_2}|}, \tag{25}$$

where $r = |\boldsymbol{r}|$. With $Z = 2$ this is the Hamiltonian for the helium atom. Interestingly although the hydride atom ($Z = 1$) exists, and its energy may be calculated, Baker et al [14] showed that for nuclear charges below $Z \approx 0.911$ the system is unbound.

Hooke's atom is an interacting system of two electrons held by a harmonic confining potential with Hamiltonian in atomic units of

$$H = -\frac{1}{2}\nabla_1^2 - \frac{1}{2}\nabla_2^2 + \frac{1}{2}\omega^2 r_1^2 + \frac{1}{2}\omega^2 r_2^2 + \frac{1}{|\boldsymbol{r_1} - \boldsymbol{r_2}|}. \tag{26}$$

Here ω is the characteristic frequency of the confining potential. The exact spatial entanglement and its approximations within LDA have previously been evaluated for Hooke's atom in Ref. [6].

5. Method of solution

We solve the many-body Schrödinger equation, using 'exact' diagonalisation (EXD) with an efficient choice of basis. For two electrons in a spherically symmetric potential the ground state can only be a function of the distance of each electron from the origin and the angle between the electron vectors. Hence we may find the ground state using exact diagonalisation with a basis in three dimensions: r_1, r_2, θ.

We use the basis

$$\phi_{ijl} = R_i(r_1)R_j(r_2)\frac{1}{\sqrt{4\pi}}\sqrt{\frac{2l+1}{4\pi}}P_l(\cos(\theta)), \tag{27}$$

where we employ the hydrogen-like wavefunction $R_i(r) = Q_i(r)e^{-\alpha r}$. The choice of α is influenced by the strength of the nuclear charge Z. Here the $Q_i(r)$ are polynomials of degree i created via the Gram-Schmidt procedure such that the R_i are orthonormal, i.e., $\int_0^\infty R_i(r)R_j(r)r^2dr = \delta_{ij}$. The Legendre polynomials P_l are orthogonal to each other and we explicitly include the normalisation factor for integration over all angles. We intend to calculate the matrix elements of the Hamiltonian with respect to the ϕ_{ijl} and $\phi_{i'j'l'}$ basis functions Eq. 27. The external potential is spherically symmetric therefore the potential integrals become just 1D integrals and are only non-zero when $l = l'$. This is propitious as they are the only integrals that have to be re-calculated each time in the iterative scheme.

The kinetic term in the integrand is

$$\nabla_i^2 = \frac{1}{r_i^2}\frac{\partial}{\partial r_i} r_i^2 \frac{\partial}{\partial r_i} - \frac{L_i^2}{r_i^2} \tag{28}$$

where L_i is the angular momentum operator. As

$$P_l(cos(\theta)) = \frac{4\pi}{(2l+1)} \sum_{m=-l}^{m=+l} Y_{lm}(\theta_1, \phi_1) Y_{lm}^*(\theta_2, \phi_2), \tag{29}$$

then

$$L_i^2 P_l = l(l+1) P_l. \tag{30}$$

Hence the kinetic terms are also 1D integrals in r_1 and r_2 when $l' = l$, otherwise the integrals are zero.

The Coulomb interaction integral is more complicated. First we expand the Coulomb interaction term using the Legendre polynomials

$$\frac{1}{|\boldsymbol{r}_1 - \boldsymbol{r}_2|} = \sum_{k=0}^{\infty} \frac{(r_<)^k}{(r_>)^{k+1}} P_k(\cos(\theta)), \tag{31}$$

where $r_< = \min\{r_1, r_2\}$ and $r_> = \max\{r_1, r_2\}$.

Next we use a result [15] for an integration of three Legendre polynomials

$$\int_{-1}^{1} P_{l'}(z) P_k(z) P_l(z) dz = 2 \begin{pmatrix} l' & k & l \\ 0 & 0 & 0 \end{pmatrix}^2. \tag{32}$$

If $l' > k + l$ then the polynomial of degree $k + l$ can be formed from Legendre polynomials each of degree less than l' hence the integral is zero, similarly for $k > l + l'$ and $l > k + l'$. Here the $3j$ symbol (see for example Ref. [16]) is zero for $l' + k + l = 2p + 1$ where p is an integer. While for $l' + k + l = 2p$

$$\begin{pmatrix} l' & k & l \\ 0 & 0 & 0 \end{pmatrix} = (-1)^p \sqrt{\Delta(l'kl)} \frac{p!}{(p-l')!(p-k)!(p-l)!}, \tag{33}$$

where

$$\Delta(abc) = \frac{(a+b-c)!(b+c-a)!(c+a-b)!}{(a+b+c+1)!}. \tag{34}$$

Hence the integral for the Coulomb interaction when switching to spherical polar integration, including the normalisation, and integrating over angles is

$$\begin{aligned} \langle \phi_{i'j'l'} | \frac{1}{|\boldsymbol{r}_1 - \boldsymbol{r}_2|} | \phi_{ijl} \rangle = \\ \sqrt{2l'+1}\sqrt{2l+1} \sum_{k=0}^{l+l'} \begin{pmatrix} l' & k & l \\ 0 & 0 & 0 \end{pmatrix}^2 \int_0^\infty dr_1 R_{i'}(r_1) R_i(r_1) \\ (\frac{1}{r_1^{k-1}} \int_0^{r_1} R_{j'}(r_2) R_j(r_2) r_2^{k+2} dr_2 + \\ r_1^{k+2} \int_{r_1}^{\infty} R_{j'}(r_2) R_j(r_2) \frac{1}{r_2^{k-1}} dr_2). \end{aligned} \tag{35}$$

As the 3j symbol is zero for $k > l + l'$ then there are at most $l + l'$ double integrals, but the other constraints then produce a zero 3j symbol for many of these. Hence efficiency is increased: we have a small sum of double integrals rather than a triple integral to compute.

We also consider the single particle KS equation (Eq. 18) where $v_{\text{ext}}(\boldsymbol{r}) = -Z/r$ for helium-like atoms and $v_{\text{ext}}(\boldsymbol{r}) = \omega r^2/2$ for Hooke's atom. For approximating v_{xc} we use $v_{\text{xc}}^{\text{LDA}} = v_{\text{x}}^{\text{LDA}} + v_{\text{c}}^{\text{LDA}}$. We employ the fit of Perdew and Wang [11] (see Eq. 23) for $v_{\text{c}}^{\text{LDA}}$ while the expression for $v_{\text{x}}^{\text{LDA}}$ is that of Wigner and Seitz [10] (see Eq. 22). We consider only spherically symmetric potentials so the KS equation reduces to a one dimensional problem. We solve this self consistently to obtain n^{target}, the LDA approximation to the ground state electron density.

This 'target' density is then used in the iterative scheme developed in Ref. [13] which allows to find the external potential for the interacting many-body system of ground state density n^{target}, in this case the i-LDA system. This scheme is based on the relation

$$v_{\text{ext}}^{i+1}(\boldsymbol{r_1}) = \frac{1}{n_i(\boldsymbol{r_1})}|E_i|[n_i(\boldsymbol{r_1}) - n^{\text{target}}(\boldsymbol{r_1})] + v_{\text{ext}}^{i}(\boldsymbol{r_1}). \tag{36}$$

Here E_i is the energy of the interacting system with external potential $\sum_j v_{\text{ext}}^i(\boldsymbol{r_j})$: this many-body system is solved at every step by using the aforementioned method to give the many-body wavefunction from which we compute the density n_i. We aid convergence for both test-bed systems by mixing v_{ext}^{i+1} with 80% of v_{ext}^i, and iterate until the error $\int d^3r\,|n_i(\boldsymbol{r}) - n^{\text{target}}(\boldsymbol{r})|/\int d^3r\, n^{\text{target}}(\boldsymbol{r})$ has reached a desired level on the chosen integration range.

We note that the inversion scheme Eq. 36 could also be used to derive the unknown potential related to a generic density n^{target} obtained, e.g., from experimental measurements.

6. Spatial entanglement calculation

For the systems we are considering, the ground state wave-function can be factorised in orbital and spin part. The spin part is a singlet which is maximally entangled and constant. We therefore ignore the spin entanglement and investigate the spatial entanglement, i.e., the particle-particle entanglement stemming from the continuous spatial degrees of freedom.

We use the two-electron wavefunction to calculate the linear entropy of the reduced density matrix

$$L = 1 - Tr\rho_A^2 \tag{37}$$

as a measure of the spatial entanglement. Here $\rho_A = Tr_B\,|\psi\rangle\,\langle\psi|$ is found by tracing out the system density matrix over the degrees of freedom of one of the two subsystems. The linear entropy was, perhaps, first put forward as a measure of purity of a quantum state by Zurek, Habib and Paz [17]. The purity of the reduced density matrix can be shown to give an indication of the number and spread of the terms in the Schmidt decomposition of the state and therefore a measure of the entanglement. Here, in the continuous case, we calculate the entanglement using

$$\rho_{\text{A}}(\boldsymbol{r_1},\boldsymbol{r_2}) = \int \Psi^*(\boldsymbol{r_1},\boldsymbol{r_3})\Psi(\boldsymbol{r_2},\boldsymbol{r_3})d\boldsymbol{r_3}, \tag{38}$$

$$\rho_{\text{A}}^2(\boldsymbol{r_1},\boldsymbol{r_2}) = \int \rho_{\text{A}}(\boldsymbol{r_1},\boldsymbol{r_3})\rho_{\text{A}}(\boldsymbol{r_3},\boldsymbol{r_2})d\boldsymbol{r_3}, \tag{39}$$

and

$$Tr\rho_{\text{A}}^2 = \int \rho_{\text{A}}^2(\boldsymbol{r},\boldsymbol{r})d\boldsymbol{r}. \tag{40}$$

We will investigate the change in spatial entanglement with the strength of the confining potential, which is characterised by ω or the nuclear charge Z for Hooke's and helium-like atoms respectively.

7. Helium-like atoms

For helium-like atoms the (numerically) exact and the i-LDA entanglement are depicted in Fig. 1. There we see that the entanglement increases as the nuclear charge decreases which may be attributed to the increase in the ratio of the electron-electron interaction to the energy from the external potential. A similar situation was observed in the spatial entanglement of Hooke's atom [6]. We note that the spatial entanglement is low for the parameters considered, when compared with the theoretical maximum of $L = 1$. The 'exact' and i-LDA entanglement are close and the gap between them increases as the weight of the particle-particle interactions increase. This fits in with the LDA performing better when many-body interactions are weak: as the confining potential increases, many-body interactions will become less important relative to it, until the system may be fairly well approximated by a non-interacting Hamiltonian thereby giving rise to a product state solution. We note that although helium-like atoms

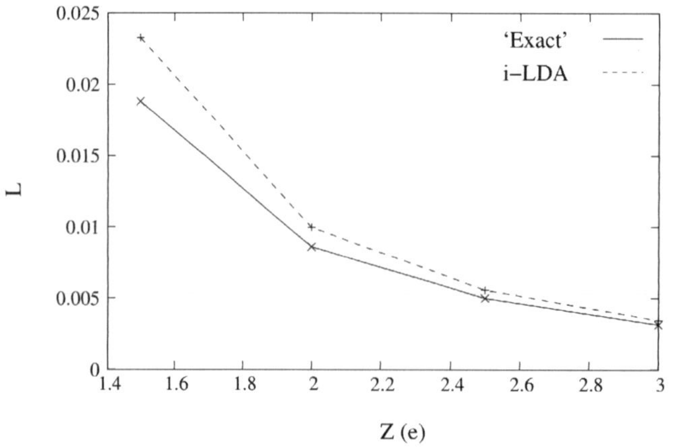

Figure 1. The linear entropy L of the reduced density matrix versus the nuclear charge Z as a measure of entanglement for helium-like atoms and the corresponding i-LDA systems.

have a bound state until $Z \approx 0.911$ we were unable to numerically solve the Kohn-Sham equations with the LDA satisfactorily for $Z < 1.5$. We hypothesise that for the exact solution with $Z \lesssim 0.911$ a transition to a resonance state—as studied recently for two electrons in a spherically symmetric square well [18] and the spherical helium atom [19]—would be expected to occur. This would result in a superposition of one electron constrained by the nucleus with the other unbound. Such a wavefunction would be expected to have a spatial entanglement of $L = 1/2$.

8. Hooke's atom

For Hooke's atom (Fig. 2) we see that the accuracy of the entanglement when using the LDA is high, but decreases a little as ω decreases and interactions, as quantified by the ratio of the electron-electron interaction to the energy from the external potential, become larger. The value of the i-LDA entanglement for $\omega = 0.00584$, lower than the value for $\omega = 0.00957$, seems anomalous as does the change from an overestimate to an underestimate of the exact entanglement. We note that $\omega = 0.00584$ has the poorest match (0.053% error) between the i-LDA estimate and the actual LDA density of the ω displayed and is the smallest ω for which we could achieve a good match with a reasonable basis size of 8^3 wavefunctions. Therefore due to a seemingly high sensitivity of the entanglement to the wavefunction yet a relative insensitivity of the density to the wavefunction, we may have crossed a threshold so the small error in reproducing the LDA density is now not small enough to prevent a large error when calculating the LDA entanglement. However we also note that 0.053% error still represents a very good accuracy in reproducing the LDA density, and that when employing an even larger basis the difference between

Journal of Physics: Conference Series **254** (2010) 012010 doi:10.1088/1742-6596/254/1/012010

old and new density estimates was just 0.089%. A basis of 8^3 was also sufficient to calculate the exact entanglement to 3 significant figures. In addition we saw in previous work [6] that the difference between exact and LDA energies exhibited a non-trivial relationship with ω, and a perturbed LDA-based approach to approximate Hooke's atom many-body wavefunction also produces a decrease of the entanglement in a similar range of ω (see Fig. 3). These observations may imply that a non-trivial relationship between ω and LDA-based estimates of the entanglement could be a real effect. We suggest that although the accuracy of the i-LDA entanglement for the two smallest values of ω may be questionable, the correct i-LDA entanglement for these values should not be too different and we are confident of the accuracy of the i-LDA entanglement for larger values of ω.

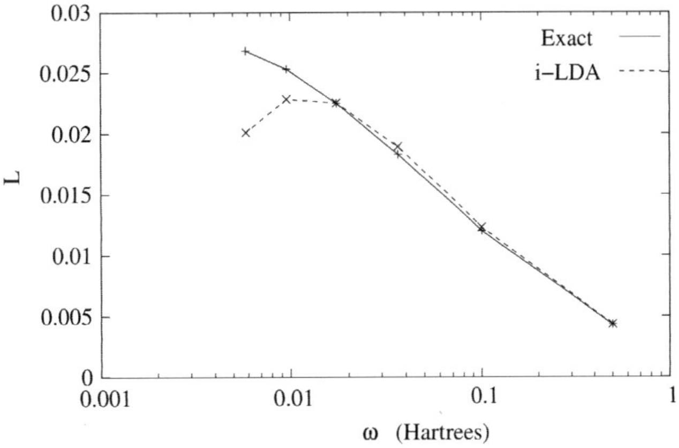

Figure 2. The linear entropy L of the reduced density matrix versus ω as a measure of entanglement for Hooke's atom and the corresponding i-LDA system.

8.1. Comparison with other LDA-based entanglement estimates

In previous work [6] we devised different ways to derive an LDA-based approximation to the entanglement.

The first is a combination of gradient descent and evolutionary algorithms to derive an approximation for the i-LDA wavefunction from which to calculate the entanglement. The main limitation of this approach is that the trial i-LDA wavefunction was based on a specific functional form. This variational-type approach included some parameters to be varied to improve performance.

The second approach to derive an LDA-based approximation for the interacting many-body wavefunction is based on a perturbative scheme whose zero-order Hamiltonian includes the Kohn-Sham effective potential (and hence, through v_{xc}, partly the many-body interaction) with the LDA approximation for the exchange-correlation term. This approach improves as higher orders are taken into account, but a first-order wavefunction would still fail to reproduce the exact entanglement when many-body interactions become dominant.

We see in Fig. 3 that, in the intermediate and strong particle-particle interaction regime, the exact i-LDA entanglement for Hooke's atom reproduces the system exact entanglement much more accurately than any of the approximations mentioned above. Approximations labelled as EA1 and EA2 were achieved by using a combination of gradient descent and evolutionary algorithms [6]. In EA1 we sought to minimise the difference between the density arising from the trial wavefunction and the exact density and then, within the wavefunctions giving an accurate enough match, we choose the wavefunction corresponding to the lowest ground state energy. The trial wavefunctions in this case are restricted to wavefunctions which can be separated in centre of mass and relative motion components. In EA2 we sought to minimise the difference between the density arising from the trial wavefunction and the exact density together with a minimisation of the energy of the trial wavefunction. The trial wavefunctions in this case used a finite number of functions in terms of r_1, r_2, and θ from a general basis for two electrons in a spherically symmetric potential. The gradient descent allows the basis coefficients for the wavefunction that gives a local minimum to be found while the evolutionary algorithm allows more of the parameter space to be explored so in principle, given a large enough basis and enough time, the combination should find the global minimum. With our finite basis and limited run-time we see that, although this method gives the trend of the i-LDA entanglement, the approximate i-LDA entanglement from the evolutionary algorithm is much higher than the actual i-LDA entanglement.

As mentioned above, in [6] we also considered a perturbative approach based on the LDA-KS equations, whose related interacting wavefunction converges toward the exact wavefunction as higher orders are considered. To first order we would hope for a good approximation to the exact entanglement that perhaps leans towards the i-LDA entanglement. We see in Fig. 3 that the entanglement results from first order becomes much greater than the i-LDA entanglement—but less so than the evolutionary algorithm—for stronger interactions, as quantified by the ratio of expectation of the interaction energy to the potential energy. However, the perturbation method allows an easier and more efficient way to approximate the entanglement within an LDA-related scheme, whose first order is accurate for low interactions; the exact i-LDA entanglement, although more accurate, is inaccessible without repeated complicated calculations: the inversion scheme requires solving the many-body equations more times than would be needed to ascertain the exact entanglement.

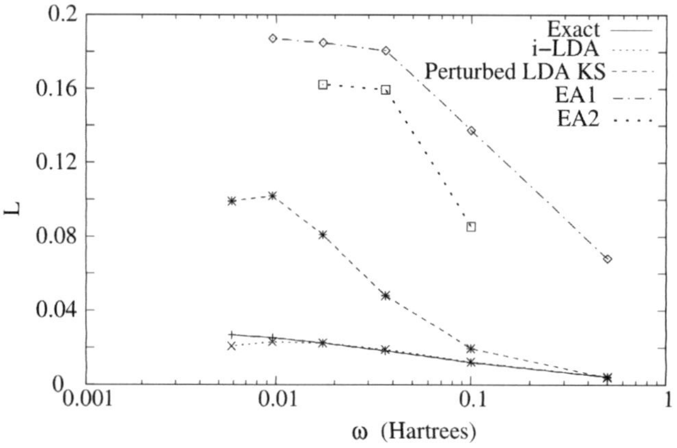

Figure 3. The linear entropy L of the reduced density matrix as a measure of the entanglement obtained using the exact, i-LDA, first order perturbation of the LDA KS equations, and evolutionary algorithm (EA1 and EA2) many-body wavefunctions.

9. Position space information entropy

We also investigate the position-space information entropy of the density

$$S_n = -\int n(\boldsymbol{r}) \ln n(\boldsymbol{r}) d\boldsymbol{r}. \tag{41}$$

as a proxy measure for the entanglement. This has been used to study the entanglement of the Moshinsky atom [20] and may also be thought of as a zeroth-order approximation to the von Neumann entropy of the reduced density matrix, where the off-diagonal terms of the reduced density matrix are set to zero [6]. As the von Neumann entropy of the reduced density matrix may be used as a measure of the entanglement and has been seen to behave similarly to L for Hooke's atom [6], it would be hoped that S_n should capture some of the behaviour of the entanglement.

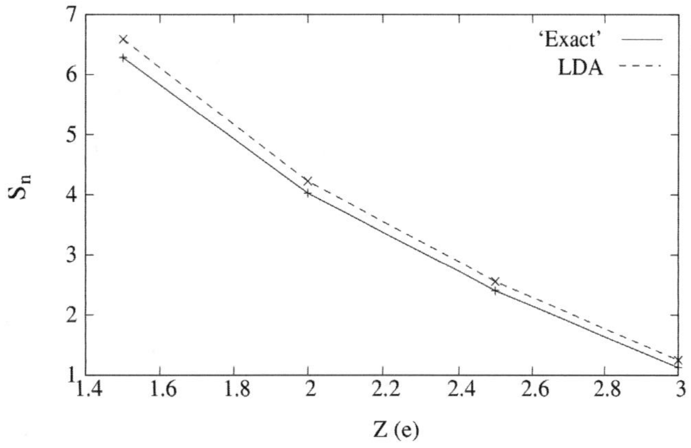

Figure 4. The position space information entropy S_n for the 'exact' and LDA densities for helium-like atoms.

We see in Fig. 4 that S_n captures the monotonically decreasing trend of the exact and i-LDA entanglement. The S_n for the LDA is greater than the exact S_n but the difference between the two does not seem to increase as Z decreases as is the case for the exact and i-LDA entanglement. The behaviour shown by S_n is also more linear with increasing Z than the entanglement.

For Hooke's atom we previously noted that S_n did not model the entanglement behaviour well over a much larger range of confining potentials [6][1]. We see in Fig. 5 that, over a smaller range of ω, S_n reproduces the general trend of the entanglement but S_n from LDA is always greater than the exact S_n which appears to not be the case for the entanglement (see Fig. 2). In addition we see that S_n is almost a straight line when plotted against ω on a logarithmic scale, this is in contrast to the entanglement which can be seen to increase less fast as ω decreases even for the small range of ω considered here.

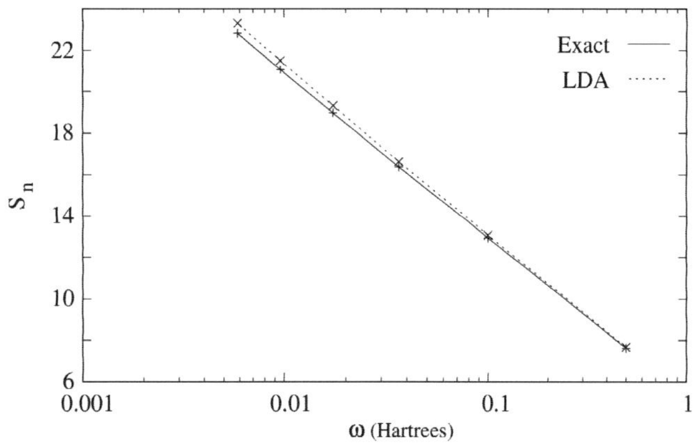

Figure 5. The position space information entropy S_n for the 'exact' and LDA densities for Hooke's atom.

[1] Label on y-axis in [6], Figs. 5 and 12, should read $S_n/\ln 2$.

10. Conclusion

We have calculated the spatial entanglement of the electrons in helium-like atoms corresponding to the exact system and to the use of the local-density approximation within density-functional theory using the concept of the interacting LDA system. The entanglement was seen to decrease with increasing nuclear charge Z. The maximum entanglement found was only a fraction $\sim 2.5\%$ of the theoretical maximum. The entanglement of the i-LDA system was seen to reproduce the qualitative and quantitative behaviour of the exact entanglement with relatively good accuracy, although the discrepancy between the measures increased as the nuclear charge diminished. We also compared the exact spatial entanglement of Hooke's atom to the entanglement of the related i-LDA system and saw that the LDA was a reasonably good approximation to the entanglement, but the accuracy decreased as the strength of the confining potential decreased and the i-LDA entanglement became non-monotonic at very small confining potential.

We also saw that previously considered approximations to the exact or i-LDA entanglement were not so reliable when the particle-particle interaction became dominant, but noted that they captured the trend of the entanglement, were successful at strong confining potentials and were often much simpler computationally. In particular the accuracy of the perturbative approach based on the LDA-KS equations could be systematically improved by going beyond first order.

We considered the position space information entropy of the density, and showed that it was able to efficiently give the general trend in the entanglement but not its detailed behaviour.

From our test-bed calculations on few-particle systems it would appear that the LDA may reproduce the entanglement behaviour well, even for systems with a highly inhomogeneous density and fairly strong interactions. More studies are necessary to ascertain if this conclusion holds for many-particle systems.

11. Acknowledgement

We acknowledge funding from EPSRC grant EP/F016719/1.

References

[1] Kohn W and Sham L J 1965 *Phys. Rev.* **140** 1133
[2] Hohenberg P and Kohn W 1964 *Phys. Rev.* **136** B864
[3] Wu L A, Sarandy M S, Lidar D A and Sham L J 2006 *Phys. Rev. A* **74** 052335
[4] Wu L A, Sarandy M S and Lidar D A 2004 *Phys. Rev. Lett.* **93** 250404
[5] Legeza O and Sólyom J 2003 *Phys. Rev. B* **68** 195116
[6] Coe J P, Sudbery A and D'Amico I 2008 *Phys. Rev. B* **77** 205122
[7] Zalka C 1998 *Fortschr. Phys.* **46** 877
[8] Nielsen M A and Chuang I L 2000 *Quantum Computation and Quantum Information* (Cambridge: Cambridge University Press)
[9] Capelle K 2006 *Brazilian J. Phys.* **36** 1318
[10] Wigner E P and Seitz F 1933 *Phys. Rev.* **43** 804
[11] Perdew J P and Wang Y 1992 *Phys. Rev. B* **45** 13244
[12] Ceperley D M and Alder B J 1980 *Phys. Rev. Lett.* **45** 566–569
[13] Coe J P, Capelle K and D'Amico I 2009 *Phys. Rev. A* **79** 032504
[14] Baker J D, Freund D E, Hill R N and Morgan III J D 1990 *Phys. Rev. A* **41** 1247
[15] Mavromatis H A and Alassar R S 1999 *Appl. Math. Lett.* **42** 101–105
[16] Messiah A 1999 *Quantum Mechanics* (Dover Publications, Inc.)
[17] Zurek W H, Habib S and Paz J P 1993 *Phys. Rev. Lett.* **70** 1187
[18] Ferrón A, Osenda O and Serra P 2009 *Phys. Rev. A* **79** 032509
[19] Osenda O and Serra P 2007 *Phys. Rev. A* **75** 042331
[20] Amovilli C and March N H 2004 *Phys. Rev. A* **69** 054302

Gaussianity bounds for quantum mixed states with a positive Wigner function

A Mandilara, E Karpov and N J Cerf

Centre for Quantum Information and Communication, École Polytechnique, CP 165, Université Libre de Bruxelles, 1050 Brussels, Belgium

E-mail: admandil@ulb.ac.be

Abstract. According to Hudson's theorem, the only quantum pure states with a positive Wigner function are Gaussian states. We summarize and compare some recent attempts at extending this theorem to the space of quantum mixed states, and complement them with new results obtained from numerical observations. Specifically, we look for upper bounds on the admissible non-Gaussianity of mixed states with a positive Wigner function.

Received: 18 November 2009

1. Introduction

The Wigner representation of quantum states [1], which is realized by joint quasi-probability distributions of canonically conjugate variables such as x and p, has a specific property that differentiates it from a true probability distribution: it can attain negative values. The existence of negative domains in the range of Wigner functions is usually interpreted as a signature of the non-classicality of the corresponding state. Several works have tried to establish theoretically a connection between this feature and other purely quantum effects such as entanglement [2], non-locality [3], and Schrödinger cats. In practice, the occurence of negative areas in the range of the Wigner function of a state can be a requirement for the success of quantum information protocols since it guarantees that the probability distribution resulting from any quadrature measurement can not be "mimicked" by a classical local model [4, 5].

It was proven by Hudson [6] that, among pure states, the only ones with a non-negative everywhere Wigner function are Gaussian states [7]. A question that arises naturally is whether Hudson's theorem can be extended to mixed states, among which not only Gaussian states may possess a positive [8] Wigner function. The first formulation and solution to this problem in the space of mixed states is due to Bröcker and Werner [9]. Working with the Fourier-Weyl transform of density matrices, they suggested a classification of mixed states according to their *Wigner spectrum.* They proved that states with a positive Wigner function are the states which include 0 and 1 in their Wigner spectrum, and they further classified this set of these states. They showed, in particular, that the set of convex mixtures of Gaussian pure states is strictly included in the set of states with a positive Wigner function. However, even though this extension of Hudson's theorem is mathematically accurate, calculating the Wigner spectrum of a state is a rather complicated procedure, rarely achievable, so these results are not operational.

In a recent work, we have suggested a different, more practical, formulation of the problem, focusing on Gaussianity [10]. Given that for pure states, Gaussianity dictates the positivity of Wigner functions, we have asked the question of how much a mixed state with positive Wigner function is allowed to deviate from a Gaussian state. We have followed two different approaches to this problem [10, 11], which we compare hereafter. Similarly as in these works, we restrict our consideration to the single-particle (single-mode) case. We then complement the results that we had obtained in [11] with a numerical search, and derive a simple criterion for the identification of mixed states possessing a Wigner function with negative areas in its range.

2. Review of our previous approaches

The first issue that arises with our suggested formulation of the extension of Hudson's theorem is the definition of *non-Gaussianity*. The problem of characterizing the non-Gaussianity of a state strongly resembles an older one, namely that of quantifying the non-classical character of a state [12]. In analogy with the common definitions of non-classicality [13, 14, 15], the non-Gaussianity of a pure state is to be defined as the minimum of the distance from the given state to all Gaussian states. The used measure of distance can be any construction that satisfies the usual requirements of a distance [16] and, in addition, that is invariant under symplectic (Gaussian) transformations. However, when looking at a computable measure of non-Gaussianity, this definition, even if correct, is not easily applicable.

In our previous works, we have chosen a recently suggested, easily evaluable, measure of non-Gaussianity [17], namely the normalized Hilbert-Schmidt distance [18] of a given state from a reference Gaussian (in general mixed) state of the same covariance matrix γ. According to this measure, the degree of non-Gaussianity δ of a state $\hat{\rho}$ from the reference Gaussian state $\hat{\rho}_G$ is quantified as,

$$\delta = \frac{1}{2\mu}\mathrm{Tr}\left((\hat{\rho}-\hat{\rho}_G)^2\right) = \frac{\mu_G + \mu - 2\mathrm{Tr}\,(\hat{\rho}\hat{\rho}_G)}{2\mu}, \tag{1}$$

where $\mu = \mathrm{Tr}\,(\hat{\rho}^2)$ is the purity of the state $\hat{\rho}$ and $\mu_G = \mathrm{Tr}\,(\hat{\rho}_G^2)$ is the purity of the corresponding Gaussian state $\hat{\rho}_G$ (having the same covariance matrix γ as the state $\hat{\rho}$). Note that the purity of the Gaussian state $\hat{\rho}_G$ can be simply expressed as $\mu_G = 1/\sqrt{\det\gamma}$. Observing that the non-Gaussianity δ is expressed in terms of purities μ and μ_G, we choose hereafter to use μ and μ_G as measures of mixedness and uncertainty, respectively. We have $0 \le \mu \le 1$ and $\mu = 1$ for pure states, while we have $\mu_G = 1/(\Delta x \Delta p)$ in the case of phase-invariant states, which makes the bridge with the Heisenberg uncertainty principle. We have $0 \le \mu_G \le 1$ and $\mu_G = 1$ for (pure Gaussian) states saturating the uncertainty principle.

Now, having chosen the measures of non-Gaussianity δ, purity μ, and uncertainty μ_G, we can formulate our problem: given the purity μ and covariance matrix γ (and therefore μ_G), what is the maximum degree of non-Gaussianity δ that a state may achieve provided that it possesses a positive Wigner function? In view of Equation (1), the search for a *maximum non-Gaussianity* can be replaced with the search for a *minimum overlap* $\mathrm{Tr}\,(\hat{\rho}\hat{\rho}_G)$ with the Gaussian reference state $\hat{\rho}_G$. In this work, we mostly employ the second option. Such a formulation of the problem implies the need for an extremization procedure. The latter may work either in the space of density matrices, or with Wigner functions, or with characteristic functions (similarly as in the work [9]). Although all these representations are equivalent, as we will see each one has its own advantages and limits.

In our first attempt [10], we have worked in Wigner *phase-space representation*. There, we have employed the method of Lagrange multipliers in order to identify the positive Wigner functions that achieve the maximum degree of non-Gaussianity δ given the constraints of μ and μ_G. Together with these constraints, which are rather simple to account for in phase-space representation, one should also require that the solution to the extremization problem is a valid Wigner function corresponding to a physical quantum state. This last condition is unfortunately hard to express in phase-space representation. In the lack of such an applicable criterion in phase space [19], we have proceeded without it. Naturally, what we have obtained then is the maximum non-Gaussianity that a classical continuous distribution in phase-space may achieve. This provides an upper bound on the maximum non-Gaussianity of quantum states with a positive Wigner function and given μ and μ_G. However this upper bound was proven by our later results to be loose, that is, not attainable by any quantum state.

In our second attempt [11], we have worked in the *state representation*, that is, using density matrices. First, we have proven that among the states that achieve the minimum permitted value of μ for given values of $\mathrm{Tr}\,(\hat{\rho}\hat{\rho}_G)$ and μ_G, there necessarily are phase-invariant states, i.e., states which can be expressed as mixtures of number states (Fock states). This observation implies that a part of the lower bound on $\mathrm{Tr}\,(\hat{\rho}\hat{\rho}_G)$ (or upper bound on the non-Gaussianity δ) as a function of μ_G and μ is realized by phase-invariant states. Then, we have identified these extremal phase-invariant states with the method of Lagrange multipliers. In this way, we were able to find the maximally non-Gaussian states in some areas of the parametric space $\{\mu_G, \mu\}$. This is what we call the *region-I* lower bound on $\mathrm{Tr}\,(\hat{\rho}\hat{\rho}_G)$ (or upper bound on the non-Gaussianity δ) in Fig. 1 ([11]). In the rest of the parametric space $\{\mu_G, \mu\}$, the lower bound on $\mathrm{Tr}\,(\hat{\rho}\hat{\rho}_G)$ corresponds to the upper bound on the purity μ when $\mathrm{Tr}\,(\hat{\rho}\hat{\rho}_G)$ and μ_G are given. This is the *region-II* lower bound on $\mathrm{Tr}\,(\hat{\rho}\hat{\rho}_G)$ (or upper bound on the non-Gaussianity δ) as exhibited in Fig. 3 in [11].

Similarly as with the extremization procedure in phase space, this procedure in density matrix representation cannot be used to account for all constraints of the problem. Indeed, we find here the maximum degree of non-Gaussianity for all mixed states and not just for states with a positive Wigner function. In fact, we could not just repeat the same procedure with the additional requirement of positivity of the Wigner function because there is no obvious way how to incorporate this constraint into the extremization problem formulated in terms of density matrices. We have proceeded numerically and have identified a part of the global bound on δ that is realized by states with positive Wigner function. Our simulation showed that the fraction of the bound where the maximum degree of non-Gaussianity for all states coincides with the maximum degree of non-Gaussianity for states with positive Wigner function is rather small, and it is mostly confined to the low purity area (see Fig. 5 in [11]).However, for the rest, big part, of the parametric space $\{\mu_G, \mu\}$ we could not draw any conclusion.

3. Numerical search

In the present work, we pursue the search for a tight upper bound on the non-Gaussianity of states with a positive Wigner function, following a numerical approach. For reasons which become obvious below, we choose to work with the overlap $\mathrm{Tr}\,(\hat{\rho}\hat{\rho}_G)$ instead of the non-Gaussianity δ. As before, we start by dividing the lower bound on $\mathrm{Tr}\,(\hat{\rho}\hat{\rho}_G)$ into two parts, namely, *region-I* where the purity μ is minimized and *region-II* where the purity μ is maximized for given values of $\mathrm{Tr}\,(\hat{\rho}\hat{\rho}_G)$ and μ_G. In Fig. 1, *region-I* and *region-II* correspond, respectively, to segment G-M_1 and M_1-P_1.

According to our previous work [11], we observe that the *region-I* of the bound for all states (segment G-M_2 in Fig. 1) can be achieved by the states with a phase-invariant Wigner function (regardless of the positivity of this Wigner function). The main step of the proof of this property is based on the Cauchy-Schwarz inequality, which holds for any pair of functions and does not depend on their positivity. Thus, this result is valid for all states, and it necessarily also holds for the subset of states with a positive Wigner function. Therefore, we can conclude that *region-I* of the bound for states with a positive Wigner function (segment G-M_1) can be realized by states with phase-invariant Wigner function, i.e., states that can be expressed as statistical mixtures of number states.

The numerical method to obtain the *region-I* of the bound for states with positive Wigner

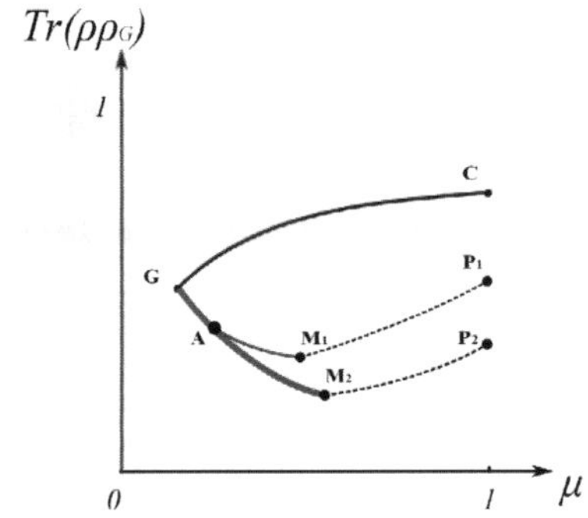

Figure 1. An explanatory depiction of the bounds on the parametric space $\{\mu_G, \mu, Tr\,(\hat{\rho}\hat{\rho}_G)\}$ for all states (index 2) and for states with positive Wigner function (index 1) employing an indicative plane of fixed μ_G. *Segment G-C* : upper bound on $Tr\,(\hat{\rho}\hat{\rho}_G)$ for all states. *Segments* G-$P_{2(1)}$: lower bounds on $Tr\,(\hat{\rho}\hat{\rho}_G)$. *Segments* G-A-$M_{2(1)}$: *region-I* of the bounds. In this part the purity μ is minimized for given values of $Tr\,(\rho_G\rho)$ and μ_G. *Segment* $M_{2(1)}$-$P_{2(1)}$: *region-II* . It stands for the part of the bound where the purity is maximized. *Point G*: Gaussian states.

function is straightforward. Among the convex combinations of number states

$$\hat{\rho} = \sum_{n=0}^{N} p_n \left|n\right\rangle \left\langle n\right| \tag{2}$$

with $0 \leq p_n \leq 1$ and $\sum_n p_n = 1$, we identify the states which possess a positive Wigner function. Then, we pick the ones which achieve the minimum purity μ given $\mathrm{Tr}\,(\hat{\rho}\hat{\rho}_G)$ and μ_G. We test states with N up to 50 and ensure that we sample over all possible combinations by choosing the mixing amplitudes $\sqrt{p_n}$ uniformly over the Haar measure (actually over the positive part of the N-dimensional sphere). Overall, we test 2.5×10^6 states and, among these, we identify $\sim 5 \times 10^4$ states with positive Wigner function. The selected states make *region-I* of the bound for states with a positive Wigner function. The latter, in turn, is composed of two parts, namely the one already identified in [11] which coincides with the global bound for all states (segment G-A in Fig. 1), and a new part which deviates from it (segment A-M_1 in Fig.1), though the deviation in terms of $\mathrm{Tr}\,(\hat{\rho}\hat{\rho}_G)$ does not exceed 0.2%. In Fig. 2, we present the *region-I* of the bounds for states with positive Wigner function and for all states projected on the planes $\{\mu_G, \delta\}$ and $\{\mu_G, \mu\}$. There, we also mark the area of coincidence for the two bounds.

Now, let us study in more details the states which belong the *region-I* of the bounds and which achieve the lowest values of $\mathrm{Tr}\,(\hat{\rho}\hat{\rho}_G)$ (points M_1 and M_2 in Fig. 1). Such states realize the boundary between *region-I* and *region-II* of the bound, and thus provide us with an ultimate (independent of the purity μ) lower bound on $\mathrm{Tr}\,(\hat{\rho}\hat{\rho}_G)$. In the case of the states with positive Wigner function, we identify this lower bound from our numerical data and represent it in Fig. 3 (thin solid line) using a polynomial function

$$M_1\,(\mu_G) = 0.0095 + 0.620\mu_G + 0.711\mu_G^2 - 0.333\mu_G^3, \tag{3}$$

fitting the numerical data in very good approximation. In Fig. 3, we also present (lower bold line) the minimum $\mathrm{Tr}\,(\hat{\rho}\hat{\rho}_G)$ for all states which, as was proven in [11], is realized by mixtures

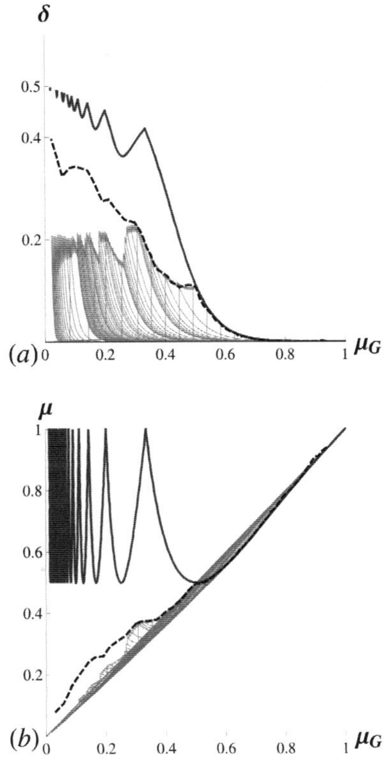

Figure 2. *Region-I* of the non-Gaussianity bound. In *(a)*, the bound for all states projected on the plane $\{\mu_G, \delta\}$ for different values of μ covers the region between the line $\delta = 0$ and the blue solid line. The allowed states lie below the surface generated by this bound for varying values of μ. The case of states with a positive Wigner function is similar, except that the blue solid line is replaced by the black dotted line. In *(b)* the bound for all states projected on the plane $\{\mu_G, \mu\}$ for different values of δ mostly covers the area between the line $\mu_G = \mu$ and the blue solid line, while it even goes slightly below $\mu_G = \mu$ as a careful inspection reveals. Note that for states below $\mu_G = \mu$, the non-Gaussianity is bounded both from below and above and a double-valued projection appears in *(b)*. Again, the case of states with a positive Wigner function is similar, except that the blue solid line is replaced by the black dotted line. Finally, we observe in *(a)* and *(b)* that a part of the bound for all states coincides with that for states with a positive Wigner function. This coinciding part of the two bounds is marked with a red grid.

of two successive number states, i.e., $p\,|n\rangle\langle n| + (1-p)\,|n+1\rangle\langle n+1|$,

$$M_2(\mu_G) = \frac{4\mu_G^2 (\mu_G - 1)^n}{(\mu_G + 1)^{n+2}} (1+n), \tag{4}$$

where $n = 0, 1, \ldots$ and $p = \left(1 - 4(1+n)\mu_G + \left(5 + 8n + 4n^2\right)\mu_G^2\right)/2\mu_G^2 \in [0, 1]$. From Fig. 3, one can see that in a certain region of μ_G, the two ultimate lower bounds, Eqs. (3) and (4), do not coincide. This clear difference suggests a sufficient condition for the existence of negative parts in the range of the Wigner function of a mixed state: *if a state characterized by a given uncertainty (i.e., a purity of the corresponding Gaussian states equal to μ_G) achieves a value of $Tr(\hat{\rho}\hat{\rho}_G)$ lower than function $M_1(\mu_G)$, then its Wigner function cannot be everywhere non-negative.* In other words, if the non-Gaussianity δ of the state exceeds a given threshold, then the state must have areas of negative values in its Wiger function.

A second question that we face, of course, is the identification of the *region-II* (segment M_1-P_1 in Fig. 1) of the bound for states with positive Wigner function. Contrary to the case of

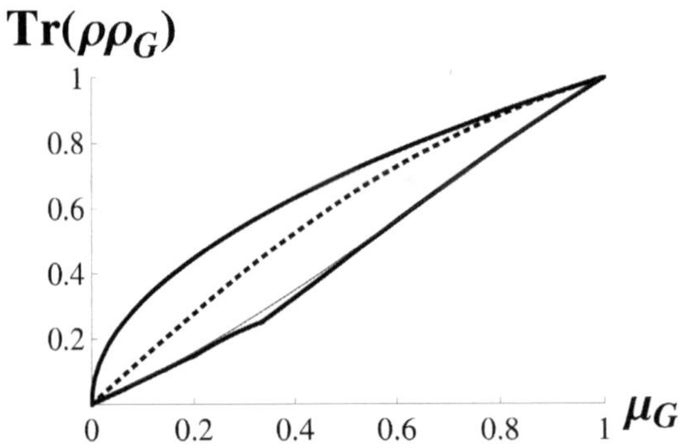

Figure 3. *Upper bold line* : global upper bound for all states derived in [10]. *Dotted line*: the lower bound, Eq.(6) for states with positive Wigner function in the limit $\mu \to 1$, Eq. (5). *Thin solid line* : ultimate lower bound, Eq.(ulti), for states with positive Wigner function. *Lower bold line* : ultimate lower bound, Eq.(4), for all states.

region-II for all states [11], we have no information about the structure of the states which realize the bound. However, for states with a positive Wigner function, one can infer the position of the bound in the limit of pure states $\mu \to 1$ (point P_1 in Fig. 1) in the following way. From Hudson's theorem, it is known that the only pure states with positive Wigner function are Gaussian states $\mu_G = 1$. Therefore, one can show that the mixed states with positive Wigner function which may approach the limit of $\mu \to 1$ (keeping $\mu_G < 1$) must be a mixture of a pure Gaussian state with some other state possessing a positive Wigner function with the mixing amplitude of the Gaussian state tending to 1. When considering the approach to the limit of pure states, one should simultaneously increase the distance (in phase space) between the two states in order to cover the whole range of values μ_G that quantifies the "spreading" of the Wigner function, otherwise we would have $\mu_G \to 1$ as $\mu \to 1$.

All the states we have tested numerically give the same result for $\mathrm{Tr}\left(\hat{\rho}\hat{\rho}_G\right)$ in the limit of pure states. The simplest one, which we choose to present here, is a mixture of two displaced and squeezed states expressed in the Wigner phase space as

$$W\left(x,p,q,s,\alpha\right) = \frac{\alpha}{\pi}\mathrm{e}^{-\frac{(x-q)^2}{s}-p^2 s} + \frac{(1-\alpha)}{\pi}\mathrm{e}^{-\frac{(x+q)^2}{s}-p^2 s}. \tag{5}$$

The limit of pure states $\mu \to 1$ is achieved, for example, when $\alpha \to 0$ while $\alpha s q^2$ is kept equal to some constant in order to have a fixed value of $\mu_G < 1$. For the state of Eq. (5), we have analytically derived the limiting value of $\mathrm{Tr}\left(\hat{\rho}\hat{\rho}_G\right)$ as a function of μ_G, namely

$$P_1\left(\mu_G\right) = \sqrt{2}\mu_G/\sqrt{1+\mu_G^2}. \tag{6}$$

We present this limiting case in Fig. 3 as a dotted line. In comparison with the ultimate lower bound, Eq. (3), this value of $\mathrm{Tr}\left(\hat{\rho}\hat{\rho}_G\right)$ is much higher. We note here that this limit $\mu \to 1$ is rather singular, as we can find states with a positive Wigner function which are almost pure, while their non-Gaussianity is far from zero (hence, they are far from saturating the uncertainty principle). Physically, these states correspond to the mixture of a Gaussian state with an infinitesimal fraction of another one, whose distance in phase space is increasingly large to keep a fixed value of the uncertainty μ_G.

Finally, for completeness, we present in Fig. 3 the ultimate upper bound on $\mathrm{Tr}\left(\hat{\rho}\hat{\rho}_G\right)$ for all states, given by $\sqrt{\mu_G}$ as that we derived in [10] employing the Cauchy-Schwartz inequality.

4. Conclusions

In conclusion, we have presented analytical and numerical estimates of the upper bound on the non-Gaussianity δ of mixed states with positive Wigner functions. From Hudson's theorem, we know that such an upper bound vanishes for pure states, but the question of estimating the highest achievable non-Gaussianity remains open in the case of mixed states, which are the states most often encountered in realistic situations. We have considered the division of the upper bound on δ into two parts, one that minimizes the purity (*region-I*) and one that maximizes the purity (*region-II*) for fixed μ_G and $Tr\,(\rho\rho_G)$. We have numerically identified *region-I* of the bound, and have shown that this part deviates slightly from the corresponding bound for all states derived in [11]. In a second time, we have numerically investigated *region-II* of the bound, and have found an indication that it moves further away from the corresponding bound for all states as the purity increases. Finally, we have numerically identified the lower bound on $\mathrm{Tr}\,(\hat{\rho}\hat{\rho}_G)$ that a state with positive Wigner function may attain given μ_G. For some interval of values of μ_G, this bound is distinct from the one for all states [11], and can be used as a sufficient criterion for the existence of negative parts in a Wigner function of a state.

Acknowledgments

AM acknowledges financial support from the Belgian National Fund for Scientific Research (FNRS). This work was carried out with the financial support of the European Commission under projects COMPAS and QAP, of the Belgian Federal program PAI under the Photonics project, and of the Brussels-Capital Region under projects CRYPTASC and Prospective Research for Brussels.

References

[1] Wigner E 1932 *Phys. Rev.* **40** 749
[2] Dahl J P, Mack H, Wolf A and Schleich W P 2006 *Phys. Rev. A* **74** 042323
[3] Bell J S 1987 *Speakable and Unspeakable in Quantum Mechanics* (Cambridge: Cambridge University Press) p 196
[4] Olivares S, Paris M G A and Bonifacio R 2003 *Phys. Rev. A* **67** 032314
[5] Garcia-Patron R, Fiurasek J, Cerf N J, Wenger J, Tualle-Brouri R and Grangier P 2004 *Phys. Rev. Lett.* **93** 130409
[6] Hudson R L 1974 *Rep. Math. Phys.* **6** 249 (1974)
[7] Soto F and Claverie P 1983 *J. Math. Phys.* **24** 97
[8] In the rest of the paper, we loosely use the term *positive Wigner* function instead of *non-negative Wigner function*. In addition, for brevity we use the term *Gaussian states* referring to the states with a Gaussian Wigner function.
[9] Bröckner T and Werner R F 1995 *J. Math. Phys.* **36**, 62
[10] Mandilara A, Karpov E and Cerf N J 2009 *Phys. Rev. A.* **79** 062302
[11] Mandilara A, Karpov E and Cerf N J 2009 Non-Gaussianity bounded uncertainty relation for mixed states *Preprint* arXiv:0910.3473
[12] Dodonov V V 2002 *J. Opt. B: Quantum Semiclass. Opt.* **4** R1
[13] Hillery M 1987 *Phys. Rev. A* **35** 725
[14] Dodonov V V, Man'ko O V, Man'ko V I, Lebedev P N and Wünsche A 2000 *J. Mod. Opt.* **47** 633
[15] Marian P, Marian T A and Scutaru A 2002 *Phys. Rev. Lett.* **88** 153601 (2002)
[16] A measure of distance should be: *(a)* positive semidefined $d\,(x_1, x_2) \geq 0$ and $d\,(x_1, x_2) = 0$ *iff* $x_1 = x_2$, *(b)* symmetric $d\,(x_1, x_2) = d\,(x_2, x_1)$ and *(c)* satisfy the triangular inequality $d\,(x_1, x_2) \leq d\,(x_1, x_3) + d\,(x_3, x_2)$.
[17] Genoni M G, Paris M G A and Banaszek K 2007 Phys. Rev. A **76** 042327
[18] Dodonov V V 2002 *J. Opt. B: Quantum Semiclass. Opt.* **4** S98-S108
[19] Tatarskiĭ V I 1983 *Usp. Fiz. Nauk.* **139** 587-619

Journal of Physics: Conference Series **254** (2010) 012012 doi:10.1088/1742-6596/254/1/012012

The 3-dimensional cube is the only periodic, connected cubic graph with perfect state transfer

Simone Severini

Department of Physics and Astronomy, University College London, WC1E 6BT London, United Kingdom

E-mail: simoseve@gmail.com

Abstract. There is *perfect state transfer* between two vertices of a graph, if a single excitation can travel with fidelity one between the corresponding sites of a spin system modeled by the graph. When the excitation is back at the initial site, for all sites at the same time, the graph is said to be *periodic*. A graph is *cubic* if each of its vertices has a neighbourhood of size exactly three. We prove that the 3-dimensional cube is the only periodic, connected cubic graph with perfect state transfer. We conjecture that this is also the only connected cubic graph with perfect state transfer.

Received: 5 January 2010

1. Introduction

1.1. State transfer

Let $G = (V, E)$ be a graph with set of vertices $V(G) = \{1, 2, ..., n\}$ and set of edges $E(G) \subseteq V(G) \times V(G) - \{\{i, i\} : i \in V(G)\}$. The *order* of G is the number of its vertices.

Let us consider a system of n spin-1/2 quantum particles with unit XY couplings. The space assigned to the entire system is $\left(\mathbb{C}^2\right)^{\otimes n}$. Each particle is attached to a vertex of G. The coupling between two particles is nonzero if and only if the particles are attached to adjacent vertices. The couplings are then specified by the adjacency matrix of the graph. The ij-th entry of the *adjacency matrix* of G is $[A(G)]_{i,j} = 1$ if $\{i, j\} \in E(G)$ and $[A(G)]_{ij} = 0$ if $\{i, j\} \notin E(G)$.

We shall work with the XY model. Let X_i and Y_i be the Pauli operators acting on the i-th particle. The Hamiltonian governing the dynamics of the spin system can be written as $H_{XY}(G) = 2^{-1} \sum_{i \neq j=1}^{n} [A(G)]_{i,j} (X_i X_j + Y_i Y_j)$. Let $\{|1\rangle \equiv \mathbf{e}_1, |2\rangle, \ldots, |n\rangle\}$ be the standard basis of the space $\mathbb{C}^n$. A vector $|i\rangle$ indicates the presence of an excitation at vertex i only. With respect to the standard basis, the ij-th entry of the Hamiltonian acting on $\mathbb{C}^n$ is $[H_{XY}(G)]_{i,j} = 2[A(G)]_{i,j}$. It follows that the Schrödinger evolution of the excitation is practically induced by a unitary matrix of the form $U_G(t) = e^{-iA(G)t}$, where $t \in \mathbb{R}^+$. We obtain a probability distribution supported by $V(G)$ by performing a projective measurement on the state $|\psi_t\rangle = U_G(t)|\psi_0\rangle$. For regular graphs, the "practically" has a much larger extension, given that, with constant couplings, any kind of interaction has an Hamiltonian proportional to $A(G)$.

Given two vertices $i, j \in V(G)$, the *fidelity* at time t between i and j is the function $f_G(i, j; t) = |\langle j|U_G(t)|i\rangle|$. We say that there is *perfect state transfer* (for short, *PST*) between the particles i and j at time t if $f_G(i,j;t) = 1$ [13]. We say that G is *periodic*, with period t, if $f_G(i,i;t) = 1$ [17]. Sometime in the physics literature periodic graphs are said to afford *perfect revival* (see, *e.g.*, [7]).

Even if the topic is not directly discussed here, it is worth noticing that in the XYZ model, the Hamiltonian restricted to $\mathbb{C}^n$ is proportional to the Laplacian matrix of G (see, *e.g.*, [10]). This fact alone is sufficient to distinguish different approaches for the two models, when the graphs considered have generic degree sequences. In this work, we consider regular graphs only. The result obtained is then also valid for the XYZ model.

The concept of PST has been introduced in [9] and [13]. The recent papers [6] and [3], even if not reviews, point out a good number of references embracing the more mathematical aspects around the notion.

1.2. Diameter

A graph $H = (W, F)$ is a *subgraph* of G if $W(H) \subseteq V(G)$ and $F(H) \subseteq E(G)$. A subgraph $H = (W, F)$ is an *induced subgraph* of G if H is a subgraph of G and, for every two vertices $i, j \in V(H)$, $\{i,j\} \in E(H)$ if and only if $\{i,j\} \in E(G)$.

The *degree* of a vertex i is the number of edges incident with i. A *path* of length l *from* vertex i *to* vertex j (if there is one) is an induced subgraph with l vertices and $l-1$ edges, such that i and j have degree one and all other vertices in the path have degree two. A graph is said to be *connected* if every two vertices are in a path.

Let $\mathcal{P}_{i,j}(G)$ be the set of all paths with end-vertices i and j. The *length* of a path with end-vertices i and j is denoted by $l(i,j)$. The (geodesic) *distance* between two vertices i and j is defined as $d(i,j) = \min_{\mathcal{P}_{i,j}(G)} l(i,j)$. The *diameter* of a connected graph G is defined as $\text{dia}(G) = \max_{i,j \in V(G)} d(i,j)$. Informally, the diameter is the longest of the shortest paths. Two vertices $i, j \in V(G)$ are said to be *antipodal* if $d(i,j) = \text{dia}(G)$.

1.3. Order/distance problems for state transfer

There is a large and growing literature concerned with the mathematics of state transfer on spin systems. Some effort towards a classification may be seen as driven by three recurrent, but essentially unstated problems:

- *General graphs:* Given $D \in \mathbb{N}$, find the graph $G = (V, E)$ with the smallest possible number of vertices $n_D = |V(G)|$ such that $d(i,j) = D$ and $f_G(i,j;t) = 1$ for some $t \in \mathbb{R}^+$. The set of these graphs is denoted by $\mathcal{G}_D$.
- *Fixed degree graphs:* Given $D, \Delta \in \mathbb{N}$, find the graph $G = (V, E)$ with the smallest possible number of vertices $n_{D,\Delta} = |V(G)|$ such that *(i)* the maximum degree of G is Δ, *(ii)* $d(i,j) = D$ and $f_G(i,j;t) = 1$ for some $t \in \mathbb{R}^+$. The set of these graphs is denoted by $\mathcal{G}_{D,\Delta}$.
- *Regular graphs:* Given $D, k \in \mathbb{N}$, find the graph $G = (V, E)$ with the smallest possible number of vertices $n_{D,k} = |V(G)|$ such that *(i)* G is k-regular, *(ii)* $d(i,j) = D$ and $f_G(i,j;t) = 1$ for some $t \in \mathbb{R}^+$. The set of these graphs is denoted by $\mathcal{G}^R_{D,k}$. A graph is *k-regular* if all of its vertices have degree k.

The requirement "smallest possible number" could be replaced with "largest possible number" to state the specular versions of the problems.

1.4. Examples

- $D = 1$: Let $P_2 = (\{1,2\},\{\{1,2\}\})$ be the path of length one. Then $[U_{P_2}(t)]_{1,2} = -i\sin(t)$ and

$$\max_{t \in \mathbb{R}^+}(|[U_{P_2}(t)]_{1,2}|) = f_{P_2}(1,2;\pi/2) = 1.$$

Hence, $P_2 \in \mathcal{G}_1$.

Journal of Physics: Conference Series **254** (2010) 012012 doi:10.1088/1742-6596/254/1/012012

- $D = 2$: Let $P_3 = (\{1,2,3\}, \{\{1,2\}, \{2,3\}\})$ be the path of length two. Then $[U_{P_3}(t)]_{1,3} = -\sin\left(t/\sqrt{2}\right)^2$ and

$$\max_{t\in\mathbb{R}^+}\left(|[U_{P_3}(t)]_{1,3}|\right) = f_{P_3}\left(1,3;\pi/\sqrt{2}\right) = 1.$$

Hence, $P_3 \in \mathcal{G}_2$.

In both cases, PST is between antipodal vertices.

- $D = 3$: Let $P_4 = (\{1,2,3,4\}, \{\{1,2\}, \{2,3\}, \{3,4\}\})$ be the path of length three. Let $a := t/2$. Then $[U_{P_4}(t)]_{1,4} = i\sqrt{5}\sin\left(a + a\sqrt{5}\right)/10 - i\sin\left(a + a\sqrt{5}\right)/2 - i\sqrt{5}\sin\left(a - a\sqrt{5}\right)/10 - i\sin\left(a - a\sqrt{5}\right)/2$ and

$$\max_{t\in\mathbb{R}^+}\left(|[U_{P_4}(t)]_{1,4}|\right) = f_{P_4}\left(1,3;2\pi/\sqrt{5}\right) = \sin\left(\pi/\sqrt{5}\right) \approx 0.986.$$

Hence, $P_4 \notin \mathcal{G}_3$.

1.5. Cartesian products

The *Cartesian product* $G = G_1 \times G_2 = (V, E)$ of two graphs $G_1 = (V_1, E_1)$ and $G_2 = (V_2, E_2)$ has set of vertices $V(G) = V(G_1) \times V(G_2)$ and $\{\{i,j\},\{k,l\}\} \in E(G)$ if *(i)* $i = k$ and $\{j,l\} \in E(G_2)$ or *(ii)* $j = l$ and $\{i,k\} \in E(G_1)$. Two facts are important: $\text{dia}(G) = \text{dia}(G_1) + \text{dia}(G_2)$; if G_1 and G_2 are k-regular and l-regular graphs, respectively, then G is $(k+l)$-regular.

The k-*dimensional cube* is the graph $P_2^{\times k}$ (in Fig. 1, $P_2^{\times 3}$). We have the following four cases:

$$[U_{P_2^{\times k}}(t)]_{1,2^k} = \begin{cases} -\sin(t)^k, & \text{if } k = 2l,\ l \text{ odd;} \\ \sin(t)^k, & \text{if } k = 2l,\ l \text{ even;} \\ i\sin(t)^k, & \text{if } k = 2l+1,\ l \text{ odd;} \\ -i\sin(t)^k, & \text{if } k = 2l+1,\ l \text{ even.} \end{cases}$$

In all these cases,

$$\max_{t\in\mathbb{R}^+}\left(\left|[U_{P_2^{\times k}}(\pi/2)]_{1,2^k}\right|\right) = f_{P_2^{\times k}}\left(1,2^k;\pi/2\right) = 1.$$

For $P_2^{\times k}$, we have $\left|V(P_2^{\times k})\right| = 2^k$ and $\text{dia}(P_2^{\times k}) = k\cdot\ \text{dia}(P_2) = 2\log_2\left|V(P_2^{\times k})\right| = k$. Thus, for the k-regular graphs in $\mathcal{G}^R_{m,m}$, $n_{m,m} \le 2^m$, where $m = D, k$.

The k-dimensional generalization of the 3×3 grid is denoted by $P_3^{\times k}$ (in Fig. 1, the grid $P_3^{\times 3}$). There are four types of matrix entries for the graph $P_3^{\times k}$:

$$[U_{P_3^{\times k}}(t)]_{1,3^k} = \begin{cases} -\sin\left(t/\sqrt{2}\right)^{2k}, & \text{if } k = 2l,\ l \text{ odd;} \\ \sin\left(t/\sqrt{2}\right)^{2k}, & \text{if } k = 2l,\ l \text{ even;} \\ i\sin\left(t/\sqrt{2}\right)^{2k}, & \text{if } k = 2l+1,\ l \text{ odd;} \\ -i\sin\left(t/\sqrt{2}\right)^{2k}, & \text{if } k = 2l+1,\ l \text{ even.} \end{cases}$$

Then,

$$\max_{t\in\mathbb{R}^+}\left(\left|[U_{P_3^{\times k}}\left(\pi/\sqrt{2}\right)]_{1,3^k}\right|\right) = f_{P_3^{\times 3}}\left(1,3^k;\pi/\sqrt{2}\right) = 1.$$

For $P_3^{\times k}$, we have $\left|V(P_3^{\times k})\right| = 3^k$ and $\text{dia}(P_3^{\times k}) = k\cdot\ \text{dia}(P_3) = 2\log_3\left|V(P_3^{\times k})\right| = 2k$.

Journal of Physics: Conference Series **254** (2010) 012012 doi:10.1088/1742-6596/254/1/012012

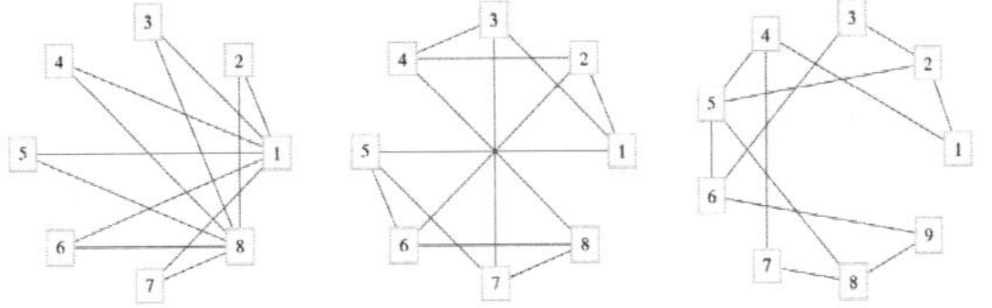

Figure 1. *(L)*: The graph W. There is PST between vertices 1 and 8. The spectrum of W is not integral. *(C)*: The 3-dimensional cube. There is PST between vertices 1 and 8. These vertices are antipodal and at distance 3. *(R)*: The graph $P_3^{\times 2}$. Note that the graph has vertices of degree two and three. There is PST between the vertices 1 and 9. These vertices are antipodal and at distance four.

For $P_2^{\times k}$ and $P_3^{\times k}$ PST is between antipodal vertices. The parameters related to PST between two vertices i and j in these graphs are given in the following tables [13]:

$P_2^{\times k}$	k	$\lvert V(P_2^{\times k})\rvert$	$d(i,j)$
	2	4	2
	3	8	3
	4	16	4
	5	32	5

;

$P_3^{\times k}$	k	$\lvert V(P_3^{\times k})\rvert$	$d(i,j)$
	2	9	4
	3	27	6
	4	81	8
	5	243	10

.

Notice that $P_3^{\times k}$ is nonregular for every k.

A square matrix M of size n consisting of unimodular entries $|M_{i,j}| = 1$ is called a *Hadamard matrix* if $HH^{\dagger} = nI$, where I is the identity matrix and † denotes the Hermitian transpose. In a *complex Hadamard matrix*, $M_{i,j} \in \mathbb{C}$ [26]. The matrix

$$U_{P_2^{\times k}}\left(\frac{\pi}{4}\right) = \frac{1}{\sqrt{2}}\begin{bmatrix} 1 & -i \\ -i & 1 \end{bmatrix}^{\otimes k}$$

is a complex Hadamard matrix.

1.6. Statement of the results

We prove that the 3-dimensional cube is the only periodic, connected cubic graph with PST. Equivalently,

Theorem. The 3-dimensional cube, $P_2^{\times 3}$, is the only periodic, connected cubic (3-regular) graph G with two different vertices i and j such that $f_G(i,j;t) = 1$, for some $t \in \mathbb{R}^+$.

The statement is verified directly in the next section. Conclusions follow. The proof is based on two known results: a periodic, connected regular graph is integral; there are only thirteen connected cubic integral graphs. The proof is technically easy, but tedious, because it goes through a seemingly unavoidable case by case analysis. Nonetheless, establishing the result is also an excuse for a further step into a systematical exploration of periodic quantum dynamics. The proof is interspersed with extra information. The broader aim would be to take a picture of periodic quantum dynamics on cubic graphs, even if here we do not state any further general result, beyond a crude analytic compilation of matrix entries.

It is an open problem to prove that $P_2^{\times 3}$ is the only connected cubic graph with PST.

Conjecture. $\mathcal{G}^R_{D,3} = \{P_2^{\times 3}\}$ if $D = 3$ and $\mathcal{G}^R_{D,3} = \emptyset$, otherwise.

Periodicity is not necessary for PST. There are examples of regular graphs that are not periodic but have PST [23].

2. Proof of the theorem

2.1. Integral graphs

The *spectrum* of a graph G is the multiset $\{\lambda_1^{[m_1]}(G), \lambda_2^{[m_2]}(G), ..., \lambda_r^{[m_r]}(G)\}$ of the eigenvalues of $A(G)$. The index $[m_i]$ in $\lambda_i^{[m_i]}(G)$ denotes the multiplicity of the eigenvalue $\lambda_i(G)$. For example, the spectrum of the complete graph on four vertices, K_4, is $\{3, -1^{[3]}\}$. A graph is said to be *integral* if its eigenvalues are integers. There is basically one survey on this area [5] . See also [15] for the relevant terminology and [1] for a nontrivial upper bounds on the total number of integral graphs with n vertices. There are a few general results that establish a relation between PST and integral graphs [17]. The next statement is directly useful to our purposes:

P0. A connected regular graph is periodic if and only if it is integral (Corollary 2.3 [17]).

The converse is not necessarily true. This fact can be observed in the examples discussed below. Moreover, integer eigenvalues have a weaker role in nonregular graphs. Let us consider, for instance, a graph W with eight vertices $\{1, 2, ..., 8\}$ and set of edges $E(W) = \{\{1, i\}, \{j, 8\} : 2 \leq i, j \leq 7;\}$ (see Fig. 1). The unitary governing the dynamics in W is defined by

$$[U_W(t)]_{i,j} = \begin{cases} -\sin\left(\sqrt{3}t\right)^2, & \text{if } i = 1 \text{ and } j = 8 \text{ or } viz.; \\ -i\sin\left(2\sqrt{3}t\right)/2\sqrt{3}, & \text{if } \{i,j\} \in E(W); \\ \cos\left(\sqrt{3}t\right)^2, & \text{If } i = j = 1, 8; \\ \left(5 + \cos\left(2\sqrt{3}t\right)\right)/6, & \text{if } i = j \neq 1, 8; \\ -\sin\left(\sqrt{3}t\right)^2, & \text{otherwise.} \end{cases}$$

There is PST between vertices 1 and 8, because $[U_W(\pi/2\sqrt{3})]_{1,8} = -\sin(3\pi/2)^2 = -1$. At the same time, the 6×6 submatrix of $U_W(\pi/2\sqrt{3})$, excluding the first/last row/column, is 2/3 in the diagonal and $-1/3$ off-diagonal. The spectrum of W is $\{\pm 2\sqrt{3}, 0^{[6]}\}$. Thus, W is not an integral graph.

We shall be interested in a special family of graphs. A *cubic* graph is a 3-regular graph. All cubic integral graphs have been classified and explicitly constructed [11, 24]. In particular,

P1. There are exactly thirteen connected cubic integral graphs.

On the light of the statements P0 and P1, a proof of the theorem can be obtained via a case by case analysis. As we have seen, and this is an already known fact, the 3-dimensional cube, $P_2^{\times 3}$, has PST between its antipodal vertices. These are vertices at distance three. We shall verify that none of the remaining twelve connected, cubic integral graphs affords PST. All graphs considered in this section are periodic.

2.2. The complete graph K_4, the complete bipartite graph $K_{3,3}$, and two connected copies of $K_{2,3}$

A *complete graph* on n vertices is a graph $K_n = (\{1, ..., n\}, E)$, where $E(K_n) = \{\{i, j\} : 1 \leq i, j \leq n\}$. The ij-entries of U_{K_4} are given by

$$[U_{K_4}(t)]_{i,j} = \begin{cases} \left(3\cos(t) + \cos(3t) + 4i\sin(t)^3\right)/4, & \text{if } i = j; \\ \cos(t)\sin(t)\left(-i\cos(t) - \sin(t)\right), & \text{if } i \neq j. \end{cases}$$

Journal of Physics: Conference Series **254** (2010) 012012 doi:10.1088/1742-6596/254/1/012012

Then $\max_{t\in\mathbb{R}^+}\left([U_{K_4}(t)]_{i,i}\right) = f_{U_{K_4}}(i,i;\pi/2) = 1$, for $1 \le i \le 4$; for every pair of vertices i and j, $\max_{t\in\mathbb{R}^+}\left([U_{K_4}(t)]_{i,j}\right) = f_{U_{K_4}}(i,j;\pi/4) = 1/2$. Note that

$$[U_{K_4}(\pi/4)]_{i,j} = \frac{1}{\sqrt{2}}\begin{cases} (1+i)/2, & \text{if } i=j; \\ -(1+i)/2, & \text{if } i \neq j. \end{cases}$$

gives a complex Hadamard matrix.

A graph $G = (V = V_1 \cup V_2, E)$ is *bipartite* if each vertex in V_1 is adjacent to vertices in V_2 only and *viz.* A *complete bipartite graph* is a bipartite graph $K_{q,p} = (V = V_1 \cup V_2, E)$ such that $|V_1| = p$, $|V_2| = q$ and $E = \{\{i,j\} : i \in V_1 \text{ and } j \in V_2\}$. The graph $K_{3,3}$ is on six vertices; $V(K_{3,3}) = \{A, B\}$, with $|A| = |B| = 3$. By definition, $\{i,j\} \in E(K_{3,3})$ if and only if $i \in A$ and $j \in B$. The spectrum of $K_{3,3}$ is $\{\pm 3, 0^{[4]}\}$ and $\mathrm{dia}(K_{3,3}) = 2$. The ij-entries of $U_{K_{3,3}}$ are as follows:

$$[U_{K_{3,3}}(t)]_{i,j} = \begin{cases} (2+\cos(3t))/3, & i=j; \\ (-1+\cos(3t))/3, & \{i,j\} \notin E(K_{3,3}); \\ -i\sin(3t)/3, & \{i,j\} \in E(K_{3,3}). \end{cases}$$

Hence, $\max_{t\in\mathbb{R}^+}\left(\left|[U_{K_{3,3}}(t)]_{i,i}\right|\right) = f_{U_{K_{3,3}}}(i,i;2\pi/3) = 1$, for $1 \le i \le 6$. For the off-diagonal entries, we need to distinguish two cases: *(i)* $\max_{t\in\mathbb{R}^+}\left(\left|[U_{K_{3,3}}(t)]_{i,j}\right|\right) = f_{U_{K_{3,3}}}(i,j;\pi/3) = 2/3$, if $i,j \in A$ or $i,j \in B$; *(ii)* $\max_{t\in\mathbb{R}^+}\left(\left|[U_{K_{3,3}}(t)]_{i,j}\right|\right) = f_{U_{K_{3,3}}}(i,j;\pi/6) = 1/3$, if $i \in A$ and $j \in B$. When $t = \pi/2$,

$$[U_{K_{3,3}}(\pi/2)]_{i,j} = \begin{cases} 2/3, & i=j; \\ -1/3, & \{i,j\} \notin E(K_{3,3}); \\ i/3, & \{i,j\} \in E(K_{3,3}). \end{cases}$$

Let $DK_{2,3}$ be the graph on ten vertices obtained from two disjoint copies of $K_{2,3}$, say $K_{2,3}^1$ and $K_{2,3}^2$, by adding three edges between the vertices of degree two in $K_{2,3}^1$ and $K_{2,3}^2$. The spectrum of $DK_{2,3}$ is $\{\pm 3, \pm 2, \pm 1^{[2]}, 0^{[2]}\}$ and $\mathrm{dia}(DK_{2,3}) = 3$. The structure of $U_{DK_{3,3}}$ consists of various kind of entries:

$$U_{DK_{2,3}}(t) = \begin{array}{cc|ccc|ccc|cc}
a_1 & a_2 & a_3 & a_3 & a_3 & a_4 & a_4 & a_4 & a_5 & a_5 \\
a_2 & a_1 & a_3 & a_3 & a_3 & a_4 & a_4 & a_4 & a_5 & a_5 \\
\hline
a_3 & a_3 & a_6 & a_7 & a_7 & a_8 & a_9 & a_9 & a_4 & a_4 \\
a_3 & a_3 & a_7 & a_6 & a_7 & a_9 & a_8 & a_9 & a_4 & a_4 \\
a_3 & a_3 & a_7 & a_7 & a_6 & a_9 & a_9 & a_8 & a_4 & a_4 \\
\hline
a_4 & a_4 & a_8 & a_9 & a_9 & a_6 & a_7 & a_7 & a_3 & a_3 \\
a_4 & a_4 & a_9 & a_8 & a_9 & a_7 & a_6 & a_7 & a_3 & a_3 \\
a_4 & a_4 & a_9 & a_9 & a_8 & a_7 & a_7 & a_6 & a_3 & a_3 \\
\hline
a_5 & a_5 & a_4 & a_4 & a_4 & a_3 & a_3 & a_3 & a_1 & a_2 \\
a_5 & a_5 & a_4 & a_4 & a_4 & a_3 & a_3 & a_3 & a_2 & a_1
\end{array},$$

where

$$\begin{aligned}
a_1 &= (5 + 3\cos(2t) + 2\cos(3t))/10, \\
a_2 &= (-5 + 3\cos(2t) + 2\cos(3t))/10, \\
a_3 &= -i(\sin(2t) + \sin(3t))/5, \\
a_4 &= (-\cos(2t) + \cos(3t))/5, \\
a_5 &= i(3\sin(2t) - 2\sin(3t))/10, \\
a_6 &= (10\cos(t) + 2\cos(2t) + 3\cos(3t))/15, \\
a_7 &= (-5\cos(t) + 2\cos(2t) + 3\cos(3t))/15, \\
a_8 &= -i(10\sin(t) - 2\sin(2t) + 3\sin(3t))/15, \\
a_9 &= i(10\sin(t) + 2\sin(2t) - 3\sin(3t))/15.
\end{aligned}$$

By considering a_1 and a_6, we can see that $\max_{t\in\mathbb{R}^+}\left(\left|[U_{DK_{2,3}}(t)]_{i,i}\right|\right) = f_{DK_{2,3}}(i,i;2\pi) = 1$, for every i. However, $\max_{1\leq j\leq 9; j\neq 1,6}\max_{t\in\mathbb{R}^+}(|a_j|) = \left(5-5\left(-1-\sqrt{5}\right)/4\right)/10 \approx 0.9$. The maximum is attained by a_2 for $t=2\pi/5$. The segments in the matrix $U_{DK_{3,3}}(t)$ help visualizing its structure and these do not have a mathematical meaning. We shall make a consistent use of this graphic tool also in the next cases. The graphs $K_{3,3}$ and $DK_{2,3}$ are in Fig. (2).

2.3. The graphs C_3+K_2 and C_6+K_2

The (*Cartesian*) *sum* $G=G_1+G_2=(V,E)$ of two graphs $G_1=(V_1,E_1)$ and $G_2=(V_2,E_2)$ has set of vertices $V(G)=V(G_1)\times V(G_2)$ and $\{\{i,j\},\{k,l\}\}\in E(G)$ if *(i)* $\{i,k\}\in E(G_1)$ or *(ii)* $\{j,l\}\in E(G_2)$ [22]. We denote by C_n the *n-cycle*: $V(C_n)=\{1,2,...,n\}$ and $\{i,(i+1) \bmod n\}\in E(C_n)$.

The graph C_3+K_2 has two cycles of length three and it can be drawn as a prism with triangular basis. It is the undirected version the Cayley digraph of the dihedral group D_6 generated by the standard set. Its spectrum is $\{3,1,0^{[2]},-2^{[2]}\}$. Given the symmetry, the unitary matrix has a neat structure:

$$U_{C_3+K_2}(t)=\left[\begin{array}{ccc|ccc} a_1 & a_2 & a_2 & a_3 & a_4 & a_4\\ a_2 & a_1 & a_2 & a_4 & a_3 & a_4\\ a_2 & a_2 & a_1 & a_4 & a_4 & a_3\\ \hline a_3 & a_4 & a_4 & a_1 & a_2 & a_2\\ a_4 & a_3 & a_4 & a_2 & a_1 & a_2\\ a_4 & a_4 & a_3 & a_2 & a_2 & a_1 \end{array}\right],$$

with

$$\begin{aligned} a_1 &= \left(2+e^{-it}+2e^{2it}+e^{-3it}\right)/6,\\ a_2 &= -ie^{-it/2}\left(\sin(t/2)+\sin(5t/2)\right)/3,\\ a_3 &= \left(2-e^{-it}-2e^{2it}+e^{-3it}\right)/6,\\ a_4 &= \left(-1-e^{-it}+e^{2it}+e^{-3it}\right)/6. \end{aligned}$$

From this, $\max_{2\leq j\leq 4}\max_{t\in\mathbb{R}^+}(|a_j|)\approx 0.9$, for $j=3$ and $t=\sqrt{3}+\pi/17$.

The graph C_6+K_2 is on twelve vertices. It has two cycles of length six and it can be drawn as a prism with hexagonal basis. In fact, in analogy with C_3+K_2, it is the undirected version the Cayley digraph of the dihedral group D_{12} generated by the standard set. As we have done for the other cases, we explicitly write down the unitary matrix. Let

$$A=\left[\begin{array}{ccc|ccc} a_1 & a_2 & a_3 & a_4 & a_3 & a_2\\ a_2 & a_1 & a_2 & a_3 & a_4 & a_3\\ a_3 & a_2 & a_1 & a_2 & a_3 & a_4\\ \hline a_4 & a_3 & a_2 & a_1 & a_2 & a_3\\ a_3 & a_4 & a_3 & a_2 & a_1 & a_2\\ a_2 & a_3 & a_4 & a_3 & a_2 & a_1 \end{array}\right] \text{ and } B=\left[\begin{array}{ccc|ccc} a_5 & a_6 & a_7 & a_8 & a_7 & a_6\\ a_6 & a_5 & a_6 & a_7 & a_8 & a_7\\ a_7 & a_6 & a_5 & a_6 & a_7 & a_8\\ \hline a_8 & a_7 & a_6 & a_5 & a_6 & a_7\\ a_7 & a_8 & a_7 & a_6 & a_5 & a_6\\ a_6 & a_7 & a_8 & a_7 & a_6 & a_5 \end{array}\right].$$

Then

$$U_{C_6+K_2}(t)=\left[\begin{array}{cc} A & B\\ B & A \end{array}\right],$$

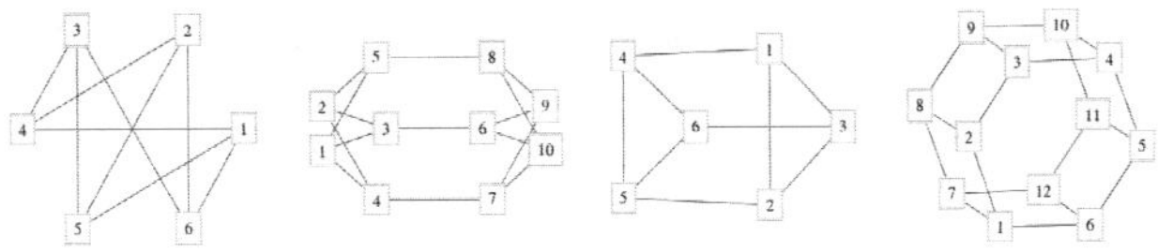

Figure 2. From the left: The complete bipartite graph $K_{3,3}$. The graph obtained by connecting together two copies of $K_{2,3}$. The graph $C_3 + K_2$. The graph $C_6 + K_2$. Although periodic, none of these graphs has PST.

and

$$\begin{aligned}
a_1 &= (2 + \cos(t) + 2\cos(2t) + \cos(3t))/6, \\
a_2 &= -i(\sin(t) + \sin(2t) + \sin(3t))/6, \\
a_3 &= (-1 + \cos(t) - \cos(2t) + \cos(3t))/6, \\
a_4 &= i(\sin(t) - 2\sin(2t) + \sin(3t))/6, \\
a_5 &= i(\sin(t) - 2\sin(2t) - \sin(3t))/6, \\
a_6 &= -(1 + 2\cos(t))\sin(t)^2/3, \\
a_7 &= -i(\sin(t) + \sin(2t) - \sin(3t))/6, \\
a_8 &= 16\left(\cos(t/2)^2 \sin(t/2)^4\right)/3.
\end{aligned}$$

For $j \neq 1$,

$$\max_{t\in\mathbb{R}^+}(|a_j|)\begin{cases}
\approx 21/50, & j = 2 \text{ and } t \approx \pi/6; \\
= 2/3, & j = 3 \text{ and } t = \pi; \\
\approx 29/50, & j = 4 \text{ and } t = 77/20 \\
\approx 9/20, & j = 5 \text{ and } t = 19/10; \\
\approx 1/2, & j = 6 \text{ and } t \approx 1.21; \\
\approx 9/25, & j = 7 \text{ and } t \approx 1.35; \\
\approx 64/81, & j = 8 \text{ and } t \approx 1.9.
\end{cases}$$

The graphs $C_3 + K_2$, and $C_6 + K_2$ are represented in Fig. (2).

2.4. The Petersen graph, a graph on ten vertices, and $L(S(K_4))$

The Petersen graph, P, illustrated in Fig. (3), is one of the best-studied single objects in the graph-theoretic literature. The Petersen graph has two cycles of length five. It is vertex-transitive but not a Cayley graph, with spectrum $\{3, 1^{[4]}, -2^{[4]}\}$ and $\text{dia}(P) = 3$. It is strongly regular with parameters $(10, 3, 0, 1)$. The symmetry appears also in $U_P(t)$, which reflects faithfully the structure of P:

$$U_P(t) = \left.\begin{array}{ccccc|ccccc}
a_1 & a_2 & a_3 & a_3 & a_2 & a_2 & a_3 & a_3 & a_3 & a_3 \\
a_2 & a_1 & a_2 & a_3 & a_3 & a_3 & a_2 & a_3 & a_3 & a_3 \\
a_3 & a_2 & a_1 & a_2 & a_3 & a_3 & a_3 & a_2 & a_3 & a_3 \\
a_3 & a_3 & a_2 & a_1 & a_2 & a_3 & a_3 & a_3 & a_2 & a_3 \\
a_2 & a_3 & a_3 & a_2 & a_1 & a_3 & a_3 & a_3 & a_3 & a_2 \\
\hline
a_2 & a_3 & a_3 & a_3 & a_3 & a_1 & a_3 & a_2 & a_2 & a_3 \\
a_3 & a_2 & a_3 & a_3 & a_3 & a_3 & a_1 & a_3 & a_2 & a_2 \\
a_3 & a_3 & a_2 & a_3 & a_3 & a_2 & a_3 & a_1 & a_3 & a_2 \\
a_3 & a_3 & a_3 & a_2 & a_3 & a_2 & a_2 & a_3 & a_1 & a_3 \\
a_3 & a_3 & a_3 & a_3 & a_2 & a_3 & a_2 & a_2 & a_3 & a_1
\end{array}\right.,$$

where

$$\begin{aligned} a_1 &= e^{-3it}\left(1+5e^{2it}+4e^{5it}\right)/10, \\ a_2 &= e^{-3it}\left(3+5e^{2it}-8e^{5it}\right)/30, \\ a_3 &= e^{-3it}\left(3-5e^{2it}+2e^{5it}\right)/30. \end{aligned}$$

Thus, $\max_{j=2,3}\max_{t\in\mathbb{R}^+}(|a_j|) = 8/15$, for $j=2$ and $t=\pi$. More generally, $[U_P(\pi)]_{i,j} = -1/5$ if $i=j$; $2/15$ if $\{i,j\}\notin E(P)$ and $-8/15$, otherwise.

There is another cubic integral graph on ten vertices, obtained by replacing with triangles two nonadjacent vertices of $K_{3,3}$. Denoted by Z, it has spectrum $\{3,2,1^{[3]},-1^{[2]},-2^{[3]}\}$. Fig. (3) contains a drawing. The unitary matrix obtained from Z is

$$U_Z(t) = \begin{array}{|cccc|cccccc|}
\hline
a_1 & a_2 & a_2 & a_2 & a_3 & a_3 & a_3 & a_3 & a_3 & a_3 \\
a_2 & a_1 & a_3 & a_3 & a_3 & a_3 & a_2 & a_3 & a_2 & a_3 \\
a_2 & a_3 & a_1 & a_3 & a_2 & a_2 & a_3 & a_3 & a_3 & a_3 \\
a_2 & a_3 & a_3 & a_1 & a_3 & a_3 & a_3 & a_2 & a_3 & a_2 \\
\hline
a_3 & a_3 & a_2 & a_3 & a_5 & a_4 & a_6 & a_6 & a_7 & a_7 \\
a_3 & a_3 & a_2 & a_3 & a_4 & a_5 & a_7 & a_7 & a_6 & a_6 \\
a_3 & a_2 & a_3 & a_3 & a_6 & a_7 & a_5 & a_6 & a_4 & a_7 \\
a_3 & a_3 & a_3 & a_2 & a_6 & a_7 & a_6 & a_5 & a_7 & a_4 \\
a_3 & a_2 & a_3 & a_3 & a_7 & a_6 & a_4 & a_7 & a_5 & a_6 \\
a_3 & a_3 & a_3 & a_2 & a_7 & a_6 & a_7 & a_4 & a_6 & a_5 \\
\hline
\end{array}.$$

The entries are

$$\begin{aligned} a_1 &= e^{-3it}\left(1+5e^{2it}+4e^{5it}\right)/10, \\ a_2 &= e^{-3it}\left(3+5e^{2it}-8e^{5it}\right)/30, \\ a_3 &= e^{-3it}\left(3-5e^{2it}+2^{5it}\right)/30, \\ a_4 &= e^{-3it}\left(e^{it}-1\right)^2\left(3+e^{it}+4e^{2it}+7e^{3it}\right)/30, \\ a_5 &= e^{-3it}\left(3+5e^{it}+5e^{2it}+10e^{4it}+7e^{5it}\right)/30, \\ a_6 &= e^{-3it}\left(3+5e^{it}-5e^{4it}-3e^{5it}\right)/30, \\ a_7 &= e^{-3it}\left(3-5e^{it}+5e^{4it}-3^{5it}\right)/30. \end{aligned}$$

We then obtain $\max_{1\leq j\leq 9; j\neq 1,5}\max_{t\in\mathbb{R}^+}(|a_j|) \approx 0.85$, for $j=4$ and $t\approx\pi-5/6$. Note that a_1, a_2, and a_3 give the same dynamics as the Petersen graph.

The graph $X = L(S(K_4))$ is constructed by replacing each vertex of K_4 with a triangle (see Fig. (3)). The triangles are then connected by independent edges. The notation indicates the line graph of the K_4 subdivision. Its spectrum is $\{3,\pm 2^{[3]},0^{[2]},-1^{[3]}\}$ and $\mathrm{dia}(X)=3$. The unitary has some symmetry:

$$U_X(t) = \begin{array}{|ccc|ccc|ccc|ccc|}
\hline
a_1 & a_2 & a_2 & a_4 & a_5 & a_6 & a_4 & \boxed{a_3} & a_4 & a_4 & a_6 & a_5 \\
a_2 & a_1 & a_2 & \boxed{a_3} & a_4 & a_4 & a_5 & a_4 & a_6 & a_4 & a_5 & a_6 \\
a_2 & a_2 & a_1 & a_4 & a_6 & a_5 & a_4 & a_4 & a_5 & \boxed{a_3} & a_4 & a_4 \\
\hline
a_4 & \boxed{a_3} & a_4 & a_1 & a_2 & a_2 & a_4 & a_5 & a_6 & a_5 & a_4 & a_6 \\
a_5 & a_4 & a_6 & a_2 & a_1 & a_2 & \boxed{a_3} & a_4 & a_4 & a_6 & a_4 & a_5 \\
a_6 & a_4 & a_5 & a_2 & a_2 & a_1 & a_4 & a_6 & a_5 & a_4 & \boxed{a_3} & a_4 \\
\hline
a_4 & a_5 & a_6 & a_4 & \boxed{a_3} & a_4 & a_1 & a_2 & a_2 & a_6 & a_5 & a_4 \\
\boxed{a_3} & a_4 & a_4 & a_5 & a_4 & a_6 & a_2 & a_1 & a_2 & a_5 & a_6 & a_4 \\
a_4 & a_6 & a_5 & a_6 & a_4 & a_5 & a_2 & a_2 & a_1 & a_4 & a_4 & \boxed{a_3} \\
\hline
a_4 & a_4 & \boxed{a_3} & a_5 & a_6 & a_4 & a_6 & a_5 & a_4 & a_1 & a_2 & a_2 \\
a_6 & a_5 & a_4 & a_4 & a_4 & \boxed{a_3} & a_5 & a_6 & a_4 & a_2 & a_1 & a_2 \\
a_5 & a_6 & a_4 & a_6 & a_5 & a_4 & a_4 & a_4 & \boxed{a_3} & a_2 & a_2 & a_1 \\
\hline
\end{array},$$

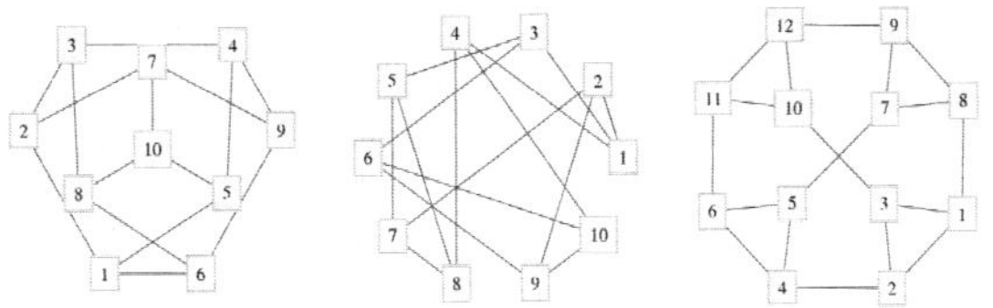

Figure 3. *(L)*: The Petersen graph. *(C)*: The graph Z on ten vertices. *(R)*: The graph $L(S(K_4))$ on twelve vertices. These graphs are periodic without PST.

with

$$\begin{aligned}
a_1 &= \left(2+3e^{it}+3e^{-2it}+3e^{2it}+e^{-3it}\right)/12,\\
a_2 &= -e^{-3it}\left(-2-5e^{it}+2e^{3it}+2e^{4it}+3e^{5it}\right)/24,\\
a_3 &= e^{-3it}\left(1+e^{it}+2e^{3it}-e^{4it}-e^{5it}\right)/12,\\
a_4 &= e^{-3it}\left(e^{it}-1\right)^2\left(2+3e^{it}+4e^{2it}+3e^{3it}\right)/24,\\
a_5 &= -e^{-3it}\left(e^{it}-1\right)^3\left(2+3e^{it}+3e^{2it}\right)/24,\\
a_6 &= -e^{-3it}\left(e^{it}-1\right)^3\left(e^{it}+1\right)/12.
\end{aligned}$$

For $j \neq 1$,

$$\max_{t\in\mathbb{R}^+}(|a_j|)\begin{cases} \approx 1/2, & j=2 \text{ and } t\approx \pi/4;\\ \approx 1/2, & j=3 \text{ and } t\approx \pi/4;\\ \approx 1/4, & j=4 \text{ and } t\approx 6/5;\\ =2/3, & j=5 \text{ and } t=\pi;\\ =\sqrt{3}/4, & j=6 \text{ and } t=2\pi/3.\end{cases}$$

2.5. The Desargues graph and its cospectral mate

The bipartite double cover of the Petersen graph is called *Desargues graph*. There are many different notations for this graph. We adopt $H_{5,2}$. The Desargues graph is on twenty vertices and its spectrum is $\left\{\pm 3, \pm 2^{[4]}, \pm 1^{[5]}\right\}$. The graph $H_{5,2}$ has a *cospectral mate*, which we will denote by $H'_{5,2}$. This is a nonisomorphic graph with the same spectrum. Two graphs G and H are said to *isomorphic* if there is a permutation matrix Q such that $QA(G)Q^T = A(H)$. It is clear that two isomorphic graphs have the same spectrum. The converse is not necessarily true. Indeed, $H_{5,2}$ and $H'_{5,2}$ are a counterexample. The graphs $H_{5,2}$ and $H'_{5,2}$ are drawn in Fig. (4). Let us define the arrays

$$A = \begin{array}{cccccccccc}
a_1 & a_2 & a_2 & a_2 & a_3 & a_2 & a_2 & a_2 & a_3 & a_3\\
a_2 & a_1 & a_2 & a_2 & a_2 & a_3 & a_3 & a_2 & a_2 & a_3\\
a_2 & a_2 & a_1 & a_3 & a_2 & a_2 & a_2 & a_3 & a_2 & a_3\\
a_2 & a_2 & a_3 & a_1 & a_2 & a_2 & a_3 & a_2 & a_3 & a_2\\
a_3 & a_2 & a_2 & a_2 & a_1 & a_2 & a_3 & a_3 & a_2 & a_2\\
a_2 & a_3 & a_2 & a_2 & a_2 & a_1 & a_2 & a_3 & a_3 & a_2\\
a_2 & a_3 & a_2 & a_3 & a_3 & a_2 & a_1 & a_2 & a_2 & a_2\\
a_2 & a_2 & a_3 & a_2 & a_3 & a_3 & a_2 & a_1 & a_2 & a_2\\
a_3 & a_2 & a_2 & a_3 & a_2 & a_3 & a_2 & a_2 & a_1 & a_2\\
a_3 & a_3 & a_3 & a_2 & a_2 & a_2 & a_2 & a_2 & a_2 & a_1
\end{array},$$

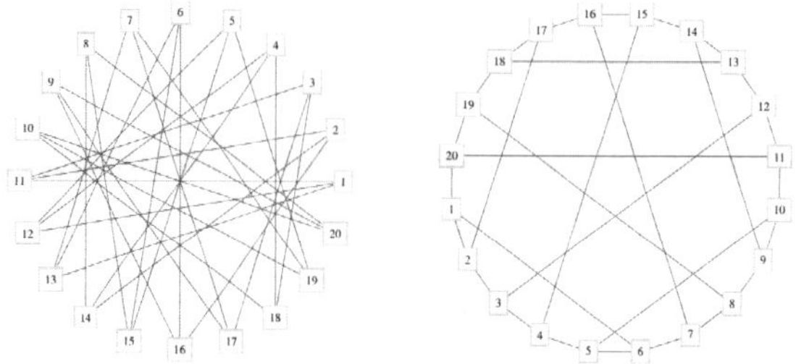

Figure 4. *(L):* The graph $H'_{5,2}$. *(R):* The Desargues graph $H_{5,2}$. This is cospectral with $H'_{5,2}$.

$$B = \begin{matrix} a_4 & a_4 & a_4 & a_8 & a_8 & a_8 & a_8 & a_8 & a_8 & a_7 \\ a_4 & a_8 & a_8 & a_4 & a_8 & a_4 & a_8 & a_8 & a_7 & a_8 \\ a_4 & a_8 & a_8 & a_8 & a_7 & a_8 & a_4 & a_4 & a_8 & a_8 \\ a_5 & a_4 & a_6 & a_4 & a_8 & a_6 & a_6 & a_4 & a_8 & a_8 \\ a_6 & a_4 & a_5 & a_8 & a_4 & a_6 & a_6 & a_8 & a_4 & a_8 \\ a_8 & a_8 & a_4 & a_8 & a_4 & a_4 & a_8 & a_7 & a_8 & a_8 \\ a_8 & a_8 & a_4 & a_7 & a_8 & a_8 & a_4 & a_8 & a_4 & a_8 \\ a_6 & a_8 & a_6 & a_4 & a_4 & a_5 & a_6 & a_8 & a_8 & a_4 \\ a_8 & a_7 & a_8 & a_8 & a_8 & a_4 & a_4 & a_8 & a_8 & a_4 \\ a_6 & a_8 & a_6 & a_8 & a_8 & a_6 & a_5 & a_4 & a_4 & a_4 \end{matrix}.$$

Let $\widetilde{M}$ be the matrix obtained by permuting the lines (rows and columns) of a square matrix M such that line i in M is line $n-j+1$ in $\widetilde{M}$. One can observe that

$$U_{H'_{5,2}}(t) = \begin{bmatrix} A & \widetilde{B} \\ B & \widetilde{A} \end{bmatrix},$$

where

$$\begin{aligned} a_1 &= e^{-3it}\left(1+4e^{it}+5e^{2it}+5e^{4it}+4e^{5it}+e^{6it}\right)/20, \\ a_2 &= e^{-3it}\left(3+2e^{it}-5e^{2it}-5e^{4it}+2e^{5it}+3e^{6it}\right)/60, \\ a_3 &= e^{-3it}\left(e^{it}-1\right)^4\left(3+4e^{it}+3e^{2it}\right)/60, \\ a_4 &= e^{-3it}\left(3+8e^{it}+5e^{2it}-5e^{4it}-8e^{5it}-3e^{6it}\right)/60, \\ a_5 &= -e^{-3it}\left(e^{it}-1\right)^3\left(1+4e^{it}+4e^{2it}+e^{3it}\right)/20, \\ a_6 &= -e^{-3it}\left(e^{it}-1\right)^3\left(3+2e^{it}+2e^{2it}+3e^{3it}\right)/60, \\ a_7 &= -e^{-3it}\left(e^{it}-1\right)^5\left(1+e^{it}\right)/20, \\ a_8 &= -e^{-3it}\left(e^{it}-1\right)^3\left(3+7e^{it}+7e^{2it}+3e^{3it}\right)/60. \end{aligned}$$

From these functions, we can see that $\max_{1\leq j\leq 8; j\neq 1}\max_{t\in\mathbb{R}^+}(|a_j|) \approx 0.83$, for $j=7$ and $t \approx 575/250$. When $t=\pi$, the probability amplitude is supported by all vertices in a class of the bipartition; in other words, the matrix $U_{H'_{5,2}}(\pi)$ is block-diagonal with two 10×10 blocks corresponding to the classes. The functions in the above equations completely specify the dynamics in $H_{5,2}$, since $H'_{5,2}$ and $H'_{5,2}$ are cospectral. One can verify that the matrix $U_{H_{5,2}}(t)$ is obtained by rearranging the entries of $U_{H'_{5,2}}(t)$.

Journal of Physics: Conference Series **254** (2010) 012012 doi:10.1088/1742-6596/254/1/012012

2.6. The Nauru graph and the Tutte-Coxeter graph

The *Nauru graph*, N_{24}, is the only cubic symmetric (arc-transitive) graph on 24 vertices. It is bipartite, with spectrum $\left\{\pm 3, \pm 2^{[6]}, \pm 1^{[3]}, 0^{[4]}\right\}$ and $\text{dia}(N_{24}) = 4$. A brief parenthesis: the Foster census of cubic symmetric graphs highlights that a large portion of periodic cubic graph is symmetric [14]. Clearly, the two sets do not coincide. There are exactly seven different kind of entries in $U_{N_{24}}(t)$. This can be seen as a 2×2 block matrix. The blocks $(1,1)$ and $(2,1)$ are below. The other blocks are just their rearrangements:

$$\begin{array}{cccccc|cccccc}
a_1 & a_2 & a_2 & a_2 & a_3 & a_2 & a_3 & a_4 & a_4 & a_3 & a_2 & a_2 \\
a_2 & a_1 & a_2 & a_2 & a_2 & a_3 & a_4 & a_2 & a_3 & a_2 & a_3 & a_4 \\
a_2 & a_2 & a_1 & a_3 & a_4 & a_2 & a_2 & a_3 & a_2 & a_2 & a_4 & a_3 \\
a_2 & a_2 & a_3 & a_1 & a_2 & a_4 & a_2 & a_3 & a_2 & a_4 & a_2 & a_3 \\
a_3 & a_2 & a_4 & a_2 & a_1 & a_2 & a_3 & a_2 & a_2 & a_3 & a_4 & a_2 \\
a_2 & a_3 & a_2 & a_4 & a_2 & a_1 & a_2 & a_2 & a_3 & a_4 & a_3 & a_2 \\
\hline
a_3 & a_4 & a_2 & a_2 & a_3 & a_2 & a_1 & a_2 & a_2 & a_3 & a_2 & a_4 \\
a_4 & a_2 & a_3 & a_3 & a_2 & a_2 & a_2 & a_1 & a_4 & a_2 & a_2 & a_3 \\
a_4 & a_3 & a_2 & a_2 & a_2 & a_3 & a_2 & a_4 & a_1 & a_2 & a_3 & a_2 \\
a_3 & a_2 & a_2 & a_4 & a_3 & a_4 & a_3 & a_2 & a_2 & a_1 & a_2 & a_2 \\
a_2 & a_3 & a_4 & a_2 & a_4 & a_3 & a_2 & a_2 & a_3 & a_2 & a_1 & a_2 \\
a_2 & a_4 & a_3 & a_3 & a_2 & a_2 & a_4 & a_3 & a_2 & a_2 & a_2 & a_1
\end{array}$$

and

$$\begin{array}{cccccc|cccccc}
a_5 & a_5 & a_6 & a_5 & a_6 & a_7 & a_7 & a_7 & a_7 & a_7 & a_6 & a_7 \\
a_5 & a_6 & a_5 & a_7 & a_7 & a_5 & a_6 & a_7 & a_7 & a_7 & a_7 & a_6 \\
a_6 & a_7 & a_7 & a_7 & a_5 & a_5 & a_7 & a_6 & a_6 & a_7 & a_7 & a_5 \\
a_7 & a_5 & a_7 & a_6 & a_5 & a_6 & a_7 & a_5 & a_7 & a_6 & a_7 & a_7 \\
a_7 & a_7 & a_6 & a_7 & a_6 & a_5 & a_5 & a_5 & a_7 & a_7 & a_6 & a_7 \\
a_7 & a_7 & a_5 & a_6 & a_7 & a_6 & a_5 & a_7 & a_5 & a_6 & a_7 & a_7 \\
\hline
a_6 & a_5 & a_5 & a_7 & a_7 & a_7 & a_7 & a_6 & a_6 & a_5 & a_7 & a_7 \\
a_7 & a_7 & a_6 & a_7 & a_6 & a_7 & a_7 & a_7 & a_5 & a_5 & a_6 & a_5 \\
a_7 & a_5 & a_5 & a_7 & a_7 & a_7 & a_6 & a_5 & a_7 & a_5 & a_5 & a_6 \\
a_5 & a_6 & a_5 & a_6 & a_7 & a_6 & a_7 & a_7 & a_7 & a_6 & a_5 & a_5 \\
a_6 & a_7 & a_5 & a_5 & a_7 & a_7 & a_5 & a_6 & a_6 & a_7 & a_5 & a_7 \\
a_7 & a_7 & a_5 & a_5 & a_5 & a_7 & a_6 & a_7 & a_5 & a_7 & a_7 & a_6
\end{array},$$

where

$$\begin{aligned}
a_1 &= (2 + 3\cos(t) + 6\cos(2t) + \cos(3t))/12, \\
a_2 &= -\left((1 + 2\cos(t))\sin(t)^2\right)/6, \\
a_3 &= 8\left(\cos(t/2)^2 \sin(t/2)^4\right)/3, \\
a_4 &= 2\left((1 + 2\cos(t))\sin(t/2)^4\right)/3, \\
a_5 &= -i(\sin(t) + 4\sin(2t) + \sin(3t))/12, \\
a_6 &= i\sin(t)^3/3, \\
a_7 &= -i(\sin(t) - 2\sin(2t) + \sin(3t))/12
\end{aligned}$$

are all the the different entries of the unitary matrix. From these functions, we can write

$$\max_{t\in\mathbb{R}^+}(|a_j|)=\begin{cases} =1, & j=1 \text{ and } t=2\pi;\\ \approx 1/2, & j=2 \text{ and } t\approx\pi/4;\\ \approx 1/4, & j=3 \text{ and } t\approx 6/5;\\ =2/3, & j=4 \text{ and } t=\pi;\\ \approx 9/20, & j=5 \text{ and } t\approx\pi/4;\\ =1/3, & j=6 \text{ and } t=\pi/2;\\ \approx 3/10, & j=7 \text{ and } t\approx 3\pi/4.\end{cases}$$

The *Tutte-Coxeter graph*, TC, is illustrated in Fig. (5). This is the largest cubic graph giving a periodic dynamics. Its eigenvalues are $\{\pm3,\pm2^{[9]},0^{[10]}\}$. Let us describe how to write down the unitary matrix U_{TC}. In Fig. (5) we have ordered the vertices anticlockwise. Each even vertex is connected to odd vertices, and *viz.* As we have seen in previous examples, TC is bipartite and its unitary has a block structure:

$$U_{TC}=\begin{bmatrix} A & B\\ B & C\end{bmatrix}.$$

The ij-th entry of the block B is

$$[B]_{i,j}=\begin{cases} -i\left(6\sin(2t)+\sin(3t)\right)/15, & \{i,j\}\in E(TC);\\ i\left(3\sin(2t)-2\sin(3t)\right)/30, & \{i,j\}\notin E(TC).\end{cases}$$

The index i runs over the odd numbers $1,3,...,29$; j over the even ones, $2,4,...,30$. For any chosen t, none of the entries of B can have unit absolute value. The blocks A and C have the same entries but rearranged:

$$\begin{matrix}
a_1 & a_2 & a_3 & a_3 & a_2 & a_3 & a_2 & a_2 & a_3 & a_2 & a_3 & a_3 & a_3 & a_3 & a_2\\
a_2 & a_1 & a_2 & a_3 & a_3 & a_2 & a_3 & a_3 & a_3 & a_2 & a_3 & a_2 & a_2 & a_3 & a_3\\
a_3 & a_2 & a_1 & a_2 & a_3 & a_2 & a_3 & a_2 & a_3 & a_3 & a_3 & a_3 & a_3 & a_2 & a_2\\
a_3 & a_3 & a_2 & a_1 & a_2 & a_3 & a_3 & a_2 & a_3 & a_2 & a_2 & a_3 & a_2 & a_3 & a_3\\
a_2 & a_3 & a_3 & a_2 & a_1 & a_2 & a_3 & a_3 & a_2 & a_3 & a_3 & a_3 & a_2 & a_3 & a_2\\
a_3 & a_2 & a_2 & a_3 & a_2 & a_1 & a_2 & a_3 & a_2 & a_3 & a_2 & a_3 & a_3 & a_3 & a_3\\
a_2 & a_3 & a_3 & a_3 & a_3 & a_2 & a_1 & a_2 & a_3 & a_3 & a_2 & a_3 & a_2 & a_3 & a_3\\
a_2 & a_3 & a_2 & a_2 & a_3 & a_3 & a_2 & a_1 & a_2 & a_3 & a_3 & a_2 & a_3 & a_2 & a_3\\
a_3 & a_3 & a_3 & a_3 & a_2 & a_2 & a_3 & a_2 & a_1 & a_2 & a_3 & a_2 & a_3 & a_3 & a_3\\
a_2 & a_2 & a_3 & a_2 & a_3 & a_3 & a_3 & a_3 & a_2 & a_1 & a_2 & a_3 & a_3 & a_2 & a_3\\
a_3 & a_3 & a_3 & a_2 & a_3 & a_2 & a_2 & a_3 & a_3 & a_2 & a_1 & a_2 & a_3 & a_2 & a_2\\
a_3 & a_2 & a_3 & a_3 & a_3 & a_3 & a_3 & a_2 & a_2 & a_3 & a_2 & a_1 & a_2 & a_3 & a_2\\
a_3 & a_2 & a_3 & a_2 & a_2 & a_3 & a_2 & a_3 & a_3 & a_3 & a_3 & a_2 & a_1 & a_3 & a_3\\
a_3 & a_3 & a_2 & a_3 & a_3 & a_3 & a_2 & a_3 & a_2 & a_2 & a_3 & a_3 & a_2 & a_1 & a_2\\
a_2 & a_3 & a_2 & a_3 & a_2 & a_3 & a_3 & a_3 & a_3 & a_3 & a_2 & a_2 & a_3 & a_2 & a_1
\end{matrix},$$

where

$$\begin{aligned} a_1 &= \left(5+9\cos(2t)+\cos(3t)\right)/15,\\ a_2 &= \left(-5+3\cos(2t)+\cos(3t)\right)/30,\\ a_3 &= 2\left(\left(7+8\cos(t)\right)\sin(t/2)^4\right)/15,\end{aligned}$$

and

$$\max_{t\in\mathbb{R}^+}(|a_j|)\begin{cases} =1, & j=1 \text{ and } t=2\pi;\\ \approx 3/10, & j=2 \text{ and } t\approx 1/4;\\ \approx 13/50, & j=3 \text{ and } t\approx 91/50.\end{cases}$$

This last equation concludes the proof of the theorem. Notice that $U_{TC}(\pi)$ is 2×2 block diagonal: the diagonal entries are $-13/15$; the off-diagonal ones $2/15$. The two blocks are Grover matrices. When $t=\pi/2$, all entries in the off-diagonal blocks are equal to $i/15$.

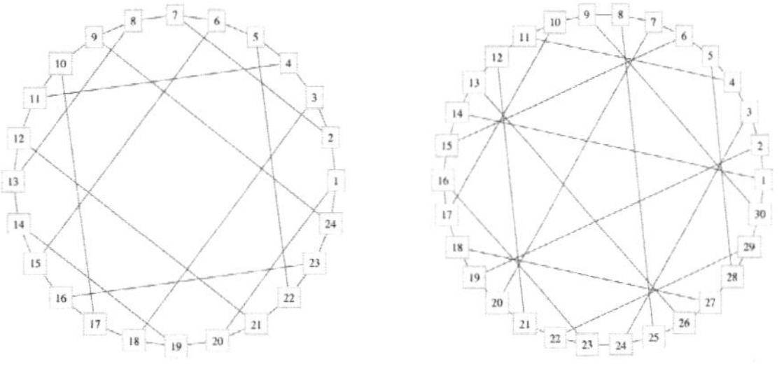

Figure 5. *(L)*: The Nauru graph. *(R)*: The Tutte-Coxeter graph. This is the largest periodic cubic graph. The Nauru graph and the Tutte-Coxeter graph do not have PST.

3. Conclusions

We have shown that the 3-dimensional cube is the only periodic, connected cubic graph with PST. The proof is based on known results of graph theory and basic definitions of quantum dynamics on graphs. Our proof goes through the relevant cases, which can be interpreted as a systematical exploration of periodic quantum dynamics on cubic graphs. A necessary and sufficient condition for PST would give a shorter, more elegant proof. However, any method for the same task requires the different cases. There is a small plus in the approach adopted here: we have written down explicitly the unitary matrices that specify the dynamics. The matrices are potentially useful in further work. The well-known mathematical scenario of state transfer together with various facts observed in this paper suggest some reflections. These may be worth a mention.

3.1. Extremality

The literature on interconnection networks for telecommunications, parallel computers and distributed systems contain many optimization scenarios dealing with order/degree/diameter [16]. The most famous one is perhaps the *degree/diameter problem* [20]: given natural numbers D and Δ, find the largest possible number of vertices $n_{\Delta,D}$ in a graph of maximum degree Δ and diameter $\leq D$. The graphs achieving the upper bound are known as *Moore graphs*. Interestingly, in the setting of PST, the original problem was somehow the opposite one [13]. In fact, the main point does not seek to maximize the number of vertices without compromising reachability, but to perform long distance communication, with a minimum of physical resources. The number of particles equals the order of the graph and it is therefore a resource. The order/distance problems for state transfer are then a new scenario, where graphs with extremal properties can be defined via some parameter associated to a dynamical process, instead of a topological condition. The landscape becomes even richer, if we consider more than a single excitation, and lift the analysis to the graph powers defined in [4]. These considerations remind that the problem of identifying extremal graphs with respect to state transfer is still a mostly unexplored area of research.

3.2. A notion of persistency

As a generalization of the rule inducing a continuous-time random walk, the operator $U_G(t) = e^{-iA(G)t}$ has been subject of intense study (see [21] and the references therein). In analogy with random walks, given the importance of these objects for constructing distributions, a principal direction has pointed uniform sampling. Additionally, still from the quantitative side, it is useful to deal with problems related to single vertices, or pairs of vertices, more than the entire graph. For example, an instance of relevant parameters is a quantum version of the hitting time. We would like to propose a notion that intends to quantify how the fidelity for state transfer between two vertices is constant in an interval, including small fluctuations. In the previous section, we have seen that half of the diagonal entries of $U_{C_6+K_2}(t)$ have the form $a_1 = (2 + \cos(t) + 2\cos(2t) + \cos(3t))/6$. When $t \in \{\pi/2, 3\pi/2\}$, $a_1 = 0$. Roughly between $\pi/2$ and $3\pi/2$ there is a plateau: the value of the function is about 1/3; it is exactly 1/3, for $t = \pi$. Starting the process from some vertex i, if at time $t = \pi$ we sample from the distribution, we are going to obtain i with probability 1/9. This probability is not spicked, but fairly stable around π. Given a graph G, the *ϵ-persistency* of a pair of vertices $i, j \in V(G)$ is the length of the longest interval $T \subset [0, 2\pi]$ such that there exists a value k for which $k - \epsilon < |[U_G(t)]_{i,j}| < k + \epsilon$, with $\epsilon \geq 0$, for every $t \in T$. By looking at all vertices, one could define the maximum persistency or the average persistency, if extending the definition in the obvious ways. A process with higher persistency requires less clock precision for sampling. A first sight reason for giving attention to persistency could arise from a connection with energy transfer problems [12]. Let us add an additional remark. Decoherence has been shown to behave as a natural smoothing mechanism on probability amplitudes [19]. Because of this fact, does decoherence increase persistency? With a similar scope but a on a different line, is the presence of decoherence compatible with PST at all?

3.3. Discrete probability transfer

The evolution governed by the Hamiltonian $H_{XY}(G)$ is driven by the exponentially smaller $U_G(t)$, when taking into account a single excitation only. While $A(G)$ faithfully represents G, the matrix $U_G(t)$ does not preserve its topological structure. In other words, $[U_G(t)]_{i,j} \neq 0$ does not imply $[A(G)]_{i,j} = 1$. A discrete evolution on G could be defined (in some cases [25]) by a unitary W_G, with $[W_G]_{i,j} \neq 0$ if and only if $[A(G)]_{i,j} = 1$. Such a process does not describe the transfer of a single excitation, but it only allows to create a probability distribution supported by $V(G)$. The process is discrete and it follows the iteration $W_G W_G^{t-1} \longmapsto W_G^t$. If there is $t \in \mathbb{N}$ such that $\left|\langle j|W_G^t|i\rangle\right| = 1$ then we have *perfect probability transfer* from vertex i to vertex j. Unless we make use of some kind of *lifting* [2] (*e.g.*, the introduction of extra degrees of freedom), this is the closest analogue to the continuous object $U_G(t)$, even if this one does not always exist. For example, we can construct on K_4 the unitary below:

$$W_{K_4} = \begin{bmatrix} 0 & -1 & 1 & 1 \\ 1 & 0 & -1 & 1 \\ 1 & -1 & 0 & -1 \\ 1 & 1 & 1 & 0 \end{bmatrix}.$$

However, it is simple to observe that each power of W_{K_4} has one of the two zero-patterns

$$\begin{bmatrix} 0 & * & * & * \\ * & 0 & * & * \\ * & * & 0 & * \\ * & * & * & 0 \end{bmatrix} \quad \text{or} \quad \begin{bmatrix} * & 0 & * & * \\ 0 & * & * & * \\ * & * & * & 0 \\ * & * & 0 & * \end{bmatrix},$$

where $*$ denotes a generic nonzero entry. This is sufficient to show that W_{K_4} does not give perfect probability transfer in K_4. In more complicated situations, the zero-patterns of matrix powers are not immediately available to imply a general statement. To verify that a graph G does not enjoy the property, we should study the spectra of unitary matrices with the same zero-pattern of W_G. Since the problem involves both spectra, zero-pattern, and an optimization procedure, its flavour reminds of the matrix analysis questions approached in [8] or various parametrizations coming from graph matrices [18]. In our context, the use of semidefinite programming techniques does not seem immediately useful, because the matrices are not stochastic, but in fact unitary.

Acknowledgments
I am supported by a Newton International Fellowship. I am grateful to Matthew Russell for finding an important error in a previous version of the paper and an anonymous referee for valuable comments. This paper is dedicated to Anthony Sudbery in the occasion of his retirement.

References

[1] Ahmadi O, Alon N, Blake I F and Shparlinski I E 2009 Graphs with Integral Spectrum, *Linear Algebra Appl.* **430** 547-552
[2] Ambainis A 2008 Quantum Random Walks – New Method for Designing Quantum Algorithms *SOFSEM 2008: Theory and Practice of Computer Science LNCS* **4910**
[3] Angeles-Canul R J, Norton R, Opperman M, Paribello C, Russell M and Tamon C 2009 On quantum perfect state transfer in weighted join graphs *Preprint* arXiv:0909.0431v1
[4] Audenaert K, Godsil C D, Royle G F and Rudolph T 2007 Symmetric squares of graphs *J. Comb. Theory, Ser. B* **97** 74-90 (*Preprint* math/0507251v1)
[5] Balińska K, Cvetković D M, Radosavljević Z, Simić S and Stevanović D 2002 A survey on integral graphs *Univ. Beograd, Publ. Elektrotehn. Fak., Ser. Mat.* **13** 42-65
[6] Bernasconi A, Godsil C and Severini S 2008 Quantum networks on cubelike graphs *Phys. Rev. A* **78** 052320 (*Preprint* arXiv:0808.0510v1)
[7] Bluhm R, Kostelecky A, Porter J and Tudose B 1997 Revivals of Quantum Wave Packets, COLBY 97-09, IUHET 372 (*Preprint* arXiv:quant-ph/9711061v1)
[8] Boyd S, Diaconis P, Sun J and Xiao L 2006 Fastest mixing Markov chain on a path *Amer. Math. Monthly* **113** 70–74
[9] Bose S 2007 Quantum Communication through Spin Chain Dynamics: an Introductory Overview *Contemp. Phys.* **48** 13-30 (*Preprint* arXiv:0802.1224v1)
[10] Bose S, Casaccino A, Mancini S and Severini S2009 Communication in XYZ All-to-All Quantum Networks with a Missing Link *Int. J. Quant. Info.* **7** (*Preprint* arXiv:0808.0748v1)
[11] Bussemaker F C and Cvetković D M 1976 There are exactly 13 connected, cubic, integral graphs *Univ. Beograd, Publ. Elektrotehn. Fak., Ser. Mat. Fiz.* **544-576** 43-48
[12] Caruso F, Chin A W, Datta, Huelga S F and Plenio M B 2009 Highly efficient energy excitation transfer in light-harvesting complexes: The fundamental role of noise-assisted transport *J. Chem. Phys.* **131** 105106 (*Preprint* arXiv:0901.4454v2)
[13] Christandl M, Datta N, Ekert A and Landahl A J 2004 Perfect state transfer in quantum spin networks *Phys. Rev. Lett.* **92**, 187902 (*Preprint* arXiv:quant-ph/0309131v2)
[14] Conder M and Dobcsányi P 2002 Trivalent Symmetric Graphs Up to *768* Vertices *J. Combin. Math. Combin. Comput.* **40** 41-63
[15] Cvetković D M, Doob M and Sachs H 1980 *Spectra of graphs – Theory and application* (Berlin, New York: Deutscher Verlag der Wissenschaften/ Academic Press)
[16] Ferrero D 1999 Introduction to interconnection network models *Publ. Mat. Urug.* 99/25
[17] Godsil C 2008 Periodic graphs *Preprint* arxiv.org/abs/0806.2074 [quant-ph]
[18] van der Holst H, Lovász L and Schrijver A 1999 The Colin de Verdiere graph parameter, Graph theory and combinatorial biology *Bolyai Soc. Math. Stud,* **7** 29-85
[19] Kendon V 2006 Decoherence in quantum walks - a review, *Math. Struct. in Comp. Sci* **17** 1169-1220 (*Preprint* arXiv:quant-ph/0606016v3)
[20] Miller M and Širáň J 2005 Moore graphs and beyond: A survey of the degree/diameter problem *Elec. J. Comb.* #DS14.
[21] Mosca M, *Quantum Algorithms*, in *Springer Encyclopedia of Complexity and Systems Science* (New York: Springer) (*Preprint* arXiv:0808.0369v1)
[22] Ore O 1962 *Theory of graphs* (American Mathematical Society)
[23] Russell M (personal communication, January 2010)
[24] Schwenk A J 1976 Exactly thirteen connected cubic graphs have integral spectra *Proc. Int. Graph Theory Conf. at Kalamazoo* ed Alavi Y and Licks D (Amsterdam: Springer)
[25] Severini S 2003 On the digraph of a unitary matrix, *SIMAX, SIAM J. Matrix Anal. Appl.* **25** 295-300 (*Preprint* math.CO/0205187)

[26] Tadej W and Życzkowski K 2006 A concise guide to complex Hamadard matrices *Open Syst. Inf. Dyn.* **13** 133-177 (*Preprint* quant-ph/0512154v2)

Journal of Physics: Conference Series **254** (2010) 012013 doi:10.1088/1742-6596/254/1/012013

The Sudbery interpretation of quantum mechanics

James Ladyman

Department of Philosophy, University of Bristol, 9 Woodland Rd, Bristol BS8 1TB, UK

E-mail: `james.ladyman@bristol.ac.uk`

Abstract. In this paper I review Tony Sudberys foundational work with a view to tracing the evolution of his ideas about the measurement problem and the interpretation of quantum mechanics.

Received: 22 June 2010

Introduction

In this paper I review Tony Sudbery's foundational work with a view to tracing the evolution of his ideas about the measurement problem and the interpretation of quantum mechanics. I also discuss briefly what I learned about the question of whether quantum mechanics is nonlocal under Sudbery's supervision.[1] In Section 1, I discuss Sudbery's critique of the application of the standard theory of quantum measurement to continuous observation. In Section 2 I briefly explain Sudbery's refutation of Karl Popper's claim that his purported experiment could be used to experimentally test the orthodox Copenhagen interpretation of quantum mechanics. In Section 3 I describe Sudbery's account of the measurement problem and the taxonomy and evaluation of interpretations of quantum mechanics in his earlier work especially in his 1986 book. In Section 4 I explain Sudbery's current defence of the Everett interpretation. Finally in Section 5 I briefly address the vexed question of whether quantum mechanics is nonlocal.

1. Discrete versus Continuous Observation

Sudbery points out at the beginning of his paper 'The Observation of Decay' [1] that time evolution in quantum mechanics is seemingly governed by two very different laws. The first is the Schroödinger equation according to which quantum systems evolve deterministically and reversibly. The second, the projection or collapse postulate is used to describe the abrupt change that occurs when a measurement is made and a system that may have been in a superposition with respect to the relevant observable is found to have a particular value. Much debate in the philosophy of quantum mechanics concerns whether one should regard collapse as a genuine

[1] I studied pure mathematics (main) and philosophy (subsidiary) at the University of York with Tony Sudbery as my supervisor graduating in 1992. I carried out a theoretical physics project with him in my final year on nonlocality in quantum mechanics and this inspired me to study philosophy of science which became my academic speciality. I am profoundly grateful to Tony for the wonderful example he set as to how to conduct oneself in academic life. He approached teaching and research in the same spirit of collective enquiry, and exhibited contagious enthusiasm and all-encompassing intellectual curiosity, as well as giving me personally great support and encouragement.

physical process (dynamical collapse theories), an expression of our limited state of knowledge (hidden variable theories), or as a perspectival illusion consequent on our experience being of only a single branch of the multiverse (Everettian theories). However, in the above paper, Sudbery discusses a fundamental problem with the collapse theory of measurement that is more rarely discussed, namely that it does not account for measurement that is passive and continuous rather than active and discrete. Consider, for example, the observation of a radioactive isotope with a Geiger counter. The system is coupled to the apparatus and simply left to decay at its own rate. There is no particular point at which a measurement is made, rather the system is continuously monitored and decay events are recorded when they happen spontaneously. The observer does not intervene actively, as in a projective measurement of the spin of an electron by passing it through a Stern-Gerlach device, but rather simply

> "waits for an event to happen" ([1], p 515).

If this kind of observation is modelled with the standard formalism one considers a number of discrete measurements with some small time interval between them and takes the limit as the time interval goes to zero. However, applying the projection postulate to each measurement in this model and then time-evolving the state to obtain the transition probability for the next measurement, leads to the conclusion that the difference in the probabilities of finding the system in a given state at different times tends to zero. Hence, the problem is that the famous 'watched-pot theorem' applies, and the model implies that no change occurs in the system at all. In the case of our radioactive element this result is of course in conflict with experience since (almost always) there will be sufficient decay events so that half the material is left after the observer has waited for the half life of the substance in question. There are various ways to resolve the paradox for many situations, for example, by supposing a finite cut-off in the limit of continual measurement, or by not modelling measurements as precise, but these do not seem to get to the heart of the problem for the case described above since the measurement is not imprecise nor is the problem well modelled by a series of measurements at discrete times. Since there is not a time-dependent interaction Hamiltonian but a time-independent one to describe continuous coupling of the system and the device a very different approach is needed.

Sudbery proposes a model of continuous observation in which the system and an apparatus are coupled and evolve in time together, hence "continuous observation is not continual measurement" ([1], p 513). He adopts a new form of the projection postulate formulated in terms of density operators. According to his Postulate A: Suppose D is an observable being continuously observed and P_k are the projection operators onto the eigenspaces of D. If ρ_0 is the density matrix of the system at time $t = 0$ its density matrix at time t is:

$$T_t(\rho_0) = \sum_k P_k\, e^{-iHt}\, \rho_0\, e^{iHt}\, P_k\,.$$

This can be considered as a supplement to the standard projection postulate to deal with continuous measurement or as a replacement for it. Sudbery's Postulate B then asserts that density operators will evolve according to the above equation for a privileged class of observables that form the relevant eigenspaces of D and the projectors onto them.

The measurement problem will not go away however. Sudbery demonstrates that although probabilities are independent of exactly where the line between system and observer is drawn, but that nonetheless some such line must be drawn (this result is due to von Neumann and gives rise to what is known as the von Neumann Chain). It is left open how we are to interpret the density matrix of a system. Sudbery points out that the problem of what kinds of physical process count as measurements remains in his approach, as does the problem of what exactly the privileged observables in postulate B are. Furthermore, with Postulate B the symmetry of

Hilbert Space is broken because some observables are regarded as special. This is analogous to classical mechanics in which there is symmetry between position and momentum under canonical transformations but an asymmetry in them when they are related to observation. Sudbery flirts with the Bohrian claim that the apparatus must be described in classical terms. He posits a sharp quantum/classical spit in return for which we only have one kind of dynamics, and we can suppose that transitions (e.g. decay) take place at definite times. In a 1986 paper Sudbery says that the above paper is part of a "general programme of showing that all interpretations of quantum mechanics are saying the same thing in different ways" ([4], p 82). This is a striking claim that comes close to positivism in so far as the seeming great metaphysical differences between dynamical collapse theories, Everettian interpretations and hidden variable interpretations are regarded as merely verbal.

This aside it is worth reiterating the conceptual distinction that Sudbery introduced in his 1984 paper. 'Measurement' is the active process of taking a system and subjecting it to an interaction that will result in the coupling of some property of the system with degrees of freedom in the measuring device. Both measurements of the first kind such as spin measurements using a Stern-Gerlach device that would result in the same outcome if repeated on the same system, and measurements of the second kind such as momentum measurements that change the momentum and hence would not result in the same outcome if repeated on the same system, are examples of 'measurement' in Sudbery's sense. On the other hand, 'observation' is when a measuring device is coupled to a system in such a way that it will not precipitate an immediate outcome, but rather will record a change in the system when it occurs such as a decay event resulting in the emission of radiation. In his most recent work Sudbery still cites the importance of recovering transition probabilities for systems that are 'passively measured', that is 'observed' not 'measured' in his 1984 terminology. Now, his explanation for the choice of basis is very different and the interpretation of the density matrix and the measurement problem are faced head on. I return to this below.

2. Sudbery falsifies Popper

In the 1982 preface to *Quantum Theory and the Schism in Physics*, Karl Popper proposed an experiment based on the EPR set-up that he claimed could be used to test the then standard Copenhagen interpretation of quantum mechanics. Popper argued that a strong form of action at a distance could be observed due to the increase in the spread of one particle's momentum due to the decrease in the spread in the other particle's position. In his paper 'Popper's variant of the EPR experiment' [3] published in *Philosophy of Science*, Sudbery shows decisively that the derivation is mistaken and that the Copenhagen interpretation cannot be tested by the experiment in the way Popper suggests.

Popper considers a pair of particles propagating in opposite directions along the x-axis such that they each have zero momentum in the y-direction, and each being passed through a slit of variable width beyond which are a semi-circle of particle detectors. He then imagines changing the sit width of particle A thereby gaining increased knowledge of its position in the y-direction, and consequently because of the entanglement of particles A and B, decreasing the spread in particle B's position too and therefore by the uncertainty relation increasing the spread in particle B's momentum. So he argues that the latter ought to be observable by the different pattern of detection in the semicircle of detectors and so the Copenhagen interpretation produces an empirically detectable form of action at a distance. However, as Sudbery shows, particles in eigenstates of y-momentum as required will not be in the entangled state of position that would lead to the prediction of a change in which counters fire because of a narrowing of the slit on the other wing.

In the commentary Sudbery endorses Hans Reichenbach's claim that the attribution of both position and momentum to a quantum particle is not a 'scientific statement' because it cannot be tested. Here again he seems to lean towards positivism as he goes on to castigate Popper for querying what can be meant by the claim that a particle cannot simultaneously have values of both position and momentum. Sudbery points out that statements about both quantities cannot be tested simultaneously, but of course this would cut no ice with Popper who thought testability was a condition for a statement being scientific but not (unlike the positivists) for it being meaningful.

In the context of discussing realism, Sudbery says there is "no more reason to think that a real particle must have a definite position and momentum than there is to think it must have a definite colour or smell" ([3], p 475). This is interesting because he here contrasts properties that are classically considered to be 'primary' with those that are classically considered to be 'secondary' in the sense defined in the scientific revolution by John Locke following Descartes, Galileo and the Greek atomists. The primary properties are defined to be those that are objectively possessed by objects as we conceive of them whereas the secondary properties are such that what the object objectively possesses is merely a configuration of primary properties that give it the stable disposition to produce in us experiences as of the secondary properties as we conceive of them. For example, it is said by the Greek atomists that honey is not really sweet but its particles are arranged in such a way as to give rise to the taste of sweetness in our perception. Sudbery is therefore suggesting that the traditional primary properties of position and momentum are on a par with paradigmatic secondary properties leaving us with the question as to what the true primary properties of particles are. Of course, mass and charge and the like may still be objectively possessed by them but they are state independent properties and so cannot be the basis for the contingent differences we observe between particles such as their apparent different spatial locations. If the quantum state is taken to be a primary property we are left with the problem of how it relates physically to the observations we make. Sudbery says that "a particle does not in general have definite values of position or momentum but it does have propensities to take up particular values of either" ([3], p 476). This replaces one mystery with another for of course we have no account of under what circumstances propensities turn into actualities.

3. The Interpretation of Quantum Mechanics

In his 1988 paper 'Testing Interpretations of Quantum Mechanics', Sudbery endorses falsificationism saying that the function of experiments is to disprove not to confirm. He goes on to distinguish three kinds of interpretation of a physical theory and how each relates to the criterion of testability.

(i) Operational interpretation This is a way of specifying the meaning of mathematical symbols or entities in terms of experimentally measurable quantities. For example, we may operationally interpret $|\psi|^2$ either as the charge density of a quantum system (Schroödinger), or as the probability of finding a particle as a given location (Born). The point about operational interpretations is that they must be testable. Without an operational interpretation we do not have a physical theory at all but only a mathematical one.

(ii) Physical interpretation Physical interpretations go further than operational ones by giving physical meaning for terms that have no direct operational interpretation. While such interpretations may be important for the development of further theoretical advances they are not themselves necessarily testable. For example, Planck interpreted his radiation law in terms of quantum oscillators with energy states in integer multiples of the product of his constant h and the frequency, and de Broglie-Bohm interpreted the phase of the wave function in terms of particles with definite trajectories and momenta.

(iii) Metaphysical interpretation This is the conceptual analysis of terms such as space and time (Einstein), probability, Bohr's complementarity and so on. According to Sudbery,

> disputes about which metaphysical interpretation of a theory to accept cannot be resolved by physics but are philosophical. Metaphysical interpretations are not testable, although they may get more or less plausibility from the role they play in the guiding the heuristics of those developing physical theories.

Sudbery goes on to summarise various interpretations of quantum mechanics that are explained in more depth in his book Quantum Mechanics and the Particles of Nature: an outline for mathematicians. The latter remains an excellent way to learn about quantum physics despite its relative age. The book goes a long way into the advanced topics of qauntum field theory, symmetry groups and particle physics for an introductory text, and so it takes one a long way to what remains the cutting edge of fundamental physics that has actually been applied to experiments[2]. One thing that makes it particularly ahead of its time is the prevalence given to foundational issues. There are four chapters that introduce quantum phenomena, quantum statics, quantum dynamics and then a selection of standard quantum systems, and then the fifth chapter is 'Quantum Metaphysics'. It is a tour de force of the philosophy of quantum mechanics into which a huge amount of important material is packed.

It is notable that the first thing in the chapter is the explanation of the distinction between pure and mixed states and the density matrix formalism. Sudbery explains and contrasts the way probability measures over classical phase space differ from the statistical operators of quantum mechanics. A table very neatly sets out the fundamental foundational differences between classical and quantum mechanics with respect to pure states, general states, the equation of motion, observables and expectation values. This forms the framework for the rest of the discussion.

When it comes to the quantum theory of measurement, Sudbery emphasises the way that measurement plays a fundamental role in the postulates of the theory, and then repeats and augments his earlier criticism of the projection postulate. His main objections to it are that:

(a) It is ill-defined since no definition of measurement is available. (It is "at worse inconsistent" ([5], p 186))
(b) It is dualistic because it splits time evolution into deterministic and probabilistic parts.
(c) It is anticausal since there is no explanation of what causes collapse and it appears to make "physical events consequences of their observation" ([5], p 212).
(d) No account of continuous measurement is possible (as we saw in 1 above).

Although Sudbery does not do so, it is possible to formulate the measurement problem in the form of three premises that are together inconsistent in the spirit of his remark quoted in (a) above.

Consider the following three propositions all of which there is reason to believe.

(I) A system has a definite value for an observable O if and only if it is an eigenstate of the corresponding hermitian operator on the Hilbert space of the system. (This is often called 'the e-state/e-value link'. It expresses the usual connection between quantum states and values of observables from right to left, and the completeness of the quantum state from left to right.)
(II) All time evolution is in accordance with the Schroödinger equation and is therefore described by a linear and unitary operator on the Hilbert space of the system.
(III) Measurements have unique outcomes, for example, a spin half system will always be found to be up or down if its spin in a particular direction is measured.

[2] It was the basis of a course given for many years by Sudbery at the University of York to third year mathematicians over three terms that managed to get those of us with very little physics to the doorstep of quantum chromodynamics.

These cannot all be true for a quantum system that is initially in a superposition of what is being measured. Consider a quantum system S described by a two-dimensional Hilbert space H which has two possible eigenstates ψ_+, ψ_- of an observable O associated with the eigenvalues +1 and -1. Suppose the measuring device M faithfully measures O for systems like S. This means that if (when M is in its ready state M_0) M interacts appropriately with S when S is in the state ψ_+, M will evolve into a state M_+. Similarly for ψ_- and M_-. The states M_+ and M_- are distinguishable by us (hence the idea of a measurement device as opposed to any other physical system that could interact with S).

The argument is as follows: Suppose S is initially in the superposition $(\psi_+ + \psi_-)/\sqrt{2}$. Since time evolution U is linear and unitary:

$$U(\frac{1}{\sqrt{2}}(\psi_+ + \psi_-) \otimes M_0) = \frac{1}{\sqrt{2}}(U(\psi_+ \otimes M_0) + U(\psi_+ \otimes M_0)) = \frac{1}{\sqrt{2}}(\psi_+ \otimes M_+ + \psi_- \otimes M_-) .$$

But the latter state is not an eigenstate of either S or M. Therefore, by the e-state/e-value link, neither S nor M have a definite value after the measurement.

The standard options for the interpretation of quantum mechanics can be classified according to which of (I) to (III) they deny. To deny (I) is usually thought to mean the positing of hidden-variables, and of course Bell's theorem tells us any such hidden variable theory must be nonlocal like de Broglie-Bohm theory. It is worth noting that quantum mechanics would also be incomplete if it did not apply to macroscopic objects and required a classical/quantum cut so that the states of the measurement device could be definite without there being corresponding quantum mechanical eigenstates. (II) would be false if the collapse of the wavefunction is a real process whether a random physical process as in dynamical collapse theories, caused by observation as in Wigner's interpretation, or caused by interaction between quantum and classical systems in line with some version of the theory of decoherence. Denying (III) gives us some version of the Everett interpretation. The other alternative is to reject the whole framework above and adopt some form of antirealism according to which the wavefunction represents only our knowledge of the system and all the probabilities in quantum mechanics are subjective.

Having explained the quantum theory of measurement, the measurement problem in the form of the cat paradox and the problem of continuous measurement and the 'watched pot theorem', Sudbery beautifully explains the EPR paradox in the context of a spin 1/2 system using to motivate the following explanation of de Broglie-Bohm theory. The latter is presented in terms of a continuity equation in terms of the probability current and density, and it is immediately apparent that the current for one particle will depend on the wavefunction of both and hence that the velocity of each particle depends on the other's position. Sudbery then shows that this manifest action at a distance in Bohm theory must be a feature of all hidden variable theories by proving a version of Bell's theorem. In his philosophical commentary, Sudbery begins by discussing indeterminism and arguing that it is straightforward to accept it but that it violates the idea that every event has a cause. This would be contested by many people; the literature in causal inference, Bayes nets and so on is replete with models of probabilistic causation (although these may not be well suited to quantum mechanics). He then says that indeterminism raises "no conceptual problems which are peculiar to quantum mechanics" ([5], p 211). I am not sure if he would now agree with this claim. It seems problematic in the light of the fact that many people argue that the principle of the common cause is false in quantum mechanics and one may connect worries about causation and determinism in quantum mechanics to the very subtle issues concerning locality discussed below. Sudbery is careful to distinguish indeterminism from indeterminacy, and notes that a system could have definite values at all times that evolve stochastically. He points out that the fact that if quantum mechanics is complete then particles cannot have definite trajectories in space over time conflicts with the "deep seated feeling that position and momentum are essential properties" of particles ([5], p 211). More generally we can ask about the physical state of a system with respect to some observable when it is not attributed an eigenstate of the corresponding operator, and the eigenstate-eigenvalue link tells us that in these circumstance the system's state with respect to the observable in question is indeterminate.

Sudbery outlines nine interpretations of quantum mechanics as follows.

(1) The *minimal interpretation* was "consistently promulgated by Bohr" ([4], p. 271) (it was notoriously not consistently promulgated consistently). It amounts to giving up on attributing states to quantum objects and on trying to say what the quantum state vector describes. Both pure states and general statistical operators refer to state preparation.
(2) The *literal interpretation* supposes that quantum states are descriptions of the properties of physical systems and there really are collapse events sometime during the measurement process as described by the projection postulate (von Neumann).
(3) The *objective interpretation* features stochastic jumps due to some physical process that is not peculiar to measurement. This is associated with Bell and with contemporary dynamical collapse theories.
(4) The *epistemic interpretation* regards quantum states as descriptions of our state of knowledge of quantum systems. This has been developed in recent years by many people working in quantum information theory, notably Carlton Caves and Chris Fuchs.
(5) The *ensemble interpretation* was favoured by Einstein and treats quantum states as descriptions of a collection of systems, and projection as shifting attention to a subensemble.
(6) The *many-worlds interpretation* due to Everett treats the state vector after projection as describing a branch in a multiverse in which collapse never objectively happens. It is now usually referred to as the 'Everett interpretation' because many people think talk of 'many worlds' is misleading.
(7) The *quantum-logical interpretation* is associated with Hilary Putnam. It is based on taking state vectors to describe propositions obeying a non-distributive logic.
(8) *Hidden-variable interpretations* deny that the quantum state is a complete description of the physical states of systems.
(9) The *stochastic interpretation* exploits the formal analogy between Schroödinger's equation and stochastic differential equations that describe processes like Brownian motion.

While (1) is subject to the challenge that it gives up on the aim of science "to understand the physical world" ([5], p 214), it is at least consistent. (2) is undermined by (a)-(d) above. The main problem with (3) is making collapse relativistically invariant. (4) is argued to be unstable between interpretations that resolve into a version of either (2) and (3). (5) is undermined by the fact that measurement involves "an objective change in the ensemble which cannot be explained away in terms of a shift of attention on the part of the observer" ([5], p 220). The Everett interpretation (6) is said to face all the same problems as the literal interpretation since the alleged splitting of the universe would occur when collapse occurs in the latter, hence we are equally owed an account of when and why it happens. (7) is allegedly based on "nothing but a mathematical pun" ([5], p 222). (8) is criticised both for requiring a distribution of hidden variables exactly matching the state vector, and for being in conflict with relativity. Finally, (9) is problematic because the motions of particles will depend on the motions of distant particles and on the form of the wavefunction at the initial time coordinate.

Sudbery recommends a choice of three (1), (3) or (8) among these nine interpretations but claims that ultimately it is a matter of individual preference since (1) is "boring", (3) is "puzzling" and (8) is "implausible" ([5], p 224). It is interesting that the interpretation that he eventually comes to advocate and does still at the time of writing is the Everett interpretation, and that it is not among those he earlier considered seriously. This perhaps reflects the fact that developments in the elucidation of the Everett interpretation seem to have made real progress with respect to the basis problem and the problem of how to interpret probability if all events happen.

4. Sudbery on Everett

By 2000 Sudbery has converted to the Everett interpretation and wants to "bring us all together to share the experience of summoning up a vanished tension and watching it relax" ([7], p.2). In fact he has in mind four tensions:

(i) The existence of spacetime versus the passage of time
(ii) Determinism versus free will
(iii) The physical description of brain states versus conscious experience
(iv) Duty versus 'why should I?'

Each of these involves the relationship between a universal third-person statement based on rational reflection and a perspectival first-person statement based on personal experience. In each case he maintains there is no genuine contradiction between them and his hope is that by treating the relationship between universal linear time evolution and the projection postulate as analogous to them he can show that there is no contradiction between them either. In short, the idea is that the projection postulate applies within a branch of the wavefunction while universal linear time evolution applies to the whole wavefunction.

His inspiration for this way of looking at the matter comes from an essay by the philosopher Thomas Nagel who distinguishes between what he calls 'internal' and 'external' questions. From an internal perspective it makes sense to ask 'why am I me?' meaning why do I have the subjective point of view that I in fact do have. From the internal perspective there is only one experiencing subject and from this perspective it is contingent who that subject is. From an external perspective it makes no sense for the question is equivalent to 'why is James Ladyman James Ladyman' and one cannot explain why a tautology is true.

In relation to (i), Sudbery argues that it is wrong to think that the perception of the flow of time is an illusion because there is no mistaken belief that we have when we experience time passing. The temporal ordering of events is no less real for being frame dependent. The way that experiences at different times seem to contradict each other is captured by thinking of the state of having a certain experience as being some quantum state that is indexed to a time. Then one cannot be in both states. The problem with the flow of time is, according to Sudbery, dissolved, when we realise that from the internal perspective time flows, while from the external perspective it does not. Internal statements are necessarily relative to a particular perspective that includes a conscious system, an instant of time and an eigenstate of experience. The latter change stochastically according to the transition probabilities given by quantum mechanics. Probability is here understood to be relative to an internal perspective, that is an Everett branch, and to be the degree of expectation of the event which is rational for a fully informed observer. From the external perspective on the other hand statements are about quantum states as functions of time where such states obey unitary dynamics. When it comes to the basis problem for the Everett interpretation which basis in Hilbert space is to be used to define the branching structure? Sudbery seems to ally himself with those like Michael Lockwood who suggest that the relevant basis is defined by the way that brain states give rise to conscious experiences. This is in contrast to those like Simon Saunders and David Wallace who think that decoherence theory suggests the emergence of a quasi-classical basis in which macroscopic objects behave quasiclassically. The latter has the disadvantage that it defines at best an approximate basis and hence an approximate branching structure, whereas the former has the disadvantage that it relies upon a promissory note about an as yet unknown science of the mind.

5. Does quantum mechanics require an Action at a Distance?

For a third year undergraduate project Sudbery gave me two papers, one by Henry Stapp and one by Karl Kraus (in *The Symposium on the Foundations of Modern Physics* edited by Lahti and Mittelstaedt), and asked me to answer the above question. The former claimed to prove that quantum mechanics is nonlocal, and the latter that it is compatible with special relativity's local causal structure based on the light cones of events. As is well known, the EPR-Bohm experiment can be carried out when the detection events on the left and right wings of the apparatus are spacelike separated. It follows that if there is a causal influence between them, such that say, the result on the left wing affects that on the right, then there are frames of reference in which the cause would occur after the effect in time. Since this seems to imply the possibility of causal loops and other problems it is usually ruled out by fiat and most physicists accept that causally connected events are in each other's light cones. Indeed, recently David Albert [2] has recently pressed the argument that quantum mechanics is incompatible with special relativity. On the other hand, physicists in general deny this and often follow Abney Shimony and talk of 'peaceful co-existence' between quantum theory and special relativity. The situation is particularly puzzling because relativistic quantum theory exists in the form of quantum field theory into which Lorentz covariance is normally explicitly built. There is a long history of controversy about action at a distance in physics. It seems fair to say that the insistence on local mechanisms to explain physical events has been extremely successful. Examples include theories that looked for intermediary carriers for the propagation of heat, sound, light, magnetic attraction and repulsion and electromagnetic induction. The field concept and its incredible fecundity may be regarded as the greatest triumph of the locality principle. While Newtonian gravitation remains a compelling counterexample, there are ways in which it respects some of the intuitions about local action. After all, the gravitational force falls off very quickly with the distance from a source, leading the conclusion that at large enough distances bodies do not affect each other gravitationally. The inverse square law can also be related to the area of an expanding sphere centred on the mass that causes the force of attraction, hence suggesting that the latter is mediated by something that propagates outwards from the source. By the time of General Relativity the idea of the finiteness of the speed of propagation of effects was fully established for all physical interactions. If it were really true that quantum mechanics was a nonlocal theory this would be a huge upset for physics.

Stapp proved a version of Bell's theorem, Krause a version of the no-signalling theorem. The task I faced was to reconcile the claims. The obvious way to do this is to disambiguate the notion of locality. If one defines locality to be the claim that experimental results are not affected by distant events except via some mechanism of contiguous intermediary processes, then I see no reason to deny that quantum systems are local, for orthodox physics posits no causal influences between spacelike separated events. While it is certainly true that the correlations exhibited by entangled systems would require action at a distance if there was a definite value for every observable for each system independently, such definite values are not a feature of the quantum formalism. I concluded in my project that some form of holism or nonseparability about the quantum world should be adopted to avoid the positing of action at a distance. Certainly one cannot prove that the quantum world requires action at a distance without assuming something like counterfactual definiteness. There is, as far as we know, complete symmetry between the two wings of the EPR-Bohm experiment. We have no more reason to believe that the left hand side result influences the right hand side result than vice versa. Hence, talk of action at a distance is misleading in this context. We have a strange kind of correlation at a distance but nothing like an ordinary interaction between the wings.

6. Conclusions

To conclude, it seems that several clear foundational commitments are constantly upheld in Sudbery's work on quantum mechanics, prior to and after his conversion to the Everett interpretation. These are as follows:

(1) There is no physical process of collapse.

(2) There is no gross nonlocality: "Quantum mechanics is not incompatible with the principle of locality, but with the conjunction of this principle and the existence of hidden variables" [1].

(3) The interpretation of quantum mechanics should be approached using the density matrix formalism and in terms of the recovery of transition probabilities, including from continuous observation.

Finally, Sudbery has always avowed and exemplified the importance and value of the philosophy of physics.

References

[1] Sudbery A S 1984 The observation of decay *Ann. Phys.* **157** 512
[2] Albert D Z and Galchen R 2009 A quantum threat to special relativity *Sci. Am.* **300** 32
[3] Sudbery A S 1985 Popper's variant of the EPR experiment *Phil. Sci.* **52** 470
[4] Sudbery A S 1986 Continuous state reduction, in *Quantum Concepts in Space and Time* ed R Penrose and C J Isham (Oxford: Oxford University Press) p 65
[5] Sudbery A S 1986 *Quantum mechanics and the particles of nature: An outline for mathematicians* (Cambridge: Cambridge University Press)
[6] Sudbery A S 1988 Testing Interpretations of Quantum Mechanics, in *Microphysical Reality and Quantum Formalism* ed A van der Merwe (Amsterdam: Kluwer) p 267
[7] Sudbery A S 2000 Why am I me and why is my world so classical? *Preprint* arXiv:quantph/0011084v1

AUTHOR INDEX

Institute of Physics
Dirac House, Temple Back
Bristol BS1 6BE UK

CURRAN ASSOCIATES INC
proceedings.com